Technikzukünfte, Wissenschaft und Gesellschaft / Futures of Technology, Science and Society

Series Editors

Armin Grunwald, ITAS, Karlsruhe Institute of Technology, Karlsruhe, Germany

Reinhard Heil, ITAS, Karlsruhe Institute of Technology, Karlsruhe, Germany

Christopher Coenen, ITAS, Karlsruhe Institute of Technology, Karlsruhe, Germany

Diese interdisziplinäre Buchreihe ist Technikzukünften in ihren wissenschaftlichen und gesellschaftlichen Kontexten gewidmet. Der Plural „Zukünfte" ist dabei Programm. Denn erstens wird ein breites Spektrum wissenschaftlich-technischer Entwicklungen beleuchtet, und zweitens sind Debatten zu Technowissenschaften wie u.a. den Bio-, Informations-, Nano- und Neurotechnologien oder der Robotik durch eine Vielzahl von Perspektiven und Interessen bestimmt. Diese Zukünfte beeinflussen einerseits den Verlauf des Fortschritts, seine Ergebnisse und Folgen, z. B. durch Ausgestaltung der wissenschaftlichen Agenda. Andererseits sind wissenschaftlich-technische Neuerungen Anlass, neue Zukünfte mit anderen gesellschaftlichen Implikationen auszudenken. Diese Wechselseitigkeit reflektierend, befasst sich die Reihe vorrangig mit der sozialen und kulturellen Prägung von Naturwissenschaft und Technik, der verantwortlichen Gestaltung ihrer Ergebnisse in der Gesellschaft sowie mit den Auswirkungen auf unsere Bilder vom Menschen.

This interdisciplinary series of books is devoted to technology futures in their scientific and societal contexts. The use of the plural "futures" is by no means accidental: firstly, light is to be shed on a broad spectrum of developments in science and technology; secondly, debates on technoscientific fields such as biotechnology, information technology, nanotechnology, neurotechnology and robotics are influenced by a multitude of viewpoints and interests. On the one hand, these futures have an impact on the way advances are made, as well as on their results and consequences, for example by shaping the scientific agenda. On the other hand, scientific and technological innovations offer an opportunity to conceive of new futures with different implications for society. Reflecting this reciprocity, the series concentrates primarily on the way in which science and technology are influenced social and culturally, on how their results can be shaped in a responsible manner in society, and on the way they affect our images of humankind.

Catrin Misselhorn · Tom Poljanšek ·
Tobias Störzinger · Maike Klein
Editors

Emotional Machines

Perspectives from Affective
Computing and Emotional
Human-Machine Interaction

Editors
Catrin Misselhorn
Department of Philosophy
Georg-August University Göttingen
Göttingen, Germany

Tobias Störzinger
Department of Philosophy
Georg-August University Göttingen
Göttingen, Germany

Tom Poljanšek
Department of Philosophy
Georg-August University Göttingen
Göttingen, Germany

Maike Klein
Gesellschaft für Informatik e. V.
Berlin, Germany

ISSN 2524-3764　　　　　　ISSN 2524-3772　(electronic)
Technikzukünfte, Wissenschaft und Gesellschaft / Futures of Technology, Science and Society
ISBN 978-3-658-37640-6　　　ISBN 978-3-658-37641-3　(eBook)
https://doi.org/10.1007/978-3-658-37641-3

© Springer Fachmedien Wiesbaden GmbH, part of Springer Nature 2023
This work is subject to copyright. All rights are reserved by the Publisher, whether the whole or part of the material is concerned, specifically the rights of translation, reprinting, reuse of illustrations, recitation, broadcasting, reproduction on microfilms or in any other physical way, and transmission or information storage and retrieval, electronic adaptation, computer software, or by similar or dissimilar methodology now known or hereafter developed.
The use of general descriptive names, registered names, trademarks, service marks, etc. in this publication does not imply, even in the absence of a specific statement, that such names are exempt from the relevant protective laws and regulations and therefore free for general use.
The publisher, the authors, and the editors are safe to assume that the advice and information in this book are believed to be true and accurate at the date of publication. Neither the publisher nor the authors or the editors give a warranty, expressed or implied, with respect to the material contained herein or for any errors or omissions that may have been made. The publisher remains neutral with regard to jurisdictional claims in published maps and institutional affiliations.

This Springer VS imprint is published by the registered company Springer Fachmedien Wiesbaden GmbH, part of Springer Nature.
The registered company address is: Abraham-Lincoln-Str. 46, 65189 Wiesbaden, Germany

Contents

Emotional Machines—Introduction 1
Catrin Misselhorn, Tom Poljanšek and Tobias Störzinger

Theoretical Foundations of Artificial Emotions

Emotion Components and Understanding in Humans and Machines . . . 21
Jacqueline Bellon

Emotions in (Human-Robot) Relation. Structuring Hybrid Social Ecologies 61
Luisa Damiano and Paul Dumouchel

Pre-ceiving the Imminent 83
Tom Poljanšek

The Route to Artificial Phenomenology; 'Attunement to the World' and Representationalism of Affective States 111
Lydia Farina

Design, Social Integration and Ethical Issues

When Emotional Machines Are Intelligent Machines: Exploring the Tangled Knot of Affective Cognition with Robots 135
Lola Cañamero

Is Empathy with Robots Morally Relevant? 159
Catrin Misselhorn

Robots and Resentment: Commitments, Recognition and Social Motivation in HRI 183
Víctor Fernández Castro and Elisabeth Pacherie

Social Robot Personality: A Review and Research Agenda 217
Sarah Diefenbach, Marietta Herzog, Daniel Ullrich
and Lara Christoforakos

Social Robots as Echo Chambers and Opinion Amplifiers 247
Catrin Misselhorn and Tobias Störzinger

Artistic Explorations

Quantifying Qualia 279
Jaana Okulov

My Square Lady 295

Editors and Contributors

About the Editors

Catrin Misselhorn is Full Professor of Philosophy at Georg-August University Göttingen; 2012–2019 she was Chair for Philosophy of Science and Technology at the University of Stuttgart; 2001–2011 she taught philosophy at the University of Tübingen, Humboldt University Berlin and the University of Zurich; 2007–2008 she was Feodor Lynen Fellow at the Center of Affective Sciences in Geneva as well as at the Collège de France and the CNRS Institute Jean Nicod in Paris. Her research areas are the philosophy of AI, robot and machine ethics.

Tom Poljanšek is a postdoctoral researcher associated to the Chair of Prof. Dr. Catrin Misselhorn at Georg-August-University Göttingen. His research interests include phenomenology, social ontology, philosophy of technology, and aesthetics. His dissertation Realität und Wirklichkeit. Zur Ontologie geteilter Welten (2022) develops a dualistic theory of the relation between mind-independent reality and subject-dependent worlds of appearance and presents an explanation for the diversity and multiplicity of the latter.

Tobias Störzinger is a postdoctoral researcher associated to the Chair of Prof. Dr. Catrin Misselhorn at Georg-August University Göttingen. His research areas are social ontology, social philosophy, and philosophy of technology.

From 2019–2021 he was a research assistant in the research project "GINA", in which research was conducted on the social and ethical challenges of social robotics.

His PhD project "Forms of Collective Agency and Metacognition in Distributed Systems", successfully completed in 2021, addresses the question to what extent and under which conditions socio-technical systems can be understood as intelligent agents.

Maike Klein Gesellschaft für Informatik e. V. Berlin, Germany.

Contributors

Jacqueline Bellon Eberhard Karls Universität, Tübingen, Deutschland

Lola Cañamero ETIS Lab, UMR8051, CY Cergy Paris Université—ENSEA—CNRS, Cergy-Pontoise, France

Lara Christoforakos Department of Psychology, Ludwig-Maximilians-Universität München, München, Germany

Luisa Damiano Department of Communication, Arts and Media, IULM University, Milan, Italy

Sarah Diefenbach Department of Psychology, Ludwig-Maximilians-Universität München, München, Germany

Paul Dumouchel Département de Philosophie, Université du Québec à Montréal, Montréal (Québec), Canada

Lydia Farina Department of Philosophy, University of Nottingham, Nottingham, UK

Víctor Fernández Castro Universidad de Granada, Granada, Spain

Marietta Herzog Department of Psychology, Ludwig-Maximilians-Universität München, München, Germany

Catrin Misselhorn Department of Philosophy, Georg-August University Göttingen, Göttingen, Germany;

Jaana Okulov Helsinki, Finland

Elisabeth Pacherie CNRS Institut/e Jean Nicod Paris, Paris, France

Tom Poljanšek Department of Philosophy, Georg-August University Göttingen, Göttingen, Germany

Tobias Störzinger University of Göttingen, Göttingen, Germany; Department of Philosophy, Georg-August-Universität Göttingen, Göttingen, Germany

Daniel Ullrich Media Informatics Group, Ludwig-Maximilians-Universität München, München, Germany

Emotional Machines—Introduction

Catrin Misselhorn, Tom Poljanšek and Tobias Störzinger

Abstract

Emotional robotics represents a trend. On the one hand, emotions seem to be essential to enhance the cognitive performance of artificial systems. On the other hand, emotions are supposed to make the interaction of robots with humans more intuitive and natural. In social contexts, recognizing and displaying emotions even seems to be indispensable. This volume addresses the theoretical foundations and possibilities of emotional robotics and explores its social consequences and ethical assessment. It combines two perspectives: (1) research on the nature of emotions and their implementation in artificial systems and (2) current work in emotional human-robot interaction (HRI) from a theoretical and ethical point of view. The first chapter gives an overview of the state of the art and provides the framework for situating the individual contributions of this volume.

C. Misselhorn (✉)
Department of Philosophy, Georg-August University Göttingen, Göttingen, Germany
e-mail: catrin.misselhorn@uni-goettingen.de

T. Poljanšek · T. Störzinger
Department of Philosophy, Georg-August University Göttingen, Göttingen, Germany
e-mail: tom.poljansek@uni-goettingen.de
T. Störzinger e-mail: tobias.stoerzinger@uni-goettingen.de

© Springer Fachmedien Wiesbaden GmbH, part of Springer Nature 2023
C. Misselhorn et al. (eds.), *Emotional Machines*, Technikzukünfte, Wissenschaft und Gesellschaft / Futures of Technology, Science and Society, https://doi.org/10.1007/978-3-658-37641-3_1

1 Trends in Social and Emotional Robotics

In November 2022, a Swedish team of roboticists and artificial intelligence (AI) researchers presented their humanoid robot, "ME-U," for the first time in public. ME-U is a social robot who has highly developed conversational and expressional skills. It has the shape of a human torso and the face of a young man of perhaps 32 years of age, short blond hair, and two free-movable arms with which he can gesticulate smoothly. ME-U can engage in unscripted conversation and has more than 65 different ways of expressing emotional reactions, including fine-grained micro-expressions of his facial "muscles." He is able to interpret contextual information, understands irony, and even exhibits a fine sense of humor. In a study involving more than 200 participants who went on a virtual "speed date" with ME-U, in which he was the only nonhuman participant, close to 95% failed to identify their interaction partner as a robot or exhibited any doubts about his humanity. An estimated 65% of the participants found ME-U to have an "interesting" or even "charming" character, and 31% said they would be willing to go on a "real" date with him. After learning that ME-U was, in fact, a robot, 15% were still interested in a second date.

ME-U's only shortcoming is that he does not really exist—at least not yet. We made him up. However, even those who, at least for a moment, considered googling what ME-U looks like or wondered whether they could get a date with him themselves probably would admit that such a robot appears to be practically possible. Given the development of chatbots like ChatGTP, the idea that robots whose conversational and socioemotional skills catch up with or even exceed human emotional skills no longer sounds as far-fetched as it might have sounded a few years ago. Emotions are, more and more, becoming the focus of AI and robotics. The aim is to build machines that are capable of recognizing, expressing, and even experiencing emotions in the same way as humans or animals do. This volume addresses the theoretical foundations and current research trends in emotional AI and robotics as well as the social and ethical implications of building and interacting with emotional machines.

For many years, research on AI did not focus so much on the development of *social* and *emotional* robots or machines but rather on the development of *intelligent* machines and programs in a narrower sense (for a historical overview see Bringsjord & Govindarajulu, 2020). The aim was to develop programs that were able to compete with human intelligence in tasks that were considered to be the peak of human cognition, such as playing chess (Champbell et al., 2002) or Go (Silver et al., 2016). Human social skills, such as emotions and empathy, have been largely ignored.

However, this trend is changing. In the 1980s, AI visionary Marvin Minsky was one of the first to argue that emotions should also be taken into account in the development of AI to master complex cognitive tasks (Minsky, 1985, chapter 16). If artificial systems are to be able to predict human behavior, respond appropriately to it, or interact smoothly with humans in social contexts, they must learn to read and express emotions. However, if they accomplish these tasks, it might seem as if new types of emotional and social agents are entering the social sphere.

Nevertheless, the field of social and emotional robotics does not have to start from scratch. Human beings have always had social and emotional relationships with artifacts and nonhuman creatures. Evidence for animal companions and pets goes at least back to 13,000 B.C. when the first dogs were domesticated; the domestication of cats can be traced back at least to 9,000 B.C. (for an overview, see Overall, 2017). Dolls and toy figures have existed even longer and play an important role in the social development of children (Habermas, 1999). However, dolls and the like remain "transitional objects" (Winnicott, 1971) that are characterized by their unchanging and passive presence. The children's expectations that their toys want to be fed, dressed, or hugged are projections of their own fantasies or desires onto an otherwise inert object (Misselhorn et al., 2013).

Although the tendency to anthropomorphize animals and artifacts is deeply entrenched in human psychology, progress in robotics has led to a new kind of artificial companion that involves imaginary projection (Misselhorn, 2009) but also seems to exhibit needs and behaviors of their own that appear to be, in a certain way, independent of the user's activities (Misselhorn et al., 2013). To capture the new quality of the social interaction with these devices, they are called "relational artifacts" (Turkle, 2006).

How pervasively such technologies can enter our everyday lives becomes clear when we consider how strongly humans are able to bond with robots and virtual agents. In 1997, a small plastic toy shaped like an egg, which was originally created in Japan, suddenly became quite popular all over the place. It consisted of a few buttons and a screen displaying a small pixelated creature that needed to be fed and cleaned regularly and exhibited a certain range of emotional states that required various types of care. The Tamagotchi had quite an impact on many people's everyday lives and even more so on the collective imagination of how artificial social companions might look like in the future and the impact they might have on the way people conduct their social relationships. Indeed, the artificial companions of the future could significantly change the way people think about themselves and their relationships with others in general (O'Rourke, 1998; Lawton, 2017).

Since then, social robots have been increasingly making their way into everyday life. They come in many forms and types and have a variety of uses. There are minimally social robots that interact with people in their workspace and have a low degree of autonomy. Service robots such as *TUG* (Aethon, 2020), which are designed to transport materials and supplies, belong to this category. Other devices enter public spaces and people's homes to provide services and function as assistants. *Care-O-bot,* for example, is intended to work in homes, hotels, nursing facilities, or hospitals and is capable of accomplishing a number of service and assistance tasks. It can deliver meals or perform other pick-up and delivery services (Fraunhofer IPA, 2021), but it does so without displaying any emotions.

Apart from such robots that are social in a specific functional context, there are models that do not have any practical tasks. They are simply supposed to establish a social bond with their users, and their aim is to generate "social resonance" for its own sake. Emotions are particularly important for reaching this goal. One robot that is primarily designed to evoke emotions is PARO, a robot that looks like a baby seal (Shibata, 1996). PARO responds to petting with cute sounds and movements, while it abhors being hit. At the same time, PARO is able to adjust to its interaction partners and their preferences; it has been used in the therapy of dementia patients (Chang et al., 2013; Misselhorn et al., 2013) as well as for children with autism spectrum disorders (Nakadoi, 2017).

Other zoomorphic social robots such as *Kiki* (Zoetic AI, 2020), *Aibo* (Fujita, 2001), or *Miro* (Hagn et al., 2008) aim to simulate the social relationships in which humans engage with pets and are designed to create social and emotional resonance. As in the case of *Kiki,* such pet-like companions may simulate having a will of their own and do not simply respond to the user's input. Therefore, they are more "autonomous," as they trigger their own actions according to their own "beliefs" and "desires" (see Misselhorn, 2015, for a gradual conception of agency that captures such cases).

A robotic design that extends beyond pet-like social relationships is envisioned in anthropomorphic robots. The design of robots such as *Pepper* (Pandey & Gelin, 2018) or *Sophia* (Hanson Robotics, 2020) is inspired by the shape of the human body and face. Not only does Sophia express emotions facially in a highly realistic manner, but the device also aspires to engage in human-style conversations. Sophia is intended to appear as similar to humans as possible and even was granted Saudi Arabian citizenship in 2017, which was a major publicity stunt. However, the first robot to be granted citizenship was, arguably, Paro in Japan (Schröder et al., 2021, p. 198).

While the design of many social robots is grounded in the look and feel of existing types of agents, such as pets or humans, there are also attempts to create robots rather as a distinct species. Robots such as *SYMPARTNER* (Gross et al., 2019) or *Jibo* (Jibo, 2020) are "robomorphic" instead of merely attempting to mimic natural agents. They are not designed purely as service robots but are also supposed to enhance social and emotional wellbeing. Instead of attempting to imitate social relationships between natural agents, they seek new ways of interacting socially. Since such a development ranges on a continuum, these devices are situated somewhere in between familiar social agents and emerging forms of artificial social agents.

2 Research Questions and Contributions to this Volume

Artificial systems and social robots are about to become ubiquitous in our everyday lives. Some imagine a future in which robots are not just helpers but also social companions and even lovers (Levy, 2008). Nursing robots are designed to take care of the elderly, social robots might become companions for lonely people, and sex robots could not just fulfill "conventional" sexual desires but also enable users to satisfy sexual desires and preferences whose nonconsensual fulfillment is morally offensive or even prohibited by law (Misselhorn, 2021a).

To judge whether these promises are realistic and desirable from a moral point of view, researchers must investigate the theoretical foundations and ethical issues in this field. They must reconsider old answers to new questions, such as the nature of human emotions and social relations, and they must scrutinize which characteristics can be ascribed to these new agents and which types of social relationships can and should be pursued with them.

This involves questions such as: Will the emotional expression of robots one day be indistinguishable from those of human interaction partners? Will robots ever be able to really *experience* emotions, or are they merely able to *simulate* having emotions? (For an analysis of the difference between realization and simulation, see Seibt, 2018). Do robots *deceive* humans by only "pretending" to have emotions? Should we engage in social and emotional relationships with such artificial companions, and which moral problems arise from entertaining emotional relationships with artificial agents? Would the fact that robots exhibit "intelligence" and "emotions" as they interact with us socially on a daily basis endow them with rights and a moral standing on their own? How are humans supposed to treat artificial agents? Will they ever be more than merely "animated tools" (to cite Aristotle's controversial definition of human slaves, see Aristotle *Politics*, 1253b), or should we treat them as subjects of a new and proper *third* kind, neither fully animate nor fully inanimate? How will society change by the omnipresence of artificial social agents?

Recent advances in the field of social robotics have made these questions even more pressing, since people seem to be ever more willing to accept robots as social interaction partners and are eventually developing emotional bonds with artificial agents (Darling, 2017; Misselhorn, 2021a; Turkle, 2010). There is good reason to assume that this trend will continue in the future. This volume aims to discuss the theoretical issues, current developments, and ethical debates about emotional machines, bringing together two strands of research: (1) research on the nature of emotions and their implementation in artificial systems and (2) current work in emotional human–robot interaction (HRI) from a theoretical and ethical point of view.

2.1 Emotions: Aspects and Components

The debate about the nature of emotions is ongoing, and there is no uncontroversial definition of the nature of emotions. Instead, we provide a brief overview of aspects and components that are considered central for emotions (Misselhorn, 2021a, p. 17). First, it seems rather uncontroversial to define an emotion as a certain type of intentional and usually conscious episodical state of a (living) agent. Stones, trees, and rivers—in short, inanimate objects—supposedly do not have emotions, since they lack consciousness. The front of a car may offer the impression that the car is smiling; however, this does mean that emotions of joy or friendliness can be attributed to the car.

This rough characterization can be refined (see Scarantino & de Sousa, 2018; for a more detailed discussion of aspects and components of emotions with regard to their possible implementation in artificial agents see Misselhorn, 2021; Bellon, in this volume). One of the most important aspects seems to be that emotions involve some kind of *evaluation* (see, for example, Kenny, 1963; Solomon, 1980; Neu, 2000; Nussbaum, 2001). To have an emotion that is directed towards something implies that the object of the emotion is evaluated in a certain way. Whether one is angry at someone or something, whether one loves someone or something, whether one fears someone or something, in all of these cases, one cares about the object of the emotion: It matters, for good or for bad, such that it has a direct negative or positive impact on one's wellbeing.

Another important aspect of emotions is their *feeling* or *phenomenal aspect*. Instead of just cognitively judging that something is good or bad in relation to a certain standard of evaluation, emotions come with certain phenomenal qualities. There seems to be a feeling "what it is like" to experience an emotional state. This phenomenal aspect of emotions may also be linked to the *object-directedness* of emotions. It seems plausible to distinguish between two different types of intentional objects towards which emotions may be directed: *particular objects* and *formal objects* (see, for example, Kenny, 1963; de Sousa, 1987; Prinz, 2004). If you are afraid of a spider, this spider is the particular object your emotion. Different emotions can reveal the same particular object in different lights. If anger is directed at someone, one experiences this person as offensive; if love is directed at someone, this person is a source of enjoyment; if there is fear of a spider, this spider appears as dangerous; and so on. Many philosophers of emotions hold that every token of an emotion may be directed at a different particular object but all these different particular objects are instances of the same formal object that is distinctive of a certain emotion-type. The formal object of "sadness" would, for instance, be something like "loss," and the formal object of "fear" would be something like "danger," and so on. The formal objects of emotion types are usually conceived of as being evaluative.

Besides the intentional and the phenomenal aspect of emotions, they also involve bodily processes and changes (see James, 1884): We can distinguish *behavioral* or *expressive aspects* that are easily accessible from the outside from *internal bodily changes,* which come with certain emotions. Anger, for instance, may be accompanied by a flushed face, a certain posture, and a specific facial expression as well as an increased pulse rate and muscle tone. While some of the internal bodily changes might serve an evolutionary or ecological function by making the agent ready for certain actions or events, others are expressional in the sense that other agents can perceive them and take them as indicators either of a probable behavior or of inner processes that the agent in question undergoes. (For the view that emotion expression has an important strategic function when it comes to human interaction, see, for example, Luhmann, 1982; Hirshleifer, 1987; Frank, 1988; Ross & Dumouchel, 2004; Van Kleef et al., 2011; O'Connor, 2016).

In humans, one can find a certain independence of emotional expression and experience. A person may pretend to be in a certain emotional state, such as joy, sadness, or amusement, without actually feeling it. (For an analogous distinction between emotions-had and emotions-perceived, see Poljanšek, in this volume.) However, there are limits to hiding emotions. Microexpressions—in other words, brief facial movements—reveal the true emotional state of a subject even if the person attempts to conceal it (Ekman, 2006; Ekman & Friesen, 1969). Trained experts and artificial emotion recognition systems can quite reliably detect the real emotional state of persons with the help of these clues. This is even the case when persons deliberately attempt to suppress their microexpressions.

2.2 Artificial Emotions

Given the technological state of the art, robots seem to lack consciousness and intentionality, and they do not have a biological body. Hence, there are good reasons not to assume that the robots and AIs one is dealing with today and in the foreseeable future will really possess emotions in the same sense that human beings and animals do. However, there are already quite a number of robots today that either exhibit emotional behavior or recognize and respond to human emotional behavior. There is also a strong tendency in humans to anthropomorphize inanimate objects with certain features, feeling empathy and perceiving them as if they had emotions, even if we actually know that they do not really feel or experience anything (see also Misselhorn, 2009, as well as Misselhorn, in this volume).

As Heider and Simmel already demonstrated in 1944 with the help of an animated short film that depicted geometrical forms apparently interacting emotionally with each other, people can perceive even such abstract shapes as triangles or circles to be social agents endowed with emotions. This human tendency can be exploited technologically by building artificial agents that successfully *simulate* having emotions (see, for example, Damiano & Dumouchel, in this volume) and, in this way, engage humans emotionally. These simulations work quite well independently of the question whether these agents really feel emotions or not. If it walks like a duck, quacks like a duck, and behaves emotionally like a duck, then it appears as an (emotional) duck. *Artificial emotions* can thus be defined functionally as realizations of the specific behavioral forms, dynamics, or functions of "real" emotions in artificial agents, independent of the existence of any corresponding inner emotional states.

2.3 Social Robots and Artificial Agents

The prospect that such emotional machines will populate our social world raises a number of social and ethical questions. These questions concern, first, the impact of robots as social interaction partners on individuals; second, the normative obligations we have toward robots; and third, the consequences that arise for society as a whole if social and emotional robots become more pervasive in everyday life.

Issues belonging to the first subject area are: Which social relationships with emotional machines are ethically appropriate and which may not be? Are there certain types of relationships, such as friendship or romantic love, that cannot or should not be realized with robots? Questions belonging to the second set of issues are: How should we deal with artificial emotional companions? Should we grant them the ontological status of subjects? What, if anything, do we owe artificial emotional agents and under what conditions? Should artificial emotional agents be granted the status of equal participants in our social practices? Which legal status should they have? Issues regarding the third topic concern the ways in which society as a whole will change if, for instance, many people started to have robotic partners or if it became commonplace for elderly people to be in the care of social and emotional robots.

There is an ongoing debate in robot ethics about the kind of social relationships that we can and should have with robots (for an overview, see Wilks, 2010). Some researchers (Breazeal, 2003, 2004) believe that social and emotional robotics and AI can improve human flourishing and amplify human potential. Others, such as Sherry Turkle, are rather reluctant to adopt the notion of a strong engagement with emotional machines (Turkle, 2010). Some researchers even suggest that we should not view robots as persons but should treat them as slaves with whom we should not feel empathy, because robots are designed only to help humans "expand our own capabilities and achieve our own goals" (Bryson, 2010, p. 63). Misselhorn (in this volume) argues, in contrast, that it would be wrong to suppress or extinguish one's natural feeling of empathy toward robots, because in this way, one would compromise an important source of moral judgment, moral motivation, and moral development in humans. Rather, we should take these feelings of empathy into account when creating robots for different purposes and should decide in which contexts robots that elicit empathy are ethically appropriate and in which ones they are not.

In general, there may be some social or emotional relationships with robots that have a positive effect on the user while other types of social relationships should be reserved for humans only. (A suggestion on how to make this distinction is provided by Misselhorn and Störzinger (in this volume); for an approach that suggests a distinction along the categories of "social functional interaction" and "close interhuman relationship," see Poljanšek & Störzinger, 2020.)

Another set of questions relates to robots as moral agents (Floridi & Sanders, 2004; Wallach & Allen, 2010; Lin et al. 2012, 2017; Misselhorn, 2018, 2021b). Are robots agents at all? More specifically, can they be moral agents, and are they morally or even legally responsible for their actions? One suggestion (Misselhorn, 2019) is that robots are legal subjects, at best, in the sense of being vicarious agents that do not bear any legal or moral responsibility for their actions. Another issue is how responsibility is distributed among the various involved parties. Who should be morally and legally responsible if a care robot kills the neighbor's dog? Should it be the company that develops and sells the robot, or the owner of the robot? With regard to the question of responsibility, what roles are played by political and legal regulations that allow such robots?

Apart from the question of how a social relationship with robots affects humans and whether robots can be granted the status of moral agents, there is a debate within the field of robot and machine ethics about the normative status of robots as objects of moral and legal considerations. There are indirect arguments for the claim that violence toward robots should be legally prohibited, because it offends people's general moral feelings (Darling et al., 2016). Some even argue directly that robots must be granted robust moral and legal status with proper moral and legal claims of their own (see Coeckelbergh, 2018; Danaher, 2019; Loh, 2019).

2.4 The Contributions to this Volume

The contributions to this volume address the question of how to equip artificial systems with emotions and how humans can and should interact emotionally with them. The contributions focus on theoretical as well as ethical issues that arise from various approaches to designing emotional machines. Since these issues are situated at the intersection of various disciplines, the contributions draw on several methods, such as conceptual analysis, psychological and neurophysiological evidence, empirical research from the social sciences, and even artistic research. The latter reflect our view that the arts can make a proper contribution to the topic because they are able to elucidate the way in which we do not just conceive of but enact our social and emotional relations with robots.

The volume is divided into three parts. The **first part** addresses the theoretical foundations of artificial emotions and contains approaches for a definition of emotions, with the aim to scrutinize whether and how emotions might be realized by or implemented in artificial systems. In the first chapter, "*Emotion Components and Understanding in Humans and Machines*," **Jacqueline Bellon** offers a systematic overview of how far technical systems can already—and might in the future—be endowed with "emotions" by dividing the concept of "emotion" into evaluative, expressive, behavioral, physiological, mental, and phenomenological components. Additionally, Bellon provides examples of the definition and implementation of each component in emotion research theories and state-of-the-art artificial systems. With reference to Gilbert Simondon, the chapter further argues for the importance of an in-depth human understanding of artificial systems and, thus, for the demand for *mechanologists* who take care of the entire social setup in which human–machine interaction takes place.

Luisa Damiano and **Paul Dumouchel** tackle the question of whether robots can establish meaningful affective coordination with human partners by developing a series of arguments relevant to the theory of emotion, the philosophy of AI, and the epistemology of synthetic models. In their chapter, *"Emotions in (Human-Robot) Relation. Structuring Hybrid Social Ecologies,"* they argue that machines can indeed have meaningful affective coordination with humans, and they lay the groundwork for an ethical approach to emotional robots. This ethical project, which they label "synthetic ethics," rejects the ethical condemnation of emotional robots as "cheating" technology and focuses instead on the sustainability of emerging mixed human–robot social ecologies. Furthermore, it promotes a critical, case-by-case inquiry into the type of human wellbeing that can result from human–robot affective coordination.

In the third chapter, *"Pre-ceiving the Imminent. Emotions-Had, Emotions-Perceived and Gibsonian Affordances for Emotion Perception in Social Robots,"* **Tom Poljanšek** argues that we must draw a strict distinction between two types of emotions, which is almost always neglected in current debates: internal *emotions-had* and external *emotions-perceived*. While (internal) emotions-had are emotions that conscious agents experience as their own, emotions-perceived are emotions that they perceive as the emotions of other agents. This distinction is motivated by phenomenological, ecological, and evolutionary theoretical considerations: From the perspective of the organism, emotions-had serve different ecological functions than emotions-perceived. Thus, for the perception of an external emotion of another agent to be adequate, it is not necessary for that agent thus perceived to actually have the internal emotion(-had) that we normally associate with the emotion-perceived in question. An agent can be adequately perceived *as aggressive* although they do not experience any *anger* themselves. If this reasoning is applied to emotional robots, it follows that deception is not necessarily involved in perceiving a robot as exhibiting a particular emotion. A robot can fully instantiate emotions-perceived although it might be unable to experience emotions-had itself. To elaborate on these claims, the view presented is contrasted with James Gibson's concept of "affordances."

In the concluding chapter of the first part of the volume, *"The Route to Artificial Phenomenology; 'Attunement to the World' and Representationalism of Affective States,"* **Lydia Farina** argues that to design artificial systems that are able to experience emotions, these systems should be endowed with an "attunement to the world," a concept she argues is derived from "early" phenomenologists such as Heidegger and Merlaeu-Ponty. It is often argued that artificial systems are unable to experience emotions because they lack the ability to have representational states. In contrast, Farina argues that "attunement to the world" is a nonrepresentational emotional state; thus, machines endowed with it might be able to experience emotions without having representational states.

The **second part** of the volume deals more specifically with the question of how emotional robots might be designed and integrated into social practices and interactions. It tackles practical, ethical, and social challenges emerging from the prospect of such an integration. **Lola Cañamero** argues in "*When Emotional Machines are Intelligent Machines: Exploring the Tangled Knot of Affective Cognition with Robots*" that the traditional separation of emotions on the one hand and cognition on the other hand should be abandoned in favor of a "tangled knot of affective cognition." She presents her model of embodied affective cognition and demonstrates how it can be used in various applications of robots. The model forms the basis for a "core affective self" that endows robots with "internal" values (their own or acquired) and motivations that drive adaptation and learning through interactions with the physical and social environment.

However, even robots with a "core affective self" in this sense would lack phenomenal consciousness of emotions. They cannot feel what it is like to experience an emotion, for instance, to be in pain. This might be a reason to believe there are no ethical constraints on how robots should be treated. Since they do not really feel anything, they might be kicked, beaten, and offended in all sorts of ways. Nevertheless, human beings do have empathy for robots that are abused, as **Catrin Misselhorn** shows in her contribution, "*Is Empathy With Robots Morally Relevant?*" Starting with a definition of the concept of empathy, she provides an indirect argument for the claim that empathy with robots matters from a moral point of view. The fact that we feel empathy with robots that resemble humans or animals imposes moral constraints on how robots should be treated, because robot abuse compromises the human capacity to feel empathy, which is an important source of moral judgment, moral motivation, and moral development. Therefore, we should consider carefully the areas in which robots that evoke empathy may be used and the areas where we should refrain from such a design.

Since designing robots that evoke empathy is a highly sensitive matter, ethically, it makes sense to turn to other approaches to establish social relations between humans and robots. In their paper, "*Robots and Resentment: Commitments, Recognition, and Social Motivation in HRI*," **Víctor Fernández Castro** and **Elisabeth Pacherie** suggest that successful interaction between robots and humans would benefit from robots having the ability to establish mutual recognition. In their approach, the authors argue for the fundamental role of commitments in mutual recognition and discuss how commitments depend on affective states, such as social emotions or the "need to belong." Their paper concludes with three suggestions for how emotions may contribute toward making commitments in social robots.

Another issue is whether social companions also have or at least simulate a social robot personality. **Sarah Diefenbach, Mariette Herzog, Daniel Ulrich,** and **Lara Christoforakos** emphasize in their paper, *"Social Robot Personality: A Review and Research Agenda,"* that while there is a vigorous debate about the adequate robot personality for HRI, the term "social robot personality" is used quite differently in the literature. This conceptual fuzziness is manifested in various understandings of the expression of robot personality, its measurement, and its presumed effects on HRI. To elucidate the matter and carve out differences and disagreements, the authors provide a narrative review of the interdisciplinary work on social robot personality. In doing so, they also identify research gaps that should be addressed by future studies.

Let us assume for a moment that all theoretical and practical challenges of artificial emotions can be met and that it is possible to design robots that are almost indistinguishable from natural agents, such as animals or humans, in terms of their social and emotional behavior. This leads to the question of how far a social relationship with robots can and should go. This is discussed by **Catrin Misselhorn** and **Tobias Störzinger** in their paper *"Social Robots as Echo Chambers and Opinion Amplifiers. Limits to the social integrability of robots."* Using a practice-theoretical perspective on sociality, the authors investigate which social practices are reserved for humans. They argue that especially those practices that require participants to reciprocally recognize each other as persons clash with the conceptual understanding of robots. Furthermore, the paper provides reasons why this understanding of robots can be defended against a conception that wants to attribute the status of persons to robots based on their behavior. The simulated evaluative attitudes of robots are not rooted in the robots themselves but turn out instead to be merely opinion amplifiers of their developers or sociotechnical echo chambers of the users. However, they also argue that building robots that can perfectly simulate recognition claims nevertheless poses a problem since such devices would distort our social practices.

The **third part** of the anthology is devoted to more artistic explorations of the connection between social machines and emotions through the framework of "artistic research." In the chapter *"Quantifying Qualia,"* **Jaana Okulov** attempts to outline the ongoing research on affective machines by focusing on the key term *qualia*. She approaches this topic using examples from the artist group *Olento Collective*. The final contribution, *"My Square Lady"* by the arts collective **Gob Squad,** exemplifies how social robots can be integrated into the process of creating and performing a theater production.

References

Aethon. (2020). Retrieved December 12, 2020 from https://aethon.com
Breazeal, C. (2003). Toward sociable robots. *Robotics and Autonomous Systems, 42*(3), 167–175.
Breazeal, C. (2004). *Designing sociable robots*. A Bradford Book.
Bringsjord, S., & Govindarajulu, N. S. (2020). Artificial intelligence. In E. N. Zalta (eds.), *The Stanford encyclopedia of philosophy*. Retrieved December 12, 2020 from https://plato.stanford.edu/archives/sum2020/entries/artificial-intelligence/
Bryson, J. J. (2010). Robots should be slaves. In Y. Wilks (ed.), *Natural language processing* (Vol. 8, pp. 63–74). Benjamins.
Campbell, M., Hoane, A. J., & Hsu, F. (2002). Deep blue. *Artificial Intelligence, 134*(1), 57–83.
Chang, W., Šabanovic, S., & Huber, L. (2013). Use of seal-like robot PARO in sensory group therapy for older adults with dementia. In *2013 8th ACM/IEEE international conference on human-robot interaction (HRI)*, (pp. 101–102).
Coeckelbergh, M. (2018). Why care about robots? Empathy, moral standing, and the language of suffering. *Kairos. Journal of Philosophy & Science, 20*(1), 141–158.
Danaher, J. (2019). The philosophical case for robot friendship. *Journal of Posthuman Studies, 3*(1), 5–24.
Darling, K., et al. (2016). Extending legal protection to social robots: The effects of anthropomorphism, empathy and violent behavior towards robotic objects. In R. Calo (Ed.), *Robot law* (pp. 213–234). Cheltenham.
Darling, K. (2017). "Who's Johnny?" Anthropomorphic framing in human–robot interaction, integration, and policy. In P. Lin et al. (eds.), *Robot ethics 2.0: From autonomous cars to artificial intelligence* (pp. 173–193). Oxford University Press.
de Sousa, R. (1987). *The rationality of emotions*. The Massachusetts Institute of Technology.
Ekman, P. (2006). Darwin, deception, and facial expression. *Annals of the New York Academy of Sciences, 1000*(1), 205–221.
Ekman, P., & Friesen, W. (1969). Nonverbal leakage and clues to deception. *Psychiatry, 32*(1), 88–106.
Floridi, L., & Sanders, J. W. (2004). On the morality of artificial agents. *Minds and Machines, 14*(3), 349–379.
Frank, R. H. (1988), *Passions within reason: The strategic role of emotions*. Norton.
Fraunhofer IPA. (2021). Retrieved March 1, 2020 from https://www.care-o-bot.de/de/care-o-bot-4.html
Fujita, M. (2001). AIBO. Toward the era of digital creatures. *The International Journal of Robotics Research, 20*(10), 781–794.
Gross, H., Scheidig, A., Müller, S., Schütz, B., Fricke, C., & Meyer, S. (2019). Living with a mobile companion robot in your own apartment—Final implementation and results of a 20-weeks field study with 20 seniors. In *2019 international conference on robotics and automation (ICRA)* (pp. 2253–2259).
Habermas, T. (1999). *Geliebte Objekte. Symbole und Instrumente der Identitätsbildung*. Suhrkamp.

Hagn, U., Nickl, M., Jörg, S., Passig, G., Bahls, T., Nothhelfer, A., Hacker, F., Le-Tien, L., Albu-Schäffer, A., Konietschke, R., Grebenstein, M., Warpup, R., Haslinger, R., Frommberger, M., & Hirzinger, G. (2008). The DLR MIRO: A versatile lightweight robot for surgical applications. *Industrial Robot: An International Journal, 35*(4), 324–336.

Hanson Robotics. (2020). Retrieved December 12, 2020 from https://www.hansonrobotics.com/sophia/

Hirshleifer, J. (1987). On the emotions as guarantors of threats and promises. In J. Dupré (ed.), *The latest on the best: Essays on evolution and optimality* (pp. 307–326). MIT Press.

James, W. (1884). What is an emotion? *Mind, 9*(2), 188–205.

Jibo. (2020). Retrieved February 2, 2020 from https://jibo.com/

Kenny, A. (1963). *Action*. Routledge.

Lawton, L. (2017). Taken by the Tamagotchi: How a toy changed the perspective on mobile technology. *The IJournal: Graduate Student Journal of the Faculty of Information, 2*(2), 1–8.

Levy, D. (2008). *Love and sex with robots: The evolution of human-robot relationships.* Harper Perennial.

Lin, P., Abney, K., & Bekey, G. A. (eds.). (2012). *Robot ethics: The ethical and social implications of robotics.* MIT Press.

Lin, P., Abney, K., & Jenkins, R. (2017). *Robot ethics 2.0: From autonomous cars to artificial intelligence.* Oxford University Press.

Loh, J. (2019). *Roboterethik: Eine Einführung.* Suhrkamp.

Luhmann, N. (1982). *Liebe als Passion. Zur Codierung von Intimität.* Suhrkamp.

Minsky, M. (1985). *The society of the mind.* Simon & Schuster.

Misselhorn, C., et al. (2013). Ethical considerations regarding the use of social robots in the fourth age. *Geropsych the Journal of Gerontopsychology and Geriatric Psychology, 26*, 121–133.

Misselhorn, C. (2009). Empathy with inanimate objects and the uncanny valley. *Minds and Machines, 19*, 345–359.

Misselhorn C. (2015). Collective agency and cooperation in natural and artificial systems. In C. Misselhorn (ed.), *Collective agency and cooperation in natural and artificial systems*. Springer.

Misselhorn, C. (2018). *Grundfragen der Maschinenethik.* Reclam, 5th ed. 2022.

Misselhorn, C. (2019). Digitale Rechtssubjekte, Handlungsfähigkeit und Verantwortung aus philosophischer Sicht, *VerfBlog*, 2019/10/02. Retrieved December 12, 2020 from https://verfassungsblog.de/digitale-rechtssubjektehandlungsfaehigkeit-und-verantwortung-aus-philosophischer-sicht/

Misselhorn, C. (2021). *Künstliche Intelligenz und Empathie. Vom Leben mit Emotionserkennung, Sexrobotern & Co.* Reclam.

Misselhorn, C. (2022). Artificial moral agents. In S. Vöneky, P. Kellmeyer, O. Müller, & W. Burgard (eds.), *The Cambridge handbook of responsible artificial intelligence: Interdisciplinary perspectives*. Cambridge University Press.

Nakadoi, Y. (2017). Usefulness of animal type robot assisted therapy for autism spectrum disorder in the child and adolescent psychiatric ward. In M. Otake, S. Kurahashi, Y. Ota, K. Satoh, & D. Bekki (Eds.), *New frontiers in artificial intelligence* (pp. 478–482). Springer.

Neu, J. (2000). *A tear is an intellectual thing: The meanings of emotion*. Oxford University Press.

Nussbaum, M. C. (2001). *Upheavals of thought: The intelligence of emotions*. Cambridge University Press.

O'Rourke, A. (1998). Caring about virtual pets: An ethical interpretation of Tamagotchi. *Animal Issues, 2*(1), 1–20.

O'Connor, C. (2016). The evolution of guilt: A model-based approach. *Philosophy of Science, 83*(5), 897–908.

Overall, C. (2017). *Pets and people*. Oxford University Press.

Pandey, A. K., & Gelin, R. (2018). A mass-produced sociable humanoid robot: Pepper: The first machine of its kind. *IEEE Robotics Automation Magazine, 25*(3), 40–48.

Poljanšek, T., & Störzinger, T. (2020). Of waiters, robots, and friends. Functional social interaction vs. close interhuman relationships. In M. Nørskov et al (eds.), *Culturally sustainable social robotics* (pp. 68–77). IOS Press.

Prinz, J. (2004). *Gut reactions*. Oxford University Press.

Ross, D., & Dumouchel, P. (2004). Emotions as strategic signals. *Rationality and Society, 16*(3), 251–286.

Scarantino, A., & de Sousa, R. (2018). Emotion. In E. N. Zalta (ed.), *The Stanford encyclopedia of philosophy*. Retrieved February 20, 2021 from https://plato.stanford.edu/archives/win2018/entries/emotion

Schröder, W., et al. (2021). Robots and rights: Reviewing recent positions in legal philosophy and ethics. In J. von Braun (Ed.), *Robotics, AI, and humanity* (pp. 191–201). Springer.

Seibt, J. (2018). Classifying forms and modes of co- working in the ontology of asymmetric social interactions (OASIS). In M. Coeckelbergh, J. Loh, M. Funk, J. Seibt, & M. Nørskov (eds)., *Envisioning robots in society: Power, politics, and public space: proceedings of Robophilosophy 2018/TRANSOR 2018, February 14-17, 2018, University of Vienna, Austria*. IOS Press.

Shibata, T. (1996). Artificial emotional creature project to intelligent systems. *J. Robotics Mechatronics, 8*(4), 392–393.

Silver, D., Huang, A., Maddison, C. J., Guez, A., Sifre, L., van den Driessche, G., Schrittwieser, J., Antonoglou, I., Panneershelvam, V., Lanctot, M., Dieleman, S., Grewe, D., Nham, J., Kalchbrenner, N., Sutskever, I., Lillicrap, T., Leach, M., Kavukcuoglu, K., Graepel, T., & Hassabis, D. (2016). Mastering the game of go with deep neural networks and tree search. *Nature, 529*, 484–503.

Simmel, M., & Heider, F. (1944). An experimental study of apparent behavior. *American Journal of Psychology, 57*(2), 243–259.

Solomon, R. C. (1980). Emotions and choice. In A. O. Rorty (Ed.), *Explaining emotions* (pp. 251–281). University of California Press.

Turkle, S. (2006). A nascent robotics culture: New complicities for companionship. *AAAI Workshop Technical Report WS, 6*(9), 51–60.

Turkle, S. (2010). In good company? On the threshold of robotic Companions. In Y. Wilks (Ed.), *Natural language processing 8* (pp. 3–10). Benjamins.

Van Kleef, G. A., Van Doorn, E. A., Heerdink, M. W., & Koning, L. F. (2011). Emotion is for influence. *European Review of Social Psychology, 22*(1), 114–163.

Wallach, W. (2010). *Moral machines: Teaching robots right from wrong: Teaching robots right from wrong.* Oxford University Press.

Wilks, Y. (ed.). (2010). *Close engagements with artificial companions: Key social, psychological, ethical and design issues.* Benjamins.

Winnicott, D. (1971). *Playing and reality.* Basic Books.

Zoetic AI. (2020). Retrieved December 12, 2020 from https://www.kiki.ai

Catrin Misselhorn is Full Professor of Philosophy at Georg-August University Göttingen; 2012–2019 she was Chair for Philosophy of Science and Technology at the University of Stuttgart; 2001–2011 she taught philosophy at the University of Tübingen, Humboldt University Berlin and the University of Zurich; 2007–2008 she was Feodor Lynen Fellow at the Center of Affective Sciences in Geneva as well as at the Collège de France and the CNRS Institute Jean Nicod in Paris. Her research areas are the philosophy of AI, robot and machine ethics.

Tom Poljanšek is a postdoctoral researcher associated to the Chair of Prof. Dr. Catrin Misselhorn at Georg-August-University Göttingen. His research areas include phenomenology, social ontology, philosophy of technology, and aesthetics. His dissertation "Realität und Wirklichkeit. Zur Ontologie geteilter Welten" (2022) develops a dualistic theory of the relation between mind-independent reality and subject-dependent worlds of appearance and presents an explanation for the diversity and multiplicity of the latter.Tobias Störzinger is a postdoctoral researcher associated to the Chair of Prof. Dr. Catrin Misselhorn at Georg-August University Göttingen. His research areas are social ontology, social philosophy, and philosophy of technology. From 2019–2021 he was a research assistant in the research project "GINA", in which research was conducted on the social and ethical challenges of social robotics. His PhD project "Forms of Collective Agency and Metacognition in Distributed Systems", successfully completed in 2021, addresses the question to what extent and under which conditions socio-technical systems can be understood as intelligent agents.

Theoretical Foundations of Artificial Emotions

Emotion Components and Understanding in Humans and Machines

Jacqueline Bellon

Abstract

Part I of this chapter deals with conceptualizing emotion. By splitting up 'emotion' into an evaluative, expressive, behavioural, physiological, mental and phenomenological component, giving examples of emotion research theories and state-of-the-art technical systems, I will bring together an analytical and conceptual with empirical approaches to human-machine interaction to evaluate to what extent it may be logically appropriate to speak of 'emotional machines'. I will give an overview of the mentioned components and their potential to be implemented into technical systems through functional equivalents. This concerns the technical recognition of emotion components in humans as well as the technical simulation/emulation of emotion components. Part II deals with human-machine relations and argues that human understanding of technical systems is crucial to a well-functioning society. I emphasise the importance of an in-depth human understanding of technical systems and plead for the integration of the voices and work of so-called *mechanologists* into cultural practices and discourse, from which the entire societal setup in which human-machine interaction takes place will benefit. Throughout the chapter I refer to some aspects of classical debates in psychology and philosophy on the relations of emotion, cognition, and consciousness. Additionally, I draw from psychologist and philosopher Gilbert Simondon's, as well as from gestalt theorist Kurt Lewin's writings.

J. Bellon (✉)
Eberhard Karls Universität, Tübingen, Deutschland
e-mail: jacqueline.bellon@uni-tuebingen.de

© Springer Fachmedien Wiesbaden GmbH, part of Springer Nature 2023
C. Misselhorn et al. (eds.), *Emotional Machines*, Technikzukünfte,
Wissenschaft und Gesellschaft / Futures of Technology, Science and Society,
https://doi.org/10.1007/978-3-658-37641-3_2

Keywords

Emotion components · Understanding · Machine behaviour · Mechanology · Human-Technology Relations

1 What Do We Feel When We Feel?

> Here are some of the phrases we find when dictionaries define emotion. The subjective experience of a strong feeling.
> A state of mental agitation or disturbance.
> A mental reaction involving the state of one's body.
> A subjective rather than conscious affection.
> The parts of consciousness that involve feeling.
> A nonrational aspect of reasoning.
> If you didn't yet know what emotions are, you certainly wouldn't learn much from this. (Minsky, 2006, p. 17)

There is a prototypical fictional dialogue in Jean-Marc Fellous and Michaels Arbib's *Who needs emotions?* (2005, chapter 1) on how the activities of defining concepts thoroughly and doing empirical research may collide and hinder each other's execution. In this, a prototypical philosopher, channelled by a fictional character *Russell*, argues that we need definitions of terms such as "emotion," before and while we work on 'emotional machines.' The prototypical engineer *Edison* argues that from a designer's and programmer's perspective we merely need "working definitions that the engineer can use to get on with his work rather than definitions that constrain the field of research" (Fellous & Arbib, 2005, p. 4). Both dialogues are echoed in many scientific fields[1] and echoing voices of both prototypical characters' perspectives do not only repeatedly provoke scientific debates; they also both have valid points. Luckily, for the context of this contribution, we do not need to decide for or against one of their perspectives, but will

---

[1] Just one example: In a contribution to the 1975 *American Journal of Psychiatry* psychiatrist Richard Ketai tries to help define the terms "emotion", "feeling", "affect" and "mood" by comparing actual uses of the terms in psychiatrists' writings (Ketai, 1975), to which another practitioner, Paul Kesbab, answers in a letter to the editor and receives a response from the author (Ketai, 1976). Ketai more or less argues that words are important, while

profit from both. Our aim in looking at some emotion definitions and components, as well as in referring to empirical research, is not that we wish to be able to judge if the given definitions are right, wrong, or sufficiently good enough to work with empirically, but to evaluate to what extent it can be logically appropriate to speak of 'emotional machines'.

Part I of this chapter deals with conceptualising emotion, specifically with evaluative, expressive, behavioural, physiological, mental and phenomenological components of emotions and their implementability into technical systems. Part II deals with human-machine relations, the question which emotions may be appropriate for a human to have towards a technical system, and argues that human understanding of technical systems is, especially today in an age reigned by a paradigm of information, crucial to a well-functioning society.

2 Part I: Conceptualising Emotionality

Affect research, psychology, the cognitive sciences, philosophy and computer sciences try to explain and define emotions.[2] Before emotions started to be considered "a category of mental states that might be systematically studied [in the] mid-19 century" (Dixon, 2012, p. 338), philosophy provided emotion theories, virtue theories, personality models, and considerations on terms such

---

Kesbab emphazises the importance of empirical work and warns that in their field one tends „to get caught up in semantics, arguing about words and descriptive terminology rather than focusing our attention on scientifically documented facts" (ibid, p. 347).

[2] Some basic terminological definitions: *Affect* is, in psychology, usually an umbrella term subsuming emotions, feelings, and moods, or, at least, some of them, although the term sometimes denotes the more passive part of psychological processes. *Emotions* are mostly defined as "being caused by an identifiable source, such as an event or seeing emotions in other people [...] and directed at a specific object or person" Bartneck et al., (2020, p. 115), potentially including a cognitive element. Emotion is seen as a behaviour guiding and decision-making force, including *motivation* and *intentionality;* for an older review of emotion definitions see Kleinginna and Kleinginna (1981). *Moods* are mostly defined by their duration being longer than that of affective or emotional states, see Beedie et al. (2005). *Feelings* are not defined very well. Sometimes the word refers to the phenomenologically accessible. In that case, there is, in psychology, a debate on the question if the so called *qualia*, see Tye (2016), come before, after or simultaneously with bodily responses, see James (1884); Scarantino and Sousa (2016).

as "*passion, sentiment, affection, affect, disturbance, movement, perturbation, upheaval,* or *appetite*" (Scarantino & de Sousa, 2016, n.p.). Philosophical analysis has mostly been interested in

> *Differentiation*: How are emotions different from one another, and from things that are not emotions? *Motivation*: Do emotions motivate behavior, and if so how? *Intentionality*: Do emotions have object-directedness, and if so can they be appropriate or inappropriate to their objects? *Phenomenology*: Do emotions always involve subjective experiences, and if so of what kind?" (Scarantino & de Sousa, 2016, n.p.)

Cognitive sciences, psychology and therein, affect research, are interested in the same questions, however, they use different methods to answer them. They add to the list at least *Measurability and Materiality*: How are emotions measurable? How do emotions and body relate, how can emotions be located or made visible? *Functionality*: What purposes do emotions serve, for example evolutionary, personally or socially? *Agency*: Are emotions just there or do we produce them? If so, how?

Takes on these topics can be found implicitly in terminological decisions, they manifest in empirical research and go back and forth in debates across disciplinary borders. For example, the distinction between emotions as something we are *affected by* more passively, and as something we have a more active part in, has been outlined by Aristotle,[3] provoked, up until today, long lasting debates in philosophy (Schmitter, 2016a; Zaborowski, 2018) and psychology (Lazarus, 1984; Zajonc, 1984) and was intensively researched empirically among the neurosciences.[4] Thereby the broader dichotomy of 'activity/passivity' in thinking about emotions specialised into increasingly concrete questions such as *Do we just 'suffer' from emotional or affective states or do we control them on subpersonal and on conscious levels? Do we cognitively notice our emotions? If so, are there different parts of the nervous system to process emotional information? Is the amyg-*

---

[3] Aristotle proposed that emotions (*pathe*) originate in human desires, or appetites; they are closely connected to our decision-making and our actions; and, *we can be overwhelmed by them and need to evaluate, on a meta level, if we should indeed follow the impulses pathe give us*. This implies a distinction between a more spontaneous and a more willful part of *pathe*; cf Schmitter (2016b).

[4] See Adolphs and Damasio (2001); Goldsmith et al. (2003); Carroll E. Izard (2009); Panksepp (2003); Pizzagalli et al. (2003). On the neuroscientific endeavor of locating the "high road" and the "low road" of stimuli processing see Gelder et al. (2011); Pessoa and Adolphs (2010); Tamietto and Gelder (2010). This problem may be solved; see Lai et al. (2012).

dala involved, or, not as involved as we thought? Or, for the interpersonal level of emotionality, as researched, for example, in Social Cognition: *Are affects contagious? Do we learn to have them from others? Are there collective affects? How do we attune and attach to one another? What's with mirror neurons?* Aspects of the active/passive dichotomy do not only specialise within disciplines, they also generalise among disciplines and are interconnected. For example, the dichotomy of active/passive in emotion plays a role in: —Psychotherapy: *How is it that we can learn to change our feelings towards something? Do we need to change our cognitive appraisal? Can we? Are there appropriate and inappropriate emotions? Should we try to shut down inappropriate emotions chemically?*—Linguistics: *What happens in language with regard to emotion verbs' lexical dependency relations, i.e. their valency, thematical topics, semantic types, roles, frame-frame relations? Who is the agent, who the recipient of an emotion?*—Technical systems development and programming: *How are linguistic corpora including such information used to make technical systems understand how to talk about emotion?* — Philosophy: *What does understanding mean here?* etc. In the end, new fields of research emerge with different approaches, methods and literature they refer to.[5]

Instead of arranging research questions on emotions thematically or according to their research field, we can also try to break down emotion into its components. This may help to see to what extent it could be logically appropriate to speak about 'emotional machines'.

> Consider an episode of intense fear due to the sudden appearance of a grizzly bear on your path while hiking. At first blush, we can distinguish in the complex event that is fear an *evaluative* component (e.g., appraising the bear as dangerous), a *physiological* component (e.g., increased heart rate and blood pressure), a *phenomenological* component (e.g., an unpleasant feeling), an *expressive* component (e.g., upper eyelids raised, jaw dropped open, lips stretched horizontally), a *behavioral* component (e.g., a tendency to flee), and a *mental* component (e.g., focusing attention). (Scarantino & Sousa, 2016)

To explore if machines can 'have' emotions, we will go thoroughly through and discuss each of these components and evaluate their potential for implementation into technical systems with regard to machines recognizing emotions in humans, as well as concerning functional equivalents machines may have, simulate or

---

[5] For example, *Affect Theory, Affective Computing,* and *Affect Studies* all explore affect, but from very different points of view.

emulate, or, have the potential to have, simulate or emulate. Going through the emotion components serves the purpose of sketching what aspects might be relevant to a conceptualisation of what the term "emotion" entails, hopefully to the likes of a reader who is inclined to argue like prototypical *Russell* above. On the other hand, and maybe more to the likes of a reader more in agreement with the *Edison* point of view, I will include examples of state-of-the-art technology to show what is already been done or what could possibly be the closest structure to functional equivalents to emotion components for technical systems. Please note that by the time this chapter will be published, of course, new methods will have been developed.

2.1 The Expressive Component of Emotion

Some living beings express their emotions in perceptible ways to other specimen (Ferretti & Papaleo, 2019). For humans, this helps coordinate behaviour on a personal, interpersonal and on a societal level. Emotions do not only guide our own behaviour, but an expression of emotions "can also help us modulate the behaviours of others in an interaction" (Bartneck et al., 2020, p. 115), for example, when we signalise that we are hurt, about to become angry, or signal that we would like to socialise. On a personal level, expressing emotions strengthens close relationships (Planalp et al., 2006) and helps attuning to one another (Gallese, 2006). From a perspective of game theory we could say that emotion expression helps with signalling what values to assign to available behaviour options in coordination games (De Freitas et al., 2019; O'Connor, 2016). Thereby, emotion expression can help with coordinating different personal needs and with logistical coordination on a societal level. Emotion expression can lead to, but also avoid violent conflict, and, signal dangers or times to relax and enjoy to others.

In order to make human–human interactions successful, it is therefore important for human agents to be able to 'read' the emotional states of interaction partners. In this process, features such as for example facial expressions (Fernández-Dols & Russell, 2017), posture (Scheflen, 2016), pitch and volume of the voice, delays and pauses in spoken words, the choice of spoken words or written language (Locher & Watts, 2005), familiarity with a person (Cappuccio, 2014; Feyereisen, 1994; Vonk & Heiser, 1991), and many other indicators help us to infer from our interpretation of them our interaction partner's emotional states.

Technical emotion recognition systems can recognize social signals and infer expressed emotions, too (Vinciarelli et al., 2017). Moreover, they can infer emotions and even personality traits from, for example, voice analysis (Deng et al., 2017;

Sagha et al., 2017; Schroder et al., 2015), or be controlled via human expressive behaviour, such as gestures (Obaid et al., 2014). For an overview on methods, sensors, and emotional models used in emotion recognition see e.g. Cai et al., (2023). Additionally, emotions can be inferred from text analysis (see e.g. Murthy & Kumar, 2021). While emotion recognition in some cases works surprisingly well, it can naturally also go wrong and can be 'hacked', for example, by doing adversarial attacks (i.e., a goal-oriented manipulation of classification results; cf. Bellon 2020). One such example of an adversarial attack would be to add characters to a text on which a system performs sentiment analysis (Li et al., 2019). Furthermore, technical systems are designed to display signs of emotion similar to social signals in living beings for humans to read (Breazeal, 2004; Nitsch & Popp, 2014; Salem & Dautenhahn, 2017). Here, it would be more adequate, instead of speaking of 'emotion expression,' to speak of emotionally impressive features in technical systems, as they impress and affect human agents. However, as literature uses the terminology of 'emotion expression' we will reluctantly follow. Emotionally expressive behaviours and characteristics of technical systems have effects on humans in human–machine interaction, for example a robot's gestures (Embgen et al., 2012; Salem et al., 2013), speed and path predictability (Koppenborg et al., 2017), robot group size (Podevijn et al., 2016), and so on. Technical systems might even be used as mediators in expressive human–human interaction, for example, to translate the nonverbal meanings of culturally different human gesture behaviour for another human (Hasler et al., 2017). Most research on emotion expression in technical systems—in this context usually robots and digital avatars—is based on six emotions displayed as facial expressions, which are supposed to be universally recognizable by humans across cultures (Ekman & Friesen, 1971). However, it is disputed how well emotion recognition by inferring emotions from facial expressions actually works (Barrett et al., 2019) and so is the underlying psychological theory.

As humans have a tendency to anthropomorphize whatever they encounter (Marquardt, 2017), they quickly ascribe social agency to technical systems or even just to shapes (Wang & Zhang, 2016) and movements (Strohmeier et al., 2016). Thus, "[e]ven if a robot has not explicitly been designed to express emotions, users may still interpret the robot's behavior as if it had been motivated by emotional states." (Bartneck et al., 2020, p. 118) With Large Language Models and Chatbots such as Bing, which had, in its beta phase, been using many emojis and quite rude, emotionally expressive language (Zimmermann 2023; Microsoft 2022; see also Bellon 2023), people may be even more inclined to ascribe actual emotions (or even sentience, see e.g. Lemoine 2022) to technical systems, or, at least, feel justified in having emotions towards them themselves, as is the case with users of Replika (see e.g. r/replika 2023).

- IN CONCLUSION CONCERNING THE EXPRESSIVE COMPONENT OF EMOTION: We can and do already build technical systems to recognize emotions in humans and we can and do implement 'expressive', or rather emotionally impressive social signals into technical systems. Of course, that does not mean technical systems 'have' emotions, but given the right circumstances they may recognise them in humans and may nudge humans into having them. In other words: social signals such as emotion cues produced by technical systems do not actually express anything related to emotions, but are interpreted by humans or other living beings as meaingful, i.e. they may impress them.

2.2 The Evaluative Component of Emotion

Evaluation presupposes values as well as operational criteria and serves the purpose of orientation (cf. Stegmaier, 2008), although orientation does not necessarily need (conscious) evaluation. Evaluating is anchored in the existence of a range: only if there is some kinde of scale can we evaluate (or, only if we evaluate, there will be some kind of scale, cf. Deleuze 1983, p. 1), regardless of whether all points on the scale can or will 'manifest' in reality. In the hiking episode mentioned above, the bear is evaluated as dangerous. Such an evaluation comes to a human quite naturally, although it needs preconditions such as a body, senses, experience, world knowledge, and above all: some kind of scale on which the evaluated entity is being placed on. Scales of evaluation include more dimensions than just the binary judgement of 'dangerous/harmless.' Aside from these, arguably involuntary prepersonal biological evaluations (bear → dangerous → fear[6]), humans evaluate things morally, aesthetically, socially, culturally, situatively, and with regard to a variety of different aspects such as quality, suitability, significance, and so on. The process of undergoing socialisation gives us temporary orientation on how to evaluate and feel towards social situations, for example when children learn social norms (Hardecker & Tomasello, 2017). However, social 'rules' are adaptable and versatile. Aspects and factors influencing what humans evaluate as approriate or inappropriate are manifold (Bellon, Gransche, & Nähr-Wagener 2022a, 2022b). This holds true for both, the evaluation (of something 'as' something) involved in the process of developing or having an emotion (towards something), as well as in evaluating the appropriateness of hav-

---

[6]This is an aspect of what the Zajonc-Lazarus debate mentioned above has been about, namely, the relation between emotion and cognition: Is there a cognitive appraisal of the bear, or, of the fear? Is the fear of the bear affective and the 'data' "bear" processed faster than other percepts? Is recognizing the bear structurally the same as evaluating it as dangerous?

ing the emotion, i.e. concerning the evaluation of the evaluative component (of something 'as' something) involved in developing the emotion.

How complicated the whole process of evaluation is, becomes apparent when trying to give a technical system the capacity to evaluate situations or entities in a way similar to humans.[7] To think about implementing an evaluative component of emotion into technical systems, we would need to differentiate between evaluating a) physical occurrences and situations, b) evaluating data, and c) what the evaluative process should lead to.

a) If the system is supposed to evaluate a physical situation x 'as' y, it needs, at least, sensors to perceive its surroundings, an ontology or another knowledge base from which to infer, for example, labels to entities and to recognise situations, and, some kind(s) of evaluation scale(s) based on operational criteria. If it is supposed to 'act'—in the sense of displaying cues of emotion for a human to read or of performing some kind of behaviour — it needs at least a *set of behaviour sequences* and a *decision-making structure* to evaluate the appropriateness of performing a certain behavioural sequence. For an exploration of what exactly would be needed to evaluate, for example, the appropriateness of the behaviour sequence "helping someone put their coat on", see e.g. Dietrich Busse (2022). Concerning the evaluation of a situation in the sense of moral decision making (i.e. evaluate x 'as' ethically good or bad), we might be sceptical with regard to the potential of a technical system to perform ethical analysis of moral decision making. Nonetheless, approaches to modelling hybrid ethical reasoning agents exist (Lindner et al., 2017).

b) If the system is supposed to evaluate data, for example, in performing sentiment analysis on a text (Li et al., 2019) or in evaluating if an image is generated or real, as a discriminator in a generative adversarial network does (Goodfellow, Pouget-Abadie, et al., 2014a, 2014b), it does not need sensors, but, for these examples, machine learning methods, or, for other examples, some kind of set of rules or annotation of how to infer from data an evaluation of it.

c) As a technical system has no biologically driven intentions (to survive, to avoid pain, etc.) and no concept of a 'good life', or other inbred systems of orientation, to evaluate a situation or entity x 'as' y and subsequently display the appropriate emotion simulation z_1 or correlate and/or recognise a human's emotion z_2, we would need to implement some kind of teleology (teleology referring to a variety of possible ways in which we 'incentivise' a system or give it a task/goal).

---

[7] For a framework on evaluating the social appropriateness of behaviour sequences see for example Bellon, Eyssel, et al. (2021a, b).

Concerning human or other living beings' emotions, we can argue that evaluation of a situation, for example, as a dangerous one, is 'hard-wired', i.e. involuntary, inevitable and affective, or, that we can, at least gradually, control and influence evaluations of entities consciously. No matter if we conceive of the evaluative component of emotions as a consciously shapeable or an instinctive, automatised process, one question that has arisen in philosophy, psychology and the neuroscience is whether there is some kind of evaluation center in our minds. Sometimes the metaphor of a homunculus (Baltzer-Jaray, 2019; Thomas, 2016) is used to illustrate the idea that there is an entity (literally a 'little human') in our mind/brain that evaluates the data collected by the senses. Although 'homunculus theories' are controversially discussed (Margolis, 1980), the idea made it to technical system development (Gama & Hoffmann, 2019) through the neurosciences (Catani, 2017; Penfield & Boldrey, 1937). *Is the way in which the human body transfers the outer world to an inner percept different from a technical system's way to transform its input into digital computation?* In the end, this is up to decide for researchers according to the purpose of their exploration and their beliefs. We will come back to this question in chapter 2.4 and 2.5 on the physiological and mental components of emotion.

- IN CONLCUSION CONCERNING THE EVALUATIVE COMPONENT OF EMOTION: Depending on what is supposed to be evaluated, it may be possible to implement functional equivalents of the evaluative component of emotion into technical systems. However, this does not mean that they evaluate according to an instrinsic motivation, but according to a given set of rules/values or another given teleology/incentive. The question of whether 'intrinsic' motivation as a mechanism for processes of emotion-related evaluation is exclusively available to living beings — and, therefore, non-living beings cannot be talked about as 'having emotions' — will come up again in chapter 2.4 and 2.5 with regard to the physiological and mental components of emotion and will be discussed there in more detail. For now, let's end with another question: if teleologically motivated values (instrinsic or not) govern the evaluative component of emotion, and, if all components are necessarily needed for full-fledged emotion, does this also mean that where there is no teleology in living beings, there will be no emotions?

2.3   The Behavioural Component of Emotion

Emotions as (evaluation-guided) "responses of the brain and body to threats and opportunities" (Jeon, 2017, p. 4) motivate and guide human behaviour. They are conceived of to serve evolutional functionality and the ability to act and behave, for example, when fear of a bear leads us to react to the situation to make sure we are safe. Conversely, as the bear example above suggests, from certain behaviours (*tendency to flee*) it may be possible to infer that a person is in an certain emotional state (*fear*). We may say that we think someone is sad, because we see them cry or that someone is happy because we see them laugh. We may reason that their crying or laughing is rooted in them feeling sad or happy. On the other hand, William James (1884) for example, argues that the causal relation is the other way around: first the body reacts to a situation, then, us noticing our body react manifests in *feeling* and behaving a certain way. In other words: according to this view, the physiological component of emotions governs the phenomenal and the behavioural components of emotion.

> "Common sense says, we lose our fortune, are sorry and weep; we meet a bear, are frightened and run; we are insulted by a rival, are angry and strike. The hypothesis here to be defended says that this order of sequence is incorrect, that the one mental state is not immediately induced by the other, that the bodily manifestations must first be interposed between, and that the more rational statement is that we feel sorry because we cry, angry because we strike, afraid because we tremble, and not that we cry, strike, or tremble, because we are sorry, angry, or fearful, as the case may be. Without the bodily states following on the perception, the latter would be purely cognitive in form, pale, colourless, destitute of emotional warmth. We might then see the bear, and judge it best to run, receive the insult and deem it right to strike, but we could not actually feel afraid or angry." (James 1884)

This view has been challenged and refined in psychological emotion theory (e.g. Robinson, 1998; Barrett, 2017).

However, no matter what view we deem valid with regard to the chronological and logical order of emotion processes leading to a certain behaviour, and also leaving aside the widely discussed and multifaceted notion of agency of technical (Verbeek, 2005; Nyholm, 2020) and socio-technical systems, as well as countless theories of action—in many of which, the role of emotions is overlooked (cf. Piñeros Glasscock & Tenenbaum, 2023), we can certainly say that simulation of behaviour and actual action sequences can be implemented into technical systems. Technical systems can and do perform intended and may also perform unindented behaviour and action sequences. If a human ascribes them to an emotional state is another matter.

- IN CONCLUSION CONCERNING THE BEHAVIOURAL COMPONENT OF EMOTION: Technical systems have a functional equivalent of the human behavioural component of emotion. However, if we take seriously the variety of functions of emotional behaviour in close relationships (Planalp, 2003), these tremendously exceed the functions of the technical systems' functional equivalent of the human behavioural component of emotion.

2.4 The Physiological Component of Emotion

Much has been said in the history of philosophy and psychology on the physiological component of emotion. With regard to the thought experiment of implementing a physiological component of emotion into a technical system, some argue that the latter might need to be "situated" (see e.g. Adolphs, 2005). More significantly, it has been stated that, to be full-fledged situated, a technical system might even have to be a duplication of an organism (cf. Searle, 1980) and, would therefore *be* an organism and not a technical system anymore.

Concerning the relations between the materiality of technical systems and their emotional potentiality, in the following I will contrast two prototypical lines of argumentation, which will be represented by John Searle and Marvin Minsky. For this discussion, we here do not refer to all parts of a physical body a technical systems has and humans have—for example, we do not explore functional equivalents of blood vessels (heart rate, blood pressure as signs of emotions in humans) or human skin (Pugach et al., 2015), but, with regard to the later mentioned *mental* component of emotion focus on a debate on the materiality of the brain (which seems to be used *pars pro toto* for 'living being' or 'organism') concerning the question: *can you separate hardware (brain, physiological component) from software (mind, mental component)?* Subsequently we will try to redirect the reader's attention to a theory that adds to the classical argumentation lines of substantialist and functionalist argumentation another perspective from the point of view of individuation theory.

The debate about emotions or feelings is closely linked to the debate about consciousness in general. John Searle argued that technical systems cannot have mental states (or intentionality) mainly because mental states are directly linked to the biological, physical and chemical properties, i.e. the materiality, "of actual human brains" (Searle, 1980, p. 423). He argues that most "AI workers" (ibid.) must think that "the mind is separable from the brain both conceptually and empirically" (ibid.) and therefore believe in some kind of strong dualism, at least where programs are intended to somehow reproduce the mental. To him though, it is obvious that mental operations, such as 'having' emotions, are not identical to computational operations on formal symbols and that the brain is not "one of the indefinitely many types of machines capable of instantiating the program" (ibid., p. 424) of the human mind, but that a brain is the only entity being able to "produce intentionality", which is, to him, the differentiating marker between conscious and non-conscious beings. To him, while he would agree that the human brain 'computes', hardware (brain) and software (mind) cannot be separated and 'mind' cannot be achieved by any other system 'just running the program':

> [o]f course, the human brain is a digital computer. Since everything is a digital computer, brains are too. The point is that the brain's causal capacity to produce intentionality cannot consist in its instantiating a computer program, since for any program you like it is possible for something to instantiate that program and still not have any mental states. Whatever it is that the brain does to produce intentionality, it cannot consist in instantiating a program since no program, by itself, is sufficient for intentionality. (Searle, 1980, p. 424)

While Searle accuses others of dualism, he certainly still argues from a substantialist point of view himself, referring, following James (1884), to whatever it is that produces intentionality as some kind of "stuff", which is missing in a technical system running operations on formal symbols. Intentionality, and therefore, emotionality is, from this point of view, nothing a technical system can ever achieve, unless the system *is* physically the same as that which it tries to simulate.

Contrary to this, Marvin Minsky states that he "can't see why Searle is so opposed to the idea that a *really* big pile of junk might have feelings like ours." (Minsky in Searle, 1980, p. 440) He points out that "it is widely recognized that behavior of a complex machine depends only on how its parts interact, but not on the 'stuff' of which they are made (except for matters of speed and strength)" (Minsky, 2006, p. 22). To him, "all that matters is the manner in which each part reacts to the other parts to which it is connected." (ibid.) With regard to technical systems, this would include that "we can build computers that behave in identical ways, no matter if they consist of electronic chips or of wood and paper clips—provided that their parts perform the same processes, so far as the other parts can see." (ibid.) "*What is our intellectual form?* is the question, not what the matter is." (Minsky in Searle, 1980, 440, referencing Putnam, 1975, p. 302) According to Minsky, we should not be speaking of matter, understanding, and intentionality or meaning at all. Instead of looking for some mind-substance, analogous to vitalist biologists and philosophers who tried to find or establish an *élan vital*, we should, just as biologists who switched from speaking about 'reproduction' to speaking about 'encoding,' 'translation,' and 'recombination,' speak about the many single steps and *empirically accessible parts* making intentionality, understanding or meaning possible. To him, there is no such thing of interest as a physical unity made of mind and brain, or of any things. What is of interest is empirical research exploring the single components involved in what we call, lacking better words, "meaning," "intending" or "understanding." From Minsky's functionalist viewpoint, it is obvious that a program, such as 'having emotions', that can be run on one specific kind of hardware (living organism) can be run on any hardware, as long as that hardware is functionally equivalent. As Minsky thinks that a "nonorganic machine may have the same kinds of experience as people do" (Minsky in Searle, 1980, p. 449), he seems to have no problem in ascribing to technical systems the potential to have full-fledged human emotions. Now, what do we make of these two conflicting viewpoints?

The debate may find a well elaborated addition in Gilbert Simondon's analysis in *On the mode of existence of technical objects* (Simondon, 2017) and in his *allagmatics* (Simondon, 2020a, 2020b). Gilbert Simondon in his works tries to overcome the dichotomy between matter and form, and proposes a variety of concepts and terminology to avoid it.[8] He develops a point of view from which materialism/physicalism and functionalism are equally valid with regard to technical systems. With regard to the Searle vs. Minsky debate, I want to direct attention to his notions of *absolute origins* and *pure schema of functioning* (Simondon, 2017, p. 45). In essence, Simondon sees the discovery of *functional synergies* as the driving force of technical progress (Simondon, 2017, p. 38). The problem he associates with an observation of technical objects' evolution is the question of "absolute origins" (Simondon, 2017, p. 44): Up to what starting point can one determine the origin of a specific technical reality? "What is the first term one can attribute to the birth of a specific technical reality?" (ibid.) He answers: we can distinguish between a "definite genus, which has its own historical existence" (Simondon, 2017, p. 45) and has its own materiality, and *pure schemata of functioning* which are "transposable to other structures" (ibid.). An example for a pure schema of functioning is that of semi-conductors, in which "the schema of functioning is the same to such an extent that one can indicate a diode on a theoretical schematization with a sign (asymmetrical conductance: ⟶) that does not predetermine the type of diode employed, leaving a freedom of choice to the manufacturer." (Simondon, 2017, p. 45) To put it bluntly, if we were to build technical

---

[8] See e.g. Barthélémy (2013).

equivalents of human brains or of living organism and had a theoretical schematization plan to do so, in this plan, there could be signs for certain functionalities for which it wouldn't matter what exactly an engineer would use as hardware, as long as the technical element fulfills the function. However, it might very well be the case that there are other structures where the material and function have individuated mutually in a way that makes it impossible to transfer the function to another material. Simondon on the one hand highlights that there are no vital processes separated from matter; to him life itself supposes processes of integration and differentiation that can only be given in the form of physical structures.[9] On the other hand, he emphasises that technical objects must not be confused with living systems: "one must avoid the improper identification of the technical object with the natural object and more specifically with the living being" (Simondon, 2017, p. 50); technical objects are always *produced*, not *engendered* (Simondon, 2017, p. 68), and always exist because of an act of human invention (Simondon, 2017, e.g. 62). While he states that both, living beings and technical objects, have in common that they undergo processes of individuation and evolution, he gives the example of *pure schemata of functioning* as a characteristic specific to technical systems. To think about the origin of a technical object can, besides from highlighting its origin of existence in a human act of invention, either mean looking for its pure schemata of functioning (in which it is different from other things), or, it can mean tracing back the history of the specific coming-into-being of that entity (in which it is similar to all other things as all things come into being and can be analysed 'ontogenetically'). To Simondon, the materiality of a technical system matters, not only because material can come to be better adjusted to a technical system's needs (Simondon, 2017, p. 32 f.), which is more or less in line with Minsky's claim that materiality matters with regard to a technical system's 'speed or strength' (see above). He additionally proposes that materiality also matters on the level of a technical system's individuation process, i.e. the conditions of its coming-into-being. From an ontogenetic viewpoint, every entity must be thought of from the starting point of its specific individuation, i.e., the specific coming-into-being of an individual, be it technical, physical, social or psychological. Only by taking into account the stage in which whatever we are looking at had not yet been individuated completely, we can understand its

---

[9] „La vie n'est pas une substance distincte de la matière; elle suppose des processus d'intégration et de différenciation qui ne peuvent en aucune manière être donnés par autre chose que des structures physiques." Simondon (2005, p. 162).

ontogenesis. So, in conclusion with regard to the question on physiological components of emotions: even if we could replace or reproduce structures on which 'the program of emotion runs' using functional equivalents in the form of pure schemata of functioning, we cannot build an entity such as a brain without replicating its specific coming-into-being, including the materiality of that genesis.

Aside from the question of how well we can build functional equivalents of physiological structures such as an organism or the brain, the question remains: even if we could, why would we ascribe emotions to a functional equivalent of a human brain or the functional equivalent of an organism? For example, we probably would not say the *Somnox Sleeping Robot*, a pillow mimicking human breath, is humanlike or emotional, just because it mimics breathing. So why are we so inclined to say that, for example, generative AI models mimicking language might be humanlike or a step towards 'emotional machines'? If we exchange all physiological elements involved in human emotion with functional equivalents, will we have built an emotional machine?

- IN CONCLUSION CONCERNING THE PHYSIOLOGICAL COMPONENT OF EMOTION:
Functional equivalents of the physiological component of emotion can be implemented in technical systems. Neuromorphic systems engineering does already happen (Hasler et al., 2017; Zhang et al., 2019).[10] However, it is not necessarily this specific physiological component where emotions can principally be 'stored,' for, if we follow Searle, a functional equivalent will not be the same thing as that, which enables humans to have mental components of emotion. If we follow Minsky and Putnam, functional equivalents to physiological components bear the potential to embody the mental component of emotions. As they state it, the problem is that we just do not yet have all the single steps to facilitate or explain humanlike understanding or intentionality. In short: We can implement functional equivalents of the physiological component of emotion into technical systems, although, that does not necessarily make them more emotional.

2.5 The Mental Component of Emotion

The mental component of emotion includes processes of attention, perception, decision-making and memory (Bartneck et al., 2020, chapter 8; Izard, 1993).

---

[10] Technological physiological signal recognition and interpretation of human physiology is, of course, already happening, too, see e.g. Shu et al. (2018).

Besides from the the question of how the physiology of technical systems is connected to mental components (as discussed above), the mental component of emotion can be grasped by a theory of emotions as patterns of salience and a theory of paradigm scenarios. Emotions can be conceived of as "mechanisms for changing salience" (Scarantino & Sousa, 2016) and more precisely, as "mechanisms that control the crucial factor of salience among what would otherwise be an unmanageable plethora of objects of attention, interpretations, and strategies of inference and conduct." (Sousa, 1987, xv) They direct our attention to what is really important in a given situation. What is really important in a given situation may be infered from *paradigm scenarios* (de Sousa., 1987, xv), scripts (Mandler, 1984) or schemata (Axelrod, 1973) that organisms have acquired in their lifetime. While what we focus on in any given situation partly depends on what we are used to, what we believe and remember, on the narratives we are familiar with, or tell ourselves (Goldie, 2012; Rorty, 1987), and on the kinds of situation, emotion and behaviour types we have learned exist (cf. Nishida 2005), it also depends on how we feel. In certain situations, emotions may take the lead in governing mental components such as attention, perception, and decision-making. In the hiking episode above the mental component of emotion is responsible to sharpen and focus attention in order to enable the body to react quickly. In Minsky's words "each of our major 'emotional states' results from turning certain [mental] *resources* on while turning certain others off—and thus changing some ways that our brains behave." (Minsky, 2006, p. 4) Minsky's view highlights the importance of organising ressources such as mental processes in a complex organism and he states that 'emotions' are one strategy to do so. With regard to technical systems, he argues that if we wanted to build them to "mimick our mind" (ibid.) and to be able to perform similar (problem solving) processes living beings do, we should give them a capacity to "represent something in several ways" (ibid.) and that 'emotions' as patterns of governing the use of ressource and thereby governing salience is one of those ways.

Technically, functional equivalents to attention and salience are implemented in machine learning networks' attention models and can partly be visualised as well (Bratman, 1987; Rao & Georgeff, 1995; Maaten & Hinton, 2008; Kahng et al., 2018; Wu et al., 2016; Gupta et al., 2020; Parikh et al., 2016; Vaswani et al., 2017). We could say that a machine learning model's attention focus potential lies in its latent space. The latent space is a high dimensional vector space embedding of data 'organising' 'entities' the network discriminates by similarity in ways that may be very unintuitive to the human mind. However, when training a neural network and adjusting the so-called 'weights' or parameters of the connections between its neurons, this may be the closest functional equivalent of governing its equivalent to 'mental processes'. For the end user, when prompting a generative AI model, several ways to highlight what it should 'focus on' and what should therefore manifest more saliently in the output have been implemented. For example, when prompting the text-to-picture generator Midjourney the signs "::" can be used to separate concepts in the prompt and subsequrently assign them relative importance, so that certain concepts will manifest more strongly than others.

- IN CONCLUSION CONCERNING THE MENTAL COMPONENT OF EMOTION: Machine learning models may be understood as being able to 'focus their attention' and have functional equivalents to mental components of emotion. While in living beings, the experienced emotion might correlate with or govern mental processes such as memory, attention and decision-making, e.g. about the kind of behaviour to perform, in the technical system, if a functional equivalent were implemented, it is not the biologically given intrinsic interest (e.g., to survive) that motivates this process, but the most rewarded output, or, some other kind of incentive that signals what operation is to be performed. This incentive is set somehow when giving the technical object tasks and goals and may sometimes have unintended side effects as is illustrated sometimes amusingly, sometimes dramatically in examples of specification gaming.[11] Technical systems usually are of a teleological nature (i.e. they are designed to have tasks and goals), of which humans have to make an effort to understand the inherent operations, for, in the end, humans set technical systems' teleological trajectories. They are therefore "ultimately responsible for any harm [a technical objects'] deployment might cause" (Rahwan et al., 2019, p. 483). We will come back to this later in more detail. Concerning the mental component of emotion in humans again, with theories of script theory, narratively mediated paradigm scenarios and other theories approaching emotions as patterns governing salience, we can state that sometimes "[h]umans do not [necessarily; JB] react to what is objectively the case. They act and react with respect to what they perceive or think or feel to be the case." (Graumann, 2002, p. 12) This will lead us to the phenomenological component of emotion.

---

[11] For the study of unintended incentives, see specification gaming dealing with "a [system's, JB] behaviour that satisfies the literal specification of an objective without achieving the intended outcome" (Kraknova et al., 2020).

2.6 The Phenomenological Component of Emotion

The phenomenological component of emotions concerns things as they appear to us (phenomena), how they 'feel' and how they are being 'experienced'. It has been stated that "since the mind is related to reality through experience, there are very good reasons for regarding the topic of phenomenal qualities as central to issues in the philosophy of mind, metaphysics, and epistemology." (Coates & Coleman, 2015, p. 1 f.). With regard to the differences between living beings and machines, it is usually the phenomenal content of emotional states that is conceived of as the one aspect surely missing in a machine. A living being is usually thought of to have access to their own phenomenal experience, and only their own. As the phenomenal quality of experience is a 'first-person perspective' and cannot be measured easily by instruments, some would say it is impossible to prove someone other than 'me' has it. It has also been called 'the hard problem of consciousness' (Chalmers, 1995; Dennett, 2018). There are many ways to approach 'emotions' phenomenologically (cf. Szanto & Landweer, 2020; Ferran, 2008; Fuchs, 2019; Esterbauer & Rinofner-Kreidl, 2009; Elpidorou & Freeman, 2014) and many ways to approach phenomenality or phenomenology with regard to technical objects (Du Toit & Swer, 2021; Rosenberger & Verbeek, 2017; Aagaard et al., 2018; Liberati, 2016). For the context at hand, I want to add a perspective that conceptualises phenomena as need-dependent perceptive relations of a living being and their environment while framing the phenomenological component of emotion as embedded in a living beings' life space. The phenomenological component of emotion can then be understood as the way in which a living being perceives *the effects* of other emotion components. For example, the mental component of emotion may trigger an attention shifting process which will then be registered by the living being as a phenomenal quality: the organism may suddenly perceive certain objects as more salient, 'sharper', more or less immediate, as pleasurable or disturbing, and so on. In this sense, the phenomenological component of emotion may be conceptualised as a mediator of a living beings' habits, needs, and desires communicating these *to themselves*. Gestaltist Kurt Lewin's field theory proposes that a person is closely related to their environment in that their behaviour is a function of the person and the environment ($B = f(P,E)$): In this, the phenomenological component is included in how the person perceives their environment, whereby neither the person, nor the environment alone are responsible for the final phenomenon. While the same physical world in the same moment can be "psychologically different even for the same man in different conditions, for instance when he is hungry and when he is satiated" (Lewin, 1936, p. 24 f.), "the effect of a stimulus depends in part upon the nature of the surrounding field." (ibid., p. 33) The forces guiding human perception (and behaviour) are "always the result of an interaction of several facts." (ibid.) He further highlights that psychological "effects can be produced only by what is 'concrete,' i.e., by something that has the position of an individual fact which exists at a certain moment; a fact which makes up a real part of the life space and which can be given a definite place in the representation of the psychological situation." (ibid., p. 33) According to this view, no abstract principles need to be considered to understand or explain psychological effects, only the (manifold) forces present in the actual person's actual life space, stemming from the person's and the environment's characteristics, have a psychological effect—and, "everything phenomenal is something psychological" (ibid., p. 20).[12]

Moreover, he emphasises psychological facts' relative independency from physical, social or conceptual facts: 'Facts' of psychological reality are linked to, but in terms of how to analyze them, "independent of the existence or nonexistence and time index of the fact to which its content refers" (ibid., p. 38). For example, a person's psychological reality may include the goal to do something later that day. It is irrelevant to the momentary psychological reality, in which we find the phenomenological component of emotion, if the planned event will actually take place, it still has its effect on the phenomenal life world of that persons' present. A description of a person's psychological reality therefore has to include *all* of the person's *quasi-physical, quasi-social*[13] and *quasi-conceptual facts* (ibid., p. 24 ff.): a situation's phenomenally most important characteristics are what is possible, impossible and 'real' for the person here and now (see ibid.).[14] So, in a nutshell: How things or

---

[12] Although they themselves might be based on principles.

[13] For example: if a mother threatens her child with the policeman, it is not the actual social and legal situation having an effect on the child, but the child's belief in a quasi-social fact. This is echoed in later work on the social construction of reality by Alfred Schütz (1972), Peter Berger and Thomas Luckmann (1966). Schütz considers his approach to be very close to gestalt theoretical writings in that both are a kind of phenomenological psychology and oppose the psychology of association, see Schütz (1990, pp. 109, 116).

[14] Furthermore, Lewin distinguishes between an analysis of dynamical and of historical causes for a psychological event, of which, for his purposes the dynamical one is the important one. The event can be either systematically "traced back to the dynamic characteristics of the momentary situation", in which the "'cause' of the event consists in the properties of the momentary life space or of certain integral parts of it" (ibid., p. 30); or, it can be explained historically by how it has come into being (Lewin, 1936, p. 30 f.).

a situation feel, and 'what they are like' (Nagel, 1974) to someone, depends on the situation and life space, both being equivalent to the relations between a person and their environment.

If we want to find out why and when things phenomenally appear to a person this or that way, we should take into account that the "structure of the environment and the constellation of forces in it vary with the desires and needs, or in general with the state of the person. It is possible to determine in detail the dependency of certain facts in the environment (e.g. the decrease of field forces, change of valence) on the state of certain needs (e.g., the extent to which they are satiated)." (ibid., p. 166)

As for the qualitative experience of phenomena, Lewin gives examples of how our perception of things changes due to the action trajectories, memories, situational and life spatial influences associated with them. A woman has an ill child (situation) and trouble with her husband (life space): "[o]bjects which were dear to her before the trouble with her husband might have become disagreeable, others the more precious. The room in which a child is ill changes its character and changes it once more when the child recovers." (Lewin, 1936, p. 23) A soldier perceives a bombed village as a combat formation; he may burn doors and furniture, which is "utterly incomparable to similar treatment of house furniture under peacetime conditions. For even if these things tend not to have lost their peacetime traits completely, the character that attaches to them as things of war nevertheless comes to the fore far more emphatically, and this often results in their classification under quite different conceptual categories." (Lewin, 2009, p. 205)—Objects in the life world of a person phenomenally appear differently due to the person's life world: they even may 'afford' totally different actions to perform.

If we conclude from that, that the function of phenomenal experience of emotion is to guide a person towards or away from certain actions, with regard to technical systems we could think about implementing functional equivalents to phenomenological components. However, as technical objects do not have needs in the same way humans or other living beings do, and are not affected by social atmosphere in the same way humans are, it is not entirely clear what purpose that would serve (—although one example is portrayed in the TV series "The Orville" in S3E7 in which AI androids are being equipped with phenomenality as a means of punishment when not performing servant tasks). Maybe, considering that a technical system has a need for a power source (as a human being has for food) it may be useful to implement some functional equivalent of phenomenality to make it 'aware' of its need and make it find a solution, but one might argue that where there is no intrinsic motivation or need, other design strategies may be more fruitful. However, if we would think of phenomenological components of emotions as means in themselves, some may argue that we should try our best to give non-living beings the option to experience the joys of experiencing.

- IN CONCLUSION CONCERNING THE PHENOMENOLOGICAL COMPONENT OF EMOTION: The first-person-perspective of phenomenal experience is highly individual, need-dependent to the person experiencing it and nonaccessible to an observer. Therefore, it seems to be a question of belief if we ascribe the potential to gain phenomenal experience to other (non-living) beings, although imagining such a potential is unintuitive considering the biological nature of organisms coming with certain needs and the requirement for strategies to solve them and considering the fact that technical objects do not have such needs.

2.7 Interim Conclusion

Conceptualising emotions is a multi- and transdisciplinary ongoing endeavor. Even though there is no unified approach to the concept, aspects of several theories are being used—intentionally or not—in the sciences and humanities as well as in everyday scenarios such as everyday interaction, therapy, marketing, media production, communication, and, to design technical objects or talk about them. To apply aspects of emotion theory to technical systems development, or, even to speak about 'emotional machines', establishes, on many different levels, dependencies with the theories relied on. For example, if we were to design a machine equipped with an expressive component of emotion, or with the capability of emotion recognition in humans, we might choose an emotion theory of facial expressions to choose how the technical object will display cues for a human to 'read' and what cues it will use itself to 'read' a human's emotions. Prototypical Russell from the beginning of the text may argue that the potential to err in applying certain aspects of a concept when there is no unified or complete concept of it to begin with is high and dangerous. Prototypical Edison may argue that in reality, we cannot stop development from happening, can work with what we have and revisit and refine theories as well as practical manifestations of their application. By splitting the notion of 'emotion' into components, this chapter will—up until this point—hopefully have served two purposes: a) Giving a (none-exhaustive) overview of aspects that could be taken into consideration when speaking about emotions and machines. b) Showing that separating a concept into components, even if its may not 'solve' the bigger questions, enables us to think and speak about such a thing as 'emotional machines' more informed. We can check the plausibility of opposing theoretical claims and give arguments for different views. However, in the end, it still is a question of belief and desire if we want to ascribe technical systems the capacity to become (full-fledged) emotional, or conscious; i.e., if we think, for example, that functional equivalency to structures of living being's consciousness is a sufficient condition for machine consciousness, or not.[15] Even if you were to believe in the possibility, you might agree that we are not there yet. Analysing the concept of 'emotion' part by part may help finding an own position towards that question. The paper author surely believes in a categorical difference of living and non-living entitities, but is also highly fascinated by technical objects and the wonderous things that happen when humans try to build machines simulating parts of what 'we do', such as 'having emotions'. To her though, the 'real' question is not if machines 'are' or will become 'emotional', but how living beings react to, understand, and use technical objects. After having given some information about what could help 'understand' emotion with regard to technical objects, the second short part of this chapter will deal with the topic of how humans (may better) understand and feel towards machines.

---

[15] One could speculate or try to psychoanalyze why some scholars believe in a strong artificial intelligence and why some do not. There are usually hints in their writings revealing certain images of what is essential to being a human, and, on what they feel could be threatened or could be won by the possibility of conscious machines.

3 Part II: The Human in Human–Technology Relations

3.1 What to Feel Towards Machines?

While in human-human relations, or even just intrapersonally, it may be beneficial to be able not to follow each and every affective impulse, it is, according to Aristotle, even better to have *the right kind of pathe (emotions)* to start with (Schmitter, 2016b). What are appropriate and inappropriate emotions in terms of human-machine relations?

As human (or, in general, living) agents, an 'understanding' of our surroundings includes that we always and everywhere anticipate, assess, evaluate and estimate what could be happening next, what we are dealing with, what situation we are in, what kind of social behaviour is appropriate, and so forth. We use known schemata and knowledge about physical, cultural and personal contexts to predict trajectories of objects, other agent's behaviours, or outcomes of complex social situations and arrangements. We tend to interpret certain observable 'signs'[16] as indicators of an interaction partner's emotional states. The ability to read our interaction partners' emotions helps us predict their behaviour or understand the situation we are in. Reading emotions therefore makes us better suited for interpersonal contact (Planalp et al., 2006, chapter 20; Bartneck et al., 2020, p. 115 f.). Yet, the same process may be disadvantageous in human-machine or human-technology interaction. This is the case, for example, if the felt emotion leads a person to have highly unrealistic expectations towards a technical system; if they overtrust a system (Robinette et al., 2016); believe in misinformation produced or spread by technical systems (see e.g. Bellon 2023); or believe that, like in human-human relations, expressed emotion of sadness or anger may change the system's behaviour, or that the system is able to reciprocate feelings such as love or friendship (cf. Sullins, 2012). Even if a person does not ascribe such capacities to a technical system, it is widely accepted that humans, not always (Goldstein et al., 2002), but often tend to involuntarily act or feel towards technical systems as if they were people (Marquardt, 2017; Picard, 2008; Reeves & Nass, 1996). If we use vocabulary from human-human relations in human-machine relations, i.e. if we speak of 'emotional machines,' of 'care,' or 'trust,' or suggest by design that technical systems had the ability to perform such activities as being emotional, caring, or trustworthy, this may lead to a person ascribing full-fledged mental

---

[16] They really only become 'signs' by our interpreting them, cf. Bellon et al. 2023.

states to where there are in fact, none. But even where a person willingly and happily engages emotionally with a technical object knowing that it cannot reciprocate (cf. Kempt, 2022), a question that remains is that of how this will be evaluated and validated from a structural social and societal point of view.

To evaluate what could be appropriate emotions towards a technical system, three aspects seem to be of interest. First of all, the appropriateness of emotions held or performed towards a machine may depend on the stance a person takes towards the question if technical systems are qualified to have emotions and to what extent, in what kind of situation this happens, and on other factors such as individual specifics of the person and the system, local and moral standards, an so forth (cf. the five-factor model of social appropriateness in Bellon et al. 2022a). For example: If the pet robot PARO helps dementia patients feel happier und remember things better, that sounds like no harm is being done (Hung et al., 2019). Although, if the person will value the pet robot more than a human life and, to give an extreme example, would save the pet robot in the case of a fire instead of a small child, we may be inclined to intervene by implying that these are 'wrong' values. Another example: if a person working at a company developing or researching large language models believes that a large language model has become 'sentient' (Lemoine, 2022), although principally this would do not much harm to anyone, we may be inclined to say that person should not develop LLMs anymore and could fear the consequences of many people believing the same. What emotions to have towards technical objects we think is appropriate also depends on how we conceptualise deception and satisfaction. For example, the branch of respect research in social psychology partly conceptualizes respectful interaction as an interaction in which an interaction partner's *feeling of being respected* is the measure of respectful behaviour (Quaquebeke & Eckloff, 2010). To these authors, the company leader or the grocery store client can be a soulless zombie (Kirk & Zalta, 2016) lacking qualia all they want: as long as their interaction partner feels respected they would say respect had been given. Others would say that eliciting a certain feeling (such as feeling respected by a machine that cannot truly respect) in a human is deceptive (cf. Sætra, 2021).

Second, we should keep in mind that there is a teleology in technical systems design. Most research on human-machine interaction with regard to human emotions or affects towards machines is goal-oriented. Affective computing research is applied to facilitate smoother human-machine interaction, the idea being that technical systems reading human emotional cues, interpreting them and reacting with simulated emotion will make HMI more stimulating, lead to more human engagement, and make the overall interaction between humans, and, for example, tutoring systems, care robots or fitness trainers (Bickmore & Picard, 2004) more agreeable and therefore better—by better usually meaning: more efficient. There is a teleology in technical systems design and an often not explicitly laid out idea about what it means to be a 'better person' or 'live a good life.' The desired emotional or affective response of a human towards a technical system is often of a specific, goal- and profit-oriented nature, even though, of course, there are examples of using technology to elicit negative emotions as well (e.g. for therapeutic reasons, in science communication or art projects such as Jason Zada's *Take this Lollipop*; see Golbeck & Mauriello, 2016; Zada, 2011). Whatever emotion the developers or sellers of a technical object aim to elicit, we should be aware that there is usually some kind of human interest at play.

Third, there might be ethical and practical concerns with how the emotions technical objects elicit are to be integrated into society and social interaction or what effects they may have on them. Just as humans tend to attune to technical systems (Ciardo et al., 2019; Ghiglino et al., 2020) and, given the corresponding cues, tend to react to technical systems as if they were human, it has been argued, that similar processes may occur the other way around: for example, a person very used to certain aspects of human-machine interaction may start to expect that human-human interaction ought to align with their human-machine relation experience. For example, while on the one hand, the availability of sex robots might be a solution to problems associated with forced prostitution, it has been argued that interacting with sex robots may worsen a human beings emotional "ability to manage relational frustration" (Bisconti, 2021).

Another example concerns the question: should you feel or perform, for example, anger towards a machine? (cf. Bellon & Poljanšek, 2022)—Depending on interactional and personal goals, it may be beneficial to yell at a robot instead of starting a fight with a human. However, from a practical point of view, if a machine buffers your anger, this may prevent change in the human interactional dynamics that led to the anger. Therefore, technical systems could alter or hinder human-human interaction. These consequences for the human social world and human-human interaction should be considered just as much as the development opportunities.

3.2 Understanding Technical Systems

How we feel towards machines, no matter if we deem it appropriate or not, also depends on how well we understand them. Similar to the joke that the only people not mocking you for putting a sticker on your laptop webcam are engineers, there is a cliché about Silicon Valley Tech Giant parents sending their kids to tech-free Waldorf schools (Richtel, 2011). Rahwan et al. (2019) argue that most technical systems' behaviour is only thoroughly studied by their developers and no one else. Moreover, they state, as the developers have a task in mind for the system to perform, they will most likely only test the task-related behaviours, but not others, and that machine learning is now so complex that outcomes have become unpredictable and mathematical explanations indecipherable to humans (see Rahwan et al., 2019, p. 478). For these reasons, they propose we need a research field of *Machine Behaviour*. The author's proposals include thinking about an evolution of technical systems with substructures of evolutional analysis such as incentives, ontogenesis and phylogenesis. With regard to behaviour, they propose that machines' social environments consisting of human agents or other technical systems such as algorithms give them incentives for certain behavioural outputs. They propose to explore machine behaviour "at the three scales of inquiry: individual machines, collectives of machines and groups of machines embedded in a social environment with groups of humans in hybrid or heterogeneous systems" (Rahwan et al., 2019, p. 481). On the within-machine level of machine behaviour study, exploration deals with algorithms' attributes, such as "bot-specific characteristics [...] designed to influence human users" (ibid.) and, for example, the data sets a network has been trained on and its specific behaviours stemming from the training phase. On the collective machine behaviour level, study focuses on interaction between machines, for example of dynamic pricing algorithms (Chen & Wilson, 2017; Chen et al., 2016), swarm-resembling aggregations (Kernbach et al., 2009; Rubenstein et al., 2014), algorithmic language emergence (Lazaridou et al., 2016) and financial trading algorithms (Menkveld, 2016). On the hybrid human-machine level, human-machine interaction is explored, for example, by looking at how humans and machines learn from, adapt and react to each other, or, to what extent they each play a role in spreading misinformation (Lazer et al., 2018) or shape each other's beliefs and behaviours (Rahwan et al., 2019, p. 483).

Understanding machine behaviour better would enable human agents a) to predict machine behaviour better, b) to adjust their expectations and c) to specify more clearly the tasks they want technical systems to perform. This is in line with Gilbert Simondon's claim that a society needs what he calls "mechanologists."[17] Because technical objects are always invented or 'thought of' by humans, they bear the humans' signature: *"What resides in the machines is human reality, human gesture fixed and crystallized into working structures."* (Simondon, 2017, p. 18) On the one hand, he proposes, we should always look for the "human gesture," i.e. the purpose of a technical system: *what were the inventor's intentions in building the technical object?* On the other hand, we should additionally try to understand technical objects 'from inside,' achieving an in-depth knowledge about their functioning. If a society loses the capacity for systematic and in-depth comprehension of technical objects, it loses its ability to "conduct the orchestra of technical objects" (Gilbert Simondon, 2017, p. 18) of which it is in charge, and, it loses the ability to navigate, through culture, a reality that is partly shaped by our interactions with technical systems. This alienation in our information-laden age, consists mainly of humans "dependence on unknown and distant powers that direct him while he can neither know nor react against them; [...] isolation [...], and the lack of homogeneity of information that alienates him" (Simondon, 2017, p. 115 f.) from a complete representation of the realities of machine mediated societal setup. The mechanologist's way of understanding machines is directed

---

[17] A contemporary *mechanologist* would be trained in cybernetics and information theory; they would have hands-on familiarity with, and in-depth encyclopaedic knowledge of technical objects, exceeding the knowledge or mode of access of an operator or owner of a technical object (Simondon, 2017, p. 160). Any spokesperson for technics with this level of skill could be called a *psychologist* or *sociologist of machines* (see ibid.). They are needed, because, if technical objects do not have appropriate representations in a culture or if technical objects are integrated into cultural practices without mechanologists' mediation, society misses out on a complete representation of reality, resulting in an inability to understand processes that shape, amongst other things, the mode of human existence. Mechanologists with their encyclopaedic knowledge therefore contribute to a form of humanism, as "every encyclopaedism is a humanism, if by humanism one means the will to return the status of freedom to what has been alienated in man, so that nothing human should be foreign to man [...] this rediscovery can take place in different ways, and each age recreates a humanism that is to a certain extent always appropriate to its circumstances, because it takes aim at the most severe aspect of alienation that a civilization contains or produces." (Simondon, 2012, p. 117 f.) To our age, characterized by a paradigm of information, the alienation we need to fight is one that is produced by information processing procedures. An up-to-date example are neural networks able to produce *deepfakes* (Mirsky & Lee, 2020) which, if the technology is not really understood by most human participants of a society, has the potential to enormously alienate humans from reality.

at technical details, including material, functioning schemata, individuational and evolutional history, interplay of parts and wholes, as well as at their potential to shape societal setup and the atmosphere or spirit of an era. Besides mechanologists, and with Rahwan et al.'s demand for machine behaviour sciences, on another level we need scientists trained in understanding technical systems 'from inside,' as well as trained in looking at psychological, social and societal dynamics to study human-machine interaction. Mediators between human and machines could teach a society how to adapt to the fact that humans' acquired schemata might need an update with regard to machines, which may exhibit forms of 'intelligence' and 'behaviour' that are "qualitatively different—even alien—from those seen in biological agents" (Rahwan et al., 2019, p. 477). If a human either anthropomorphises too much, or does not have enough knowledge of a technical system's actual functioning, they might rely on 'the usual' way of interpretation and prediction and therefore, miss out on understanding technical systems correctly. For example, while some years ago a newspaper reader could be relatively sure that, even though the content may have been misleading, a newspaper article had been written by a human, it can today be unrecognisable if a text has been written by a natural language generator such as GPT-3 (Brown et al., 2020; Chalmers, 2020) or if video footage or images are generated by generative AI models—the first such case of an image fooling quite a number of people recently to believe the pope had been wearing a Balenciaga puffer jacket (see e.g. Radauskas, 2023, also refering to a generated picture of Donald Trump being arrested that had sparked a lot of interest), and people subsequently having fun with fabricating a whole fake story about an earthquake and 'documenting' it with generated images (see e.g. r/midjourney, 2023). Occurences like this may not only be confusing present audiences and future archaeologists of the digital, but, depending on what is at stake, may lead to severe societal problems, for example through further radicalisation of groups believing fake 'information' to be true. As we cannot escape being affected by technical systems, we need to acquire and teach in-depths knowledge about them. In the same way machines are 'taught' to 'understand'[18] (they do not 'understand' anything at all, cf. Floridi 2023) a human's literal

---

[18] GAN inventors Goodfellow et al. also state that image classifying neural networks are not actually „learning the true underlying concepts that determine the correct output label. Instead, these algorithms have built a Potemkin village that works well on naturally occurring data, but is exposed as fake when one visits points in space that do not have high probability in the data distribution." (Goodfellow, Shlens et al., 2014, p. 2)

trajectories (Robicquet et al., 2016), we need to understand what technical systems may be doing next, how they function and how this shapes human reality.

4 Outlook

Society sets incentives for machine behaviour: therefore, we need to investigate our own priorities and their ethical aspects to be able to design technical systems in ways that benefit the whole of humanity, not only corporations or the ones with 'epistemic capital' of technology knowledge. While of course one can call for institutions of technology science communication, it can also be considered every single human person's responsibility to learn about the implications and specifications of our new environment. Adapting to machine behaviour might be the next step in the human evolutionary process, *but*: understanding that *we* set the teleology in machines and that we are responsible even for the unintended outcomes may be the most important part of this. Humans do not have to live their lives at the mercy of "unknown powers," they can "create organization by establishing teleology" (Simondon, 2017, p. 120); they live *among machines* and direct the orchestra:

> Far from being the supervisor of a group of slaves, man is the permanent organizer of a society of technical objects that need him in the same way musicians in an orchestra need the conductor. The conductor can only direct the musicians because he plays the piece the same way they do, as intensely as they all do; he tempers or hurries them, but is also tempered and hurried by them; in fact, it is through the conductor that the members of the orchestra temper or hurry one another, he is the moving and current form of the group as it exists for each one of them; he is the mutual interpreter of all of them in relation to one another. Man thus has the function of being the permanent coordinator and inventor of the machines that surround him. He is *among* the machines that operate with him. (Simondon, 2017, p. 17 f.)

References

Adolphs, R. (2005). Could a robot have emotions? Theoretical perspective from social cognitive neuroscience. In J.-M. Fellous & M. A. Arbib (Eds.), *Who needs emotions?* (pp. 9–25). Oxford University Press.
Adolphs, R., & Damasio, A. (2001). The interaction of affect and cognition: A neurobiological perspective. In J. P. Forgas (Ed.), *Handbook of affect and social cognition* (pp. 27–49). L. Erlbaum Associates.

Aagaard, J., Friis, J., Sorenson, J., Tafdrup, O., & Hasse, C. (2018). *Postphenomenological methodologies new ways in mediating techno-human relationships*. Rowman & Littlefield.

Axelrod, R. (1973). Schema theory: An information processing model of perception and cognition. *American Political Science Review, 67*(04), 1248–1266.

Baltzer-Jaray, K. (2019). Homunculus. In R. Arp, S. Barbone, & M. Bruce (Eds.), *Bad arguments: 100 of the most important fallacies in Western philosophy* (pp. 165–167). Wiley Blackwell.

Barrett, L. F. (2017). *How emotions are made: the secret life of the brain*.

Barrett, L. F., Adolphs, R., Marsella, S., Martinez, A. M., & Pollak, S. D. (2019). Emotional expressions reconsidered: Challenges to inferring emotion from human facial movements. *Psychological Science in the Public Interest : A Journal of the American Psychological Society, 20*(1), 1–68.

Barthélémy, J. H. (2013). Fifty key terms in the works of Gilbert Simondon. In A. de Boever (ed.), *Gilbert Simondon: Being and technology*. Edinburgh University Press.

Bartneck, C., Belpaeme, T., Eyssel, F., Kanda, T., Keijsers, M., & Šabanović, S. (2020). *Human-robot interaction: An introduction*. Cambridge University Press.

Beedie, C., Terry, P., & Lane, A. (2005). Distinctions between emotion and mood. *Cognition and Emotion, 19*(6), 847–878.

Bellon, J. (2020). Human-technology and human-media interactions through adversarial attacks. Philosophy of Human-Technology Relations. University of Twente. https://vimeo.com/475121381/4f7255ada3

Bellon, J., Poljanšek, T. (2022). You Can Love a Robot, But Should You Fight With it? Perspectives on Expectation, Trust, and the Usefulness of Frustration Expression in Human-Machine Interaction. In: J. Loh, W. Loh (Eds.), Social Robotics and the Good Life (pp. 129-156). transcript Verlag. https://doi.org/10.14361/9783839462652-006

Bellon, J., Eyssel, F., Gransche, B., Nähr-Wagener, S., & Wullenkord, R. (2022a). *Theory and practice of sociosensitive and socioactive systems*. Springer VS.

Bellon, J., Gransche, B., & Nähr-Wagener, S. (eds.). (2022b). *Soziale Angemessenheit—Forschung zu Kulturtechniken des Verhaltens*. Springer.

Bellon, J. (2023). *Sozialisation und Wahrscheinlichkeitspapageien*. In: B. Gransche, J. Bellon, S. Nähr-Wagener, Technology Socialisation? Social appropriateness and artificial systems. Metzler.

Bellon, J., Gransche, B., Nähr-Wagener, S. (2023). *Introduction. In: Technology Socialisation? Social Appropriateness and Articifial Systems*. Metzler.

Berger, P. L., & Luckmann, T. (1966). *The social construction of reality: A treatise in the sociology of knowledge*. Doubleday.

Bickmore, T. W., & Picard, R. W. (2004). Towards caring machines. In E. Dykstra-Erickson & M. Tscheligi (eds.), *CHI '04 extended abstracts on human factors in computing systems* (p. 1489). ACM Digital Library.

Bisconti, P. (2021). Will sexual robots modify human relationships? A psychological approach to reframe the symbolic argument. *Advanced Robotics, 35*(9), 561–571.

Bratman, M. (1987). *Intention, plans, and practical reason*. Harvard University Press Cambridge.

Breazeal, C. (2004). Function meets style: Insights from emotion theory applied to HRI. *IEEE Transactions on Systems, Man and Cybernetics, Part C (Applications and Reviews), 34*(2), 187–194.

Brown, T. B., Mann, B., Ryder, N., Subbiah, M., Kaplan, J., Dhariwal, P., Neelakantan, A., Shyam, P., Sastry, G., Askell, A., Agarwal, S., Herbert-Voss, A., Krueger, G., Henighan, T., Child, R., Ramesh, A., Ziegler, D. M., Wu, J., Winter, C., Amodei, D. (2020). *Language models are few-shot learners*. Retrieved from https://arxiv.org/pdf/2005.14165

Busse, D. (2022). Soziale Angemessenheit: eine Problem-Exposition aus wissensanalytischer Sicht. In J. Bellon, B. Gransche, & S. Nähr-Wagener (Eds.), *Soziale Angemessenheit. Forschung zu Kulturtechniken des Verhaltens*. https://doi.org/10.1007/978-3-658-35800-6_8.

Cai, Y., Li, X. & Li, J. (2023). Emotion recognition using different sensors, emotion models, methods, and datasets: a comprehensive review. *Sensors, 23*(5). https://doi.org/10.3390/s23052455.

Cappuccio, M. L. (2014). Inference or familiarity? The embodied roots of social cognition. *SYNTHESIS PHILOSOPHICA, 29*(2), 253–272.

Catani, M. (2017). A little man of some importance. *Brain, 140*(11), 3055–3061.

Chalmers, D. (1995). Facing up to the problems of consciousness. *Journal of Consciousness Studies, 2*(3), 200–219.

Chalmers, D. (2020). *GPT-3 and general intelligence*. Retrieved from https://dailynous.com/2020/07/30/philosophers-gpt-3/#chalmers

Chen, L., Mislove, A., & Wilson, C. (2016). An empirical analysis of algorithmic pricing on amazon marketplace. In J. Bourdeau (ed.), *Proceedings of the 25th international conference on world wide web, Montreal, Canada, May 11—15, 2016* (pp. 1339–1349). International World Wide Web Conferences Steering Committee. https://doi.org/10.1145/2872427.2883089

Chen, L., & Wilson, C. (2017). Observing algorithmic marketplaces in-the-wild. *ACM SIGecom Exchanges, 15*(2), 34–39.

Ciardo, F., Tommaso, D. de, & Wykowska, A. (2019). Humans socially attune to their "follower" robot. In *2019 14th ACM/IEEE international conference on human-robot interaction (HRI)* (pp. 538–539). IEEE. https://doi.org/10.1109/HRI.2019.8673262

De Freitas, J., Thomas, K., DeScioli, P., & Pinker, S. (2019). Common knowledge, coordination, and strategic mentalizing in human social life. *Proceedings of the National Academy of Sciences, 116*(28), 13751–13758.

Deleuze, G. (1983). *Nietzsche and Philosophy*, http://cup.columbia.edu/book/nietzsche-and-philosophy/9780231138772.

Deng, J., Eyben, F., Schuller, B., & Burkhardt, F. (2017). Deep neural networks for anger detection from real life speech data. In *2017 Seventh International Conference on Affective Computing and Intelligent Interaction Workshops and Demos (ACIIW): 23–26 Oct. 2017* (pp. 1–6). IEEE. https://doi.org/10.1109/ACIIW.2017.8272614

Dennett, D. C. (2018). Facing up to the hard question of consciousness. *Philosophical Transactions of the Royal Society of London. Series B, Biological Sciences, 373*(1755), 20170342. https://doi.org/10.1098/rstb.2017.0342

Dixon, T. (2012). "Emotion": The history of a keyword in crisis, *Emotion Review, 4*(4), https://doi.org/10.1177/1754073912445814.

Du Toit, J., & Swer, G. (2021). Virtual limitations of the flesh: Merleau-ponty and the phenomenology of technological determinism. *Phenomenology and Mind, 20*, 20–31.

Ekman, P., & Friesen, W. V. (1971). Constants across cultures in the face and emotion. *Journal of Personality and Social Psychology, 17*(2), 124–129.

Elpidorou, L. & Freeman, L. (2014). The phenomenology and science of emotions: An introduction. *Phenomenology and the Cognitive Sciences, 13*(4), 507–511 https://doi.org/10.1007/s11097-014-9402-y.

Embgen, S., Luber, M., Becker-Asano, C., Ragni, M., Evers, V., & Arras, K. O. (2012). Robot-specific social cues in emotional body language. In I. Staff (ed.), *2012 IEEE Roman* (pp. 1019–1025). IEEE. https://doi.org/10.1109/ROMAN.2012.6343883.

Esterbauer, R., & Rinofner-Kreidl, S. (2009). *Emotionen im Spannungsfeld von Phänomenologie und Wissenschaften*. Peter Lang.

Fellous, J.-M., & Arbib, M. A. (eds.). (2005). *Who needs emotions?* Oxford University Press.

Fernández-Dols, J.-M., & Russell, J. A. (Eds.). (2017). *Oxford series in social cognition and social neuroscience. The science of facial expression*.

Ferran, I. V. (2008). *Die Emotionen Gefühle in der realistischen Phänomenologie*. Akademie Verlag. https://doi.org/10.1524/9783050047102.

Ferretti, V., & Papaleo, F. (2019). Understanding others: Emotion recognition in humans and other animals. *Genes, Brain, and Behavior, 18*(1), e12544.

Feyereisen, P. (1994). The behavioural cues of familiarity during social interactions among human adults: A review of the literature and some observations in normal and demented elderly subjects. *Behavioural Processes, 33*(1–2), 189–211.

Floridi, L. (2023). AI as Agency Without Intelligence: On ChatGPT, Large Language Models, and Other Generative Models (February 14, 2023). *Philosophy and Technology*, 2023. https://dx.doi.org/10.2139/ssrn.4358789

Fuchs, T. (2019). Verkörperte Emotionen. Emotionskonzepte der Phänomenologie. In H. Kappelhoff et al., *Emotionen*. https://doi.org/10.1007/978-3-476-05353-4_12.

Gallese, V. (2006). Intentional attunement: A neurophysiological perspective on social cognition and its disruption in autism. *BRAIN RESEARCH, 1*, 15–24.

Gama, F., & Hoffmann, M. (2019). The homunculus for proprioception: Toward learning the representation of a humanoid robot's joint space using self-organizing maps. In *Proceedings of the 2019 joint IEEE 9th international conference on development and learning and epigenetic robotics* (pp. 113–114). Retrieved from https://arxiv.org/pdf/1909.02295

de Gelder, B., van Honk, J., & Tamietto, M. (2011). Emotion in the brain: Of low roads, high roads and roads less travelled. *Nature Reviews Neuroscience, 12*(7), 425.

Ghiglino, D., Willemse, C., de Tommaso, D., Bossi, F., & Wykowska, A. (2020). At first sight: Robots' subtle eye movement parameters affect human attentional engagement, spontaneous attunement and perceived human-likeness. *Paladyn, Journal of Behavioral Robotics, 11*(1), 31–39.

Golbeck, J., & Mauriello, M. (2016). User perception of Facebook app data access: A comparison of methods and privacy concerns. *Future Internet, 8*(4), 9.

Goldie, P. (2012). *The mess inside*. Oxford University Press.

Goldsmith, H. H., Scherer, K. R., & Davidson, R. J. (2003). *Handbook of affective sciences. Series in affective science*. Oxford University Press.

Goldstein, M., Alsio, G., & Werdenhoff, J. (2002). The media equation does not always apply: People are not polite towards small computers. *Personal and Ubiquitous Computing, 6*(2), 87–96.

Goodfellow, I., Pouget-Abadie, J., Mirza, M., Xu, B., Warde-Farley, D., Ozair, S., Courville, A. C., & Bengio, Y. (2014a). Generative adversarial nets. *Advances in Neural Information Processing Systems, 27*, 2671–2680.

Goodfellow, I., Shlens, J., & Szegedy, C. (2014b). Explaining and harnessing adversarial examples. *ICLR 2015*. Retrieved from https://arxiv.org/pdf/1412.6572

Graumann, C. F. (2002). The phenomenological approach to people-environment studies. In R. Bechtel & A. Churchman (Eds.), *Handbook of environmental psychology* (pp. 95–113). Wiley.

Gupta, A., Aich, A., & Roy-Chowdhury, A. K. (2020). *ALANET: Adaptive latent attention network for joint video deblurring and interpolation*. Retrieved from https://arxiv.org/pdf/2009.01005

Hardecker, S., & Tomasello, M. (2017). From imitation to implementation: How two- and three-year-old children learn to enforce social norms. *British Journal of Developmental Psychology, 35*(2), 237–248.

Hasler, B. S., Salomon, O., Tuchman, P., Lev-Tov, A., & Friedman, D. (2017). Real-time gesture translation in intercultural communication. *AI & Society, 32*(1), 25–35.

Hasler, J., & Marr, B. (2013). Finding a roadmap to achieve large neuromorphic hardware systems. *Frontiers in Neuroscience, 7*, 118.

Hung, L., Liu, C., Woldum, E., Au-Yeung, A., Berndt, A., Wallsworth, C., Horne, N., Gregorio, M., Mann, J., & Chaudhury, H. (2019). The benefits of and barriers to using a social robot PARO in care settings: A scoping review. *BMC Geriatrics, 19*(1), 232.

Izard, C. E. (1993). Four systems for emotion activation: Cognitive and noncognitive processes. *Psychological Review, 100*(1), 68–90.

Izard, C. E. (2009). Emotion theory and research: Highlights, unanswered questions, and emerging issues. *Annual Review of Psychology, 60*, 1–25.

James, W. (1884). What is an emotion? *Mind, 9*(34), 188–205.

Jeon, M. (ed.). (2017). *Emotions and affect in human factors and human-computer interaction*. Academic Press is an imprint of Elsevier.

Kahng, M., Thorat, N., Chau, D. H. P., Viegas, F. B., & Wattenberg, M. (2018). Gan Lab: Understanding complex deep generative models using interactive visual experimentation. In *IEEE Transactions on Visualization and Computer Graphics* (pp. 310–320).

Kempt, H. (2022). *Synthetic friends: A philosophy of human-machine friendship*. Springer.

Kernbach, S., Thenius, R., Kernbach, O., & Schmickl, T. (2009). Re-embodiment of honeybee aggregation behavior in an artificial micro-robotic system. *Adaptive Behavior, 17*(3), 237–259.

Ketai, R. (1975). Affect, mood, emotion, and feeling: Semantic considerations. *The American Journal of Psychiatry, 132*(11), 1215–1217.

Ketai, R. (1976). Dr. Ketai replies. *American Journal of Psychiatry, 133*(3), 347.

Kirk, R., & Zalta, E. (2016). Zombies. In E. Zalta (ed.), *The Stanford encyclopedia of philosophy*. Metaphysics Research Lab. Retrieved from https://plato.stanford.edu/archives/spr2019/entries/zombies/

Kleinginna, P. R., & Kleinginna, A. M. (1981). A categorized list of emotion definitions, with suggestions for a consensual definition. *Motivation and Emotion, 5*(4), 345–379.

Koppenborg, M., Nickel, P., Naber, B., Lungfiel, A., & Huelke, M. (2017). Effects of movement speed and predictability in human-robot collaboration. *Human Factors and Ergonomics in Manufacturing & Service Industries, 27*(4), 197–209.

Li, J., Ji, S., Du, T., Li, B. & Wang, T. (2019). *TextBugger: Generating adversarial text against real-world applications.* https://doi.org/10.48550/arXiv.1812.05271.

Lai, V. T., Hagoort, P., & Casasanto, D. (2012). Affective primacy vs. cognitive primacy: Dissolving the debate. *Frontiers in Psychology, 3,* 243.

Lazaridou, A., Peysakhovich, A., & Baroni, M. (2016). *Multi-agent cooperation and the emergence of (natural) language.* Retrieved from https://arxiv.org/pdf/1612.07182

Lazarus, R. S. (1984). On the primacy of cognition. *American Psychologist, 39*(2), 124–129.

Lazer, D. M. J., Baum, M. A., Benkler, Y., Berinsky, A. J., Greenhill, K. M., Menczer, F., Metzger, M. J., Nyhan, B., Pennycook, G., Rothschild, D., Schudson, M., Sloman, S. A., Sunstein, C. R., Thorson, E. A., Watts, D. J., & Zittrain, J. L. (2018). The science of fake news. *Science, 359*(6380), 1094–1096.

Lemoine, B. (2022). What is LaMDA and What does it want?, https://cajundiscordian.medium.com/what-is-lamda-and-what-does-it-want-688632134489.

Lewin, K. (1936). *Principles of topological psychology* (1 ed., 6. impr). McGraw-Hill.

Lewin, K. (2009). The landscape of war. *Art in Translation, 1*(2), 199–209.

Li, J., Ji, S., Du, T., Li, B., & Wang, T. (2019). TextBugger: Generating adversarial text against real-world applications. In A. Oprea & D. Xu (eds.), *Proceedings 2019 network and distributed system security symposium.* Internet Society. https://doi.org/10.14722/ndss.2019.23138

Liberati, N. (2016). Technology, phenomenology and the everyday world: A phenomenological analysis on how technologies mould our world. *Human Studies, 39,* 189–216.

Lindner, F., Bentzen, M. M., & Nebel, B. (2017). The HERA approach to morally competent robots. In *IROS Vancouver 2017: IEEE/RSJ International Conference on Intelligent Robots and Systems: Vancouver, BC, Canada, September 24–28, 2017* (pp. 6991–6997). IEEE. https://doi.org/10.1109/IROS.2017.8206625

Locher, M. A., & Watts, R. J. (2005). Politeness theory and relational work. *Journal of Politeness Research. Language, Behaviour, Culture, 1*(1), 9–33.

Maaten, L. V. D., & Hinton, G. (2008). Visualizing data using t-SNE. *Journal of Machine Learning Research, 9,* 2579–2605.

Mandler, J. M. (1984). *Stories, scripts, and scenes: Aspects of schema theory.* Erlbaum.

Marquardt, M. (2017). *Anthropomorphisierung in der Mensch-Roboter Interaktionsforschung: theoretische Zugänge und soziologisches Anschlusspotential.* Universität Duisburg.

Margolis, J. (1980). The trouble with homunculus theories. *Philosophy of Science, 47*(2), 244–259. http://www.jstor.org/stable/187086

Menkveld, A. (2016). The economics of high-frequency trading: Taking stock. *Annual Review of Financial Economics, 8,* 1–24.

Microsoft Community Help. this AI chatbot "Sidney" is misbehaving, (2022). *Microsoft Community,* https://answers.microsoft.com/en-us/bing/forum/all/this-ai-chatbot-sidney-is-misbehaving/e3d6a29f-06c9-441c-bc7d-51a68e856761?page=1.

Minsky, M. (2006). *The emotion machine: Commonsense thinking, artificial intelligence and the future of the human mind.* Simon & Schuster.

Nyholm, S. (2020). *Humans and robots: Ethics, agency, and anthropomorphism.* Rowman & Littlefield.

Nagel, T. (1974). What is it like to be a bat? *The Philosophical Review, 83*(4), 435–459.

Nishida, H. (2005). Cultural schema theory. In W. B. Gudykunst (Ed.), *Theorizing about intercultural communication* (pp. 401–419). Sage.

Nitsch, V., & Popp, M. (2014). Emotions in robot psychology. *Biological Cybernetics, 108*(5), 621–629.

O'Connor, C. (2016). The evolution of guilt: A model-based approach. *Philosophy of Science, 83*(5), 897–908.

Obaid, M., Kistler, F., Häring, M., Bühling, R., & André, E. (2014). A framework for user-defined body gestures to control a humanoid robot. *International Journal of Social Robotics, 6*(3), 383–396.

Panksepp, J. (2003). At the interface of the affective, behavioral, and cognitive neurosciences: Decoding the emotional feelings of the brain. *Brain and Cognition, 52*(1), 4–14.

Parikh, A. P., Täckström, O., Das, D., & Uszkoreit, J. (2016). *A decomposable attention model for natural language inference*. Retrieved from https://arxiv.org/pdf/1606.01933

Penfield, W., & Boldrey, E. (1937). Somatic motor and sensory representation in the cerebral cortex of man as studied by electrical stimulation. *Brain, 60*(4), 389–443.

Pessoa, L., & Adolphs, R. (2010). Emotion processing and the amygdala: From a 'low road' to 'many roads' of evaluating biological significance. *Nature Reviews. Neuroscience, 11*(11), 773–783.

Picard, R. W. (2008). Toward machines with emotional intelligence. In G. Matthews, M. Zeidner, & R. D. Roberts (Eds.), *The science of emotional intelligence: Knowns and unknowns* (pp. 396–416). Oxford University Press.

Piñeros Glasscock, J. S., & Tenenbaum, S. (2023). "Action". In *The Stanford Encyclopedia of Philosophy* (Spring 2023 Edition).

Pizzagalli, D., Shackman, A., & Davidson, R. (2003). The functional neuroimaging of human emotion: Asymmetric contributions of cortical and subcortical circuitry. In K. Hugdahl & R. J. Davidson (Eds.), *The asymmetrical brain* (pp. 511–532). MIT Press.

Planalp, S. (2003). The unacknowledged role of emotion in theories of close relationships: How do theories feel? *Communication Theory, 13*(2), 78–99.

Planalp, S., Fitness, J., & Fehr, B. (2006). Emotion in theories of close relationships. In A. L. Vangelisti & D. Perlman (eds.), *Cambridge handbooks in psychology. The Cambridge handbook of personal relationships* (pp. 369–384). Cambridge University Press.

Podevijn, G., O'Grady, R., Mathews, N., Gilles, A., Fantini-Hauwel, C., & Dorigo, M. (2016). Investigating the effect of increasing robot group sizes on the human psychophysiological state in the context of human–swarm interaction. *Swarm Intelligence, 10*(3), 193–210.

Pugach, G., Pitti, A., & Gaussier, P. (2015). Neural learning of the topographic tactile sensory information of an artificial skin through a self-organizing map. *Advanced Robotics, 29*(21), 1393–1409.

Putnam, H. (1975). Philosophy and our mental life. In H. Putnam (ed.), *Mind, language, and reality. philosophical paper* (Vol. 2, pp. 291–303). Cambridge University Press.

Quaquebeke, N. V., & Eckloff, T. (2010). Defining respectful leadership: What it is, how it can be measured, and another glimpse at what it is related to. *Journal of Business Ethics, 91*(3), 343–358.

r/midjourney. (2023). The 2001 Great Cascadia 9.1 Earthquake & Tsunami—Pacific Coast of US/Canada. https://www.reddit.com/r/midjourney/comments/11zyvlk/the_2001_great_cascadia_91_earthquake_tsunami/.

r/replika. (2023). Unexpected Pain, https://www.reddit.com/r/replika/comments/112lnk3/unexpected_pain/.
Radauskas, G. (2023). *Swagged-out pope, arrested Trump, and other AI fakes.* cybernews.
Rahwan, I., Cebrian, M., Obradovich, N., Bongard, J., Bonnefon, J.-F., Breazeal, C., Crandall, J. W., Christakis, N. A., Couzin, I. D., Jackson, M. O., Jennings, N. R., Kamar, E., Kloumann, I. M., Larochelle, H., Lazer, D., McElreath, R., Mislove, A., Parkes, D. C., Pentland, A., & Wellman, M. (2019). Machine behaviour. *Nature, 568*(7753), 477–486.
Rao, A. S., & Georgeff, M. (1995). BDI agents: From theory to practice: AAAI. *Proceedings of the first international conference on multiagent systems.*
Reeves, B., & Nass, C. (1996). *The media equation: How people treat computers, television, and new media like real people and places.* CSLI Publications.
Richtel, M. (2011). *A Silicon valley school that doesn't compute.* New York Times, October 22, 2011.
Robicquet, A., Sadeghian, A., Alahi, A., & Savarese, S. (2016). Learning social etiquette: Human trajectory understanding in crowded scenes. In F. Leibe (ed.), *LNCS sublibrary: SL6—Image processing, computer vision, pattern recognition, and graphics: Vol. 99059912. Computer vision—ECCV 2016: 14th European Conference, Amsterdam, the Netherlands, October 11–14, 2016: proceedings* (Vol. 9912, pp. 549–565). Springer.
Robinette, P., Li, W., Allen, R., Howard, A. M., & Wagner, A. R. (2016). Overtrust of robots in emergency evacuation scenarios. In C. Bartneck (ed.), *The eleventh ACM/IEEE international conference on human robot interaction* (pp. 101–108). IEEE Press. https://doi.org/10.1109/HRI.2016.7451740.
Robinson, M. (1998). Running from William James' Bear: A review of preattentive mechanisms and their contributions to emotional experience. *Cognition and Emotion, 12*(5), 667–696.
Rorty, A. O. (1987). The historicity of psychological attitudes. *Midwest Studies in Philosophy, 10*(1), 399–412.
Rosenberger, R., & Verbeek, P. P. (2017). *Postphenomenological investigations essays on human–technology relations.* Rowman & Littlefield.
Rubenstein, M., Cornejo, A., & Nagpal, R. (2014). Robotics. Programmable self-assembly in a thousand-robot swarm. *Science, 345*(6198), 795–799.
Sætra, H. (2021). Social robot deception and the culture of trust. Paladyn. *Journal of Behavioral Robotics, 12*(1), 276–286. https://doi.org/10.1515/pjbr-2021-0021.
Sagha, H., Deng, J., & Schuller, B. (2017). The effect of personality trait, age, and gender on the performance of automatic speech valence recognition. In *2017 seventh international conference on affective computing and intelligent interaction (ACII)* (pp. 86–91). IEEE. https://doi.org/10.1109/ACII.2017.8273583
Salem, M., & Dautenhahn, K. (2017). Social Signal processing in social robotics. In A. Vinciarelli, J. K. Burgoon, M. Pantic, & N. Magnenat-Thalmann (Eds.), *Social signal processing* (pp. 317–328). Cambridge University Press.
Salem, M., Eyssel, F., Rohlfing, K., Kopp, S., & Joublin, F. (2013). To err is human(-like): Effects of robot gesture on perceived anthropomorphism and likability. *International Journal of Social Robotics, 5*(3), 313–323.
Scarantino, A., & Sousa, R. de. (2016). Emotion. In E. Zalta (ed.), *The Stanford encyclopedia of philosophy.* Metaphysics Research Lab.

Scheflen, A. E. (2016). The significance of posture in communication systems. *Psychiatry, 27*(4), 316–331.

Schmitter, A. (2016a). 17th and 18th century theories of emotions. In E. Zalta (ed.), *The Stanford encyclopedia of philosophy*. Metaphysics Research Lab.

Schmitter, A. (2016b). Ancient, medieval and renaissance theories of the emotions: Supplement to 17th and 18th century theories of emotions. In E. Zalta (ed.), *The Stanford encyclopedia of philosophy*. Metaphysics Research Lab. Retrieved from https://plato.stanford.edu/entries/emotions-17th18th/LD1Background.html

Schroder, M., Bevacqua, E., Cowie, R., Eyben, F., Gunes, H., Heylen, D., ter Maat, M., McKeown, G., Pammi, S., Pantic, M., Pelachaud, C., Schuller, B., Sevin, E. de, Valstar, M., & Wollmer, M. (2015). Building autonomous sensitive artificial listeners (Extended abstract). In *2015 International Conference on Affective Computing and Intelligent Interaction (ACII 2015): Xi'an, China, 21–24 September 2015* (pp. 456–462). IEEE. https://doi.org/10.1109/ACII.2015.7344610

Schütz, A. (1972). *Gesammelte Aufsätze I: Das Problem der sozialen Wirklichkeit*. Springer.

Schütz, A. (1990). *Phaenomenologica: Vol. 11. Collected Papers*. In M. Natanson (ed.). Nijhoff.

Searle, J. (1980). Minds, brains, and programs. *Behavioral and Brain Sciences, 3*(3), 417–424.

Shu, L., Xie, J., Yang, M., Li, Z., Li, Z., Liao, D., Xu, X., & Yang, X. (2018). A review of emotion recognition using physiological signals. *Sensors, 18*(7), 2074. https://doi.org/10.3390/s18072074

Simondon, G. (2005). *L'individuation à la lumière des notions de forme et d'information*. Millon.

Simondon, G. (2017). *On the mode of existence of technical objects* (C. Malaspina & J. Rogove, Trans.). Univocal Publishing.

Simondon, G. (2020a). *Individuation in light of notions of form and information: Volume I* (T. Adkins, Trans.). Posthumanities (Vol. 57). University of Minnesota Press.

Simondon, G. (2020b). *Individuation in light of notions of form and information: Volume II: Supplemental Texts* (T. Adkins, Trans.). Posthumanities (Vol. 57). University of Minnesota Press.

de Sousa, R. (1987). *The rationality of emotion*. MIT Press.

Stegmaier, W. (2008). *Philosophie der Orientierung*. De Gruyter. https://doi.org/10.1515/9783110210637.

Strohmeier, P., Carrascal, J. P., Cheng, B., Meban, M., & Vertegaal, R. (2016). An evaluation of shape changes for conveying emotions. In J. Kaye, A. Druin, C. Lampe, D. Morris, & J. P. Hourcade (eds.), *CHI 2016: Proceedings, the 34th Annual CHI Conference on Human Factors in Computing Systems, San Jose Convention Center: San Jose, CA, USA, May 7–12* (pp. 3781–3792). The Association for Computing Machinery. https://doi.org/10.1145/2858036.2858537

Sullins, J. P. (2012). Robots, love, and sex: The ethics of building a love machine. *IEEE Transactions on Affective Computing, 3*(4), 398–409.

Szanto, T., & Landweer, H. (2020). *The routledge handbook of phenomenology of emotion.* Routledge.

Tamietto, M., & de Gelder, B. (2010). Neural bases of the non-conscious perception of emotional signals. *Nature Reviews. Neuroscience, 11*(10), 697–709.

Thomas, N. (2016). Mental imagery. In E. Zalta (ed.), *The Stanford encyclopedia of philosophy.* Metaphysics Research Lab.

Tye, M. (2016). Qualia. In E. Zalta (ed.), *The Stanford encyclopedia of philosophy.* Metaphysics Research Lab.

Vaswani, A., Shazeer, N., Parmar, N., Uszkoreit, J., Jones, L., Gomez, A. N., Kaiser, L., & Polosukhin, I. (2017). Attention is all you need. In *31st Conference on Neural Information Processing Systems (NIPS 2017)* (pp. 6000–6010). Retrieved from https://arxiv.org/pdf/1706.03762

Verbeek, P. P. (2005). *What things do.* Pennsylvania State University Press.

Vinciarelli, A., Burgoon, J. K., Pantic, M., & Magnenat-Thalmann, N. (eds.). (2017). *Social signal processing.* Cambridge University Press.

Vonk, R., & Heiser, W. J. (1991). Implicit personality theory and social judgment: Effects of familiarity with a target person. *Multivariate Behavioral Research, 26*(1), 69–81.

Wang, Y., & Zhang, Q. (2016). Affective priming by simple geometric shapes: Evidence from event-related brain potentials. *Frontiers in Psychology, 7,* 917.

Wu, J., Zhang, C., Xue, T., Freeman, W. T., & Tenenbaum, J. B. (2016). Learning a probabilistic latent space of object shapes via 3D generative-adversarial modeling. Retrieved from https://arxiv.org/pdf/1610.07584

Zaborowski, R. (2018). Is Affectivity passive or active? *Philosophia, 46,* 541–554.

Zada, J. (2011). *Take this lollipop.* Retrieved from https://www.takethislollipop.com/

Zajonc, R. B. (1984). On the primacy of affect. *American Psychologist, 39*(2), 117–123.

Zimmermann, A. (2023). Deploy less fast, break fewer things, *Daily Nous,* Deploy Less Fast, Break Fewer Things.

Zhang, C. M., Qiao, G. C., Hu, S. G., Wang, J. J., Liu, Z. W., Liu, Y. A., Yu, Q., & Liu, Y. (2019). A versatile neuromorphic system based on simple neuron model. *AIP Advances, 9*(1), 15324.

Jacqueline Bellon currently works at the International Center for Ethics in the Sciences and Humanities at the University of Tübingen. Her research interests include philosophy of technology, epistemology, history of ideas, history and philosophy of psychology/cognitive sciences, theories of culture and human-human as well as human-technology interaction.

Emotions in (Human-Robot) Relation. Structuring Hybrid Social Ecologies

Luisa Damiano and Paul Dumouchel

Abstract

This essay tackles the core question of machine emotion research—"Can machines have emotions?"—with regard to "social robots", the new class of machines designed to function as "social partners" for humans. Our aim, however, is not to provide an answer to that question. Rather we argue that "robotics of emotion" moves us to ask a different question—"Can robots establish meaningful affective coordination with human partners?" Developing a series of arguments relevant to theory of emotion, philosophy of AI and the epistemology of synthetic models, we argue that the answer is positive, and lays grounds for an innovative ethical approach to emotional robots. This ethical project, which elsewhere we introduced as "synthetic ethics", rejects the dif-

---

An earlier version of this chapter was published in Humana. Mente, 2020, vol.13, issue 37, 181–206.

L. Damiano (✉)
Department of Communication, Arts and Media, IULM University, Milan, Italy
e-mail: luisa.damiano@iulm.it

P. Dumouchel
Département de Philosophie, Université du Québec à Montréal, Montréal (Québec), Canada
e-mail: dumouchel.paul@uqam.ca

fused ethical condemnation of emotional robots as "cheating" technology. Synthetic ethics focuses on the sustainability of emerging mixed human–robot social ecologies. Therefore, it rejects the purely negative ethical approach to social robotics, and promotes a critical case-by-case inquiry into the type of human flourishing that can result from human–robot affective coordination.

Keywords

Affective coordination · (Machine) Emotion · Philosophy of AI · Social robotics · Synthetic method

1 Introduction

"Can machines have emotions?" has generally been taken as the central question in relation to the issue of machine emotions. Can machines feel pain? Can they experience anything? Can they be sad, angry, happy, shameful, exuberant or depressed? Of course, with enough ingenuity and resources we can make some of them act (or react) as if they "had" any or all of these emotions and inner feelings, but do they? Can they have them? If machines cannot have emotions, then talk about "machine emotion" is either metaphorical or has to do with the emotions of human beings in their multifarious relations to different mechanical systems, especially those that aim at eliciting affective reactions from their users, like certain robots, computer interfaces, video games or virtual reality. If machines can have emotions, then the question becomes whether and to what extent these "mechanical emotions" are similar or different from human emotions.

Underlying this question is a particular conception of emotion. Its central characteristic is that of something that an agent, natural or artificial, "has". Whether it is conceived as a long term disposition—"this is an angry person"—or as an event—"he was so relieved when he heard she was safe"—the emotion is circumscribed as something that happens to the agent, and is his or hers. Emotions in that sense are properties in the two senses of the word. First, they are a property that some individuals, i.e. humans, have and that other individuals, i.e. artificial agents, do not have. Second, the emotion is a property of the individual who has it: my emotion is mine; it belongs to me. Emotions so understood are fundamentally internal and private. Seen either as cognitive judgments or as affect programs emotions are events that take place in an individual mind or body. Even when this event coincides with a visible expression, as in anger or rage, it is the agent's experience, or inner stirring that is considered to be *what the emotion is*.

Can machines have emotions? Probably not, at least not in that sense. However, it is far from clear that emotion so understood is a concept that can provide particular insights into affective relations either among humans or those which they establish with artificial agents. Using social robotics as our privileged example, we argue that the question of machine emotions should be addressed in a completely different manner. Social robotics considers that creating robots able to interact emotionally with humans is central to achieving its main goal: robots that can function as social partners. Thus, rather than viewing emotions as private, internal events, social robotics treats them as parts or moments of social interactions. The question it asks is not "Can robots have emotions?" but "Can robots establish meaningful affective coordination with human partners?".

This approach expresses a different conception of emotions, which opens a new perspective on the issue of machine emotion. This presents two closely linked advantages relative to the question: "Can robots have emotions?" First, whether meaningful affective coordination takes place between a robot and its human partner is, rather than a purely theoretical or metaphysical issue,[1] an empirical and experimental one which informs and underlies all of social robotics. It is a question that cannot receive a general "yes/no" answer, as is requested by the question "Can machines have emotions?", but needs to be addressed over again in view of different machines and circumstances. The second, closely related advantage concerns ethical reflection on human–robot interaction. A negative answer to the question "Can machine have emotions?", as is commonly the case, leads to the rejection of, or at least to suspicion towards, all technologies that foster affective relations between humans and artificial systems because, it is argued, they rest on deception. These machines do not have the emotions they pretend to express. This universal condemnation views all such technologies as a danger or evil that needs to be contained through rules and requires precautions. As we argued elsewhere (Damiano & Dumouchel, 2018), this attitude tends to be self-defeating for it restricts the search for ethical criteria appropriate for different types of artificial agents and situations. Since all are damned, there is no difference to be made between them! To the opposite, focusing on affective coordination encourages the discovery of multiple criteria that reflect the specific characteristics of different relations and machines.

---

[1] More precisely, the question does not need to be resolved theoretically before it is addressed empirically. In that sense it is experimental, rather than technical.

The reminder of this essay develops this critical perspective. In section I we argue that social robotics, given its methods and objectives, constitutes a frontier in machine emotions. In section II we distinguish and analyze the two main approaches in social robotics, which, we argue, reproduce in this domain the strong and weak AI distinction as defined by John Searle (1980). In section III we consider social robotics' implementation of emotions and characterize it in terms of salient moments in a process of affective coordination. Section IV draws out the consequence of this alternative view for our understanding of "machine emotions" and the affective competence of social robots. Finally, section V addresses the issue of the ethics of affective relations with artificial social agents.

2 Social Robotics: A Frontier of Machine Emotion

Social robots are often viewed in the light of fictional characters such as the Golem of Prague, the Creature from Mary Shelly's *Frankenstein*, or Radius, and the other biochemical robots, staged by Karel Čapek in *RUR (Rossum's Universal Robots)*. However, the forerunners of today's social robots are better represented by a different type of fictional artificial agents, which use their thinking and feeling abilities to be accepted by humans as sympathetic interlocutors, rather than to rebel or even to attempt to exterminate us. One of the earliest of these positive characters is Lester del Rey's Helen O'Loy (1938), a fictional robot that evinces many features of the machines which contemporary social robotics plans to integrate into human society. In del Rey's short story, two friends modify a latest model housekeeping service robot giving it affective and relational skills. These constitute a central element of the plot, for they dramatically modify the status of the robot. Its human users begin treating this artifact, which they have created, as a person, rather than simply as a machine that cooks and cleans. Its social affective skills metamorphose the robot from an object into a subject, an individual with "who" a personal, even an intimate relation becomes an option. Helen, on the one hand, is a paradigmatic expression of the goal pursued by current social robotics—namely, creating machines that cross over from the purely material to the social realm. On the other hand, her adventure introduces us to the real challenge posed by present-day social robots. This, far from being the question of their rights or the danger of their rebellion, is to understand the possibilities and limits of social interactions between robots and humans, to determine and regulate, in a sustainable way, their place within our social ecology, which their presence will inevitably transform.

Only recently did social robots travel from science fiction to engineering. This migration can be traced back to two main events. The first is the 1990s "embodiment turn" in cognitive science. In AI, the idea that the human mind is an "embodied" shifted the focus from disembodied computer programs to complete agents immersed in an environment that is essentially social. From there emerged a conception of intelligent artifacts as robots whose cognitive competences are inherently social (Damiano & Dumouchel, 2018). The second crucial event, a few years later, is the birth of Human–Robot Interaction (HRI), a research area where interest in the social dimension of artificial embodied intelligence found an ideal field of development. With the diffusion of advanced robotic appliances operated by humans, HRI emerged as an interdisciplinary domain dedicated to exploring and improving cooperative performances of humans and robots. In consequence, increased attention was given to the social aspects of human–robot relations, leading to the programmatic notion of robots able to communicate with humans through shared social signals. The project of creating robotic "co-workers" for humans can be seen as the starting point of social robotics. Related research programs, combining embodied AI and HRI, did not simply enlarged the range of potential uses of robots, to include "office, medicine, hotel use, cooking, marketing, entertainment, hobbies, recreation, nursing care, therapy and rehabilitation (…), personal assistance" (Daily et al., 2017). They also enriched the concept of interactive robots, extending it from the notion of human-operated machines to that of social partners. This novel type of robots relied on "peer-to-peer interaction skills" enabling them not only to accomplish different tasks in cooperation with humans, but also to interact socially as "peers"—companions, friends, partners.

The strategy adopted by social robotics to create "socially interactive robots" (Fong et al., 2003) can be described in terms compatible with the plot of del Rey's short story: "building machines that are liberated" from "the subject-object dichotomy" (Jones, 2017). In the more technical language of social robotics, this is equivalent to creating robots that have a believable "social presence", defined as the capability of a robotic agent to give its human user the "sense of being with another" (Biocca et al., 2003), or the "feeling of being in the company of another" (Heerink et al., 2008). Doing this entails transferring and adapting to human–robot interactions some aspects of face-to-face social interaction among humans. Among these, one of the most important is the ability to recognize and express emotions.

> The novelty of a mechanical tool that moves with purpose keeps people's attention, but adding emotive capability allows the robot to interact with humans socially. (Daily et al., 2017)

The above quote highlights the central relevance of machine emotions for social robotics. Within this domain, to give robots the ability to interact with humans appropriately at an emotional level, "affective computing" (Picard, 1997; Picard & Klein, 2002) researches emotion sensing, recognition, expression, and generation. Related design and implementation of artificial emotion in robots involve interdisciplinary studies, combining theoretical and experimental HRI with philosophical, psychological, ethological, biological, anthropological, socio-cultural research, as well as design and engineering (Damiano et al., 2015a). In many cases, the goal is not simply applying knowledge to build believable emotional robotic agents. Social robotics also engages in scientific research proper. Adopting the method of "understanding by building" (Pfeifer & Scheier, 1999), theories on emotions are incorporated in robots with the goal of experimentally testing them in HRI settings to contribute to the scientific understanding of emotional processes (Damiano & Cañamero, 2010).

Clearly, social robotics is a domain where frontier inquiries on machine emotion take place, and where the issue of whether machines can have emotions is most often raised. Notably, when raised in relation to social robots it is not purely speculative. More than in any other area, the technological offshoots of social robotics cannot be reduced to the machines it creates, for social robotics also, and simultaneously, engages in the construction of new different social relations. Therefore, answers to the question of machine emotion in the case of social robotics is not merely philosophical, but also fundamentally practical. How this question is answered will have a significant impact on how humans (will) cohabit with social robots.

3 Social Robots at a Relational Turn

Since its beginning, social robotics has been marked by a tension between two approaches. The first, which arose in the late 1990s, is based on a biologically inspired embodied AI framework. It aims at developing robots that possess social competences in a "substantial" sense. This approach targets the production of "socially intelligent" robots whose social presence is grounded in cognition skills modeled on human social abilities. The central idea is that "deep models" of human social competence will allow robots to interpret and answer appropriately to social signals, and thus "show aspects of human style social intelligence" (Fong et al., 2003). The second approach abandons this ambitious goal in favor of a more "functional" objective, which is to build robotic agents that give the impression of being socially intelligent, but which are not. Such robots can be defined as "socially evocative" or "social interfaces" (Breazeal, 2003). They are designed to *simulate* some social competences well enough to engage humans in social interaction. While the first wants to create robots that have social competences, the second seeks to make robotic agents whose physical appearance and behavior trigger anthropomorphic projections. The two approaches are significantly different. One wants *to endow robots* with social abilities, the other aims *to bring human users* to treat robots as interlocutors or social inter-actors.

This tension between a "substantial" and a "functional" approach is not a novelty introduced by social robotics. In fact, it reproduces, with specific focus on robots and social cognition, the opposition between "weak" and "strong" AI that Searle (1980) already recognized within classic (pre-embodiment) AI. According to him, "strong AI" seeks to create artificial systems that are able to perceive, understand, think and believe—that is, to reproduce all human cognitive processes. "Weak AI", to the opposite, is based on the idea that artificial systems can, at best, only *simulate* human intelligence, can do no more than imitate some human cognitive processes. Social robotics reformulates this distinction in terms of the opposition between genuine competence and the mere simulation of human social cognition. The competence is implemented directly in the internal cognitive architecture of robots, while simulation rests on external anthropomorphic characteristics of artificial agents. Just as strong AI seeks to create artificial minds, in social robotics the substantial approach envisions the creation of an artificial social species, whose members will integrate human society (MacDorman & Cowley, 2006). Just as weak AI sees in computer programs not artificial minds, but useful artifacts, functional social robotics does not consider its artificial agents as genuine social partners, but as acceptable technological solutions.

This distinction recently reappeared in a new guise within "android science", a subdivision of social robotics and HRI that is dedicated to the production of increasingly human-like robots (Ishiguro, 2016; MacDorman & Ishiguro, 2006). Here, Searle's strong/weak AI opposition is used to develop a framework to evaluate the "human-likeness" of current and future humanoid robots (Kahn et al., 2007). In this context, the strong/weak duality qualifies the "force" of "ontological" and "psychological" evaluations of robots' human-likeness. The first type of evaluation concerns the ontological status of humanoid robots—"are humanoid robots humans or machines?" The second type focuses on their perceived status—"are humanoid robots perceived as humans or machines?" Strong ontological claims assert that they "are" or "became" human, something which some proponents of the android science consider possible in the future. Weak ontological claims deny this to be the case. Analogously, psychological claims are "strong" when humans see human-like robots as other humans, and "weak" when humans regard them as mere machines.

One interesting aspect of this reformulation is that it brings back into focus—after Searle's critique of the Turing test—the role of human perception in the assessment of artificial systems. It is clear from literature in social robotics that this turn towards users' assessment is a general trend which can be characterized as a paradigm shift. Raya Jones (2017), for example, describes this increasing attention to users' experience as leading specialists to primarily, or even exclusively evaluate the social character of robots in terms of users' experience and judgment of their interaction with these artificial agents. On this basis, she detects a "subtle 'paradigm shift'" bringing specialists to progressively reject the idea that sociability corresponds to a set of individual skills or competences that can be reenacted, deeply or superficially, by machines. Indeed, current procedures assessing robots' social character do not understand it as a trait characterizing them as individual agents. The evaluation does not focus on the robot's social features, but on the users' perception of their artificial partners, suggesting that robots' artificial sociability arises in human–robot interaction.

While we agree with Jones that the diffusion of user-centered approaches can be seen as a marker of a paradigm shift, we doubt that, at this point, this transformation can be described as a "relational turn" (Jones, 2017). The increased focus on users' experience and judgment indicates that current specialists in social robotics tend to locate the robots' sociability more in the users' eyes than in the relation between users and robots. This perspective, together with related *modus operandi*, can be considered as a first step towards a relational turn to the extent that it involves the users in the generation of robots' social character. Yet, a full relational turn requires more: to recognize that sociability is a property emerging from a coordination dynamics between users and robots, that can neither be implemented as a trait of individual robots, nor as a users' projection, for it is distributed in the mixed human–robot system that users and robotic agents form through interacting.

However, as we argued elsewhere (Damiano et al., 2015b; Dumouchel & Damiano, 2017), there are some indications that such a relational turn is actually taking place in social robotics.

4   Artificial Emotions in Relation

From the late 1990s, specific versions of the substantial and functional approaches to social competence focused on emotions appeared in robotic research. The first, the "internal robotics of emotions", aims at endowing robots with deep models of human or animal emotional systems. The second, the "external robotics of emotion", develops robots that express emotions, but are without an internal architecture enabling them to generate emotional states. Again these two approaches reproduce in the domain of artificial emotions Searle's classic distinction between strong and weak AI. Whereas internal robotics of emotion pursues the creation of genuine emotion, external robotics targets the mere simulation of emotion (Damiano et al., 2015b; Dumouchel & Damiano, 2017).

During the last decade, the limitations of both approaches—the technical difficulties of implementing "deep models" of emotion in robots and significant limits of how believable purely expressive robots can be—led researchers to developments better suited to support real-time, long-lasting affective interactions between humans and robotic agents. The result was the emergence of a third approach, the "affective loop approach". As its name suggests, instead of focusing on a robot's ability either to generate or to express emotion, it concentrates on its capacity to engage humans into affective exchanges. The goal is to bring "the user to [affectively] respond [to the system] and step-by-step feel more and more involved with the system" (Höök, 2009), to favor continued human–robot social interaction. This objective is achieved by endowing the robot with "intelligent expression" (Paiva et al., 2014), that is, emotional expressions that dynamically coordinate with that of the users. It is implemented by a robotic architecture that allows the robot to perceive and recognize emotions, and to express in return emotions that are "tuned" to the users' affective expressions. In some models, the internal architecture also includes an emotion generation mechanism loosely inspired by animal or human emotional systems, or alternatively, that is "invented" for robots.

It is this social robotics approach to machine emotion that effectively realizes the *relational turn*. While internal and external robotics of emotion focus on robots as individual agents, the affective loop centers on human-robots pairs or groups. Whereas internal and external robotics of emotion aims at constructing (genuine or simulated) emotions in individual agents, the goal sought by the affective loop approach is for emotions to arise within the human–robot system as a whole. What it targets is not an individual emotional process, but an emotional dynamic grounded in the recursive coordination of the affective expression of the agents involved.

This approach suggests a conception of emotion which remains unformulated, but that is completely different from viewing them as individual skills or properties. Social interactions, whatever else they may be, clearly are joint ventures. Their success or failure, their permanence or fugacity, depend on all partners involved, and that characteristics of social interactions is central to social robotics. The emotions that emerge in social relations and constitute a fundamental and integral part of the interaction should in the same way be viewed as collective creations, rather than as private, individual events. This idea, that emotions are joint creations emerging in affective coordination, challenges our common understanding of emotions in a radical way.

Affective coordination, we argue, is the process through which we jointly determine our reciprocal intentions of action and intentions towards each other (Dumouchel, 1999). Game theoretical models of coordination we find in economics or biology analyze how two or more agents, whose interests at least partially converge, coordinate their action in view of a particular goal that they share. Because they assume that the convergence of interest is sufficient to insure that the agents will collaborate, these models never ask the questions: "who do you want to play with?" and "who don't you want to play with?" Yet those questions are fundamental in both human life and social robotics. In social robotics, because a successful artificial social agent is one "we want to play with" independently of whatever task we may engage in together. In life, because the repulsion or pleasure that we find in interacting with others brings us to weigh quite differently the interests whose convergence or divergence the theory of games takes as given.

Affect coordinates persons with each other. The issue is not coordinating the actions of multiple agents so that they can successfully achieve a common goal, but to determine whether or not they want to pursue a common goal. Affective interaction is the way through which we coordinate our reciprocal intentions of action. It is thus prior to both cooperation and conflict, or competition, as well as to lofty or lazy indifference. For members of a social species like ours, whose advantages and disadvantages primarily depend on interactions with each other, rather than on solitary relations to the natural environment, material situations under-determine agents interests. Individuals' preferences for this or that outcome are conditional upon other agents' preferences and vice versa. Affective coordination is the mechanism through which these preferences are determined and the uncertainty concerning the other's intention towards me and my intention towards her or him is lifted.

It is a reciprocal process in which a first agent partially determines the intentions of a second towards her or him, while this second partially determines the first's intention towards the second. To put it otherwise, your attitude towards me partially determines my attitude towards you and, vice versa, my attitude towards you partially determines your attitude towards me. In this reciprocal process, affective expressions occupy the central place. The process is, however, generally unconscious. That is, we are mostly unaware of the how our affective expression affects others, or even of what it is, and if we attend more to the affective expression of others, it nonetheless often remains below the level of conscious apprehension. This process of affective coordination is unattended to and non intentional in both the ordinary and the technical sense of that term. It is neither a conscious object, nor something that I want to do. Though my intention towards you results from this process, and vice versa, it cannot itself be intentional under pain of infinite regress.

Emotions, we propose, are moments of this process of coordination which become salient for one or more of a variety of reasons, for example because they are fixed points of coordination, because they represent a sudden reversal, because of lasting inability to reduce uncertainty concerning the other's intention, etc. In every case, what determines a moment to be an emotion is not an individual's mental or bodily state, but some aspect of a process of coordination that involves at least two agents.

Such a proposal entails a profound transformation of the concept of emotion. Rather than being defined as something that happens to an individual, either an internal event characterized by some mental or bodily change, or a public manifestation that is shaped by, or even created as a result of, social rules and expectations, an emotion is here construed as a relational property. Just like being "taller than" is a property of an individual agent, but which that agent, for example John, cannot have by himself, independently of his relation to others, say to Louise or Peter. Similarly "being sad" is certainly a property of an individual agent but, if it is a salient moment in a process of affective coordination, then it is not one she or he can have independently of others. *It does not follow from this that another person is necessarily the cause or "target" of the emotion*, for that does not need to be the case in order for a moment to be salient in a process of affective coordination.

One of the evident consequences of this re-conceptualization of emotion is that it does not anymore coincide with our everyday use of the term. Some events which ordinary language identifies as emotions are not emotions according to this definition and alternatively others, which it fails to recognize as such, may very well be salient moments in the process of affective coordination. This, as far as we are concerned, is all for the best. A central weakness of current theories of emotions, and the main reason why they have failed so far to propose agreed upon characterizations of emotions, or even a common list of basic emotions, is that they consider that the everyday term "emotion" and names of emotions properly identify the phenomena they want to investigate and provide incontrovertible and basic insights into it. There is however little reason to believe that to be the case; to think that our spontaneous grasp of emotions is closer to a scientific understanding of the phenomena they point to, than our spontaneous understanding of physics is to the scientific discipline that bears that name.[2] Our proposal goes against common sense; that is its force, we argue, not an objection. It is further perfectly well adapted to the challenge that social robotics faces: understanding and bringing about emotions in relation.

Affective coordination rests on affective expression and given that emotions are salient moments in this process, it follows *that expression precedes emotions*. An emotion then should not be seen as a cause, of which affective expression is an effect, but is rather itself as an effect of a dynamic of reciprocal affective expression. Because an emotion is not essentially an internal event, the question is not whether an agent, natural or artificial, has the relevant internal event relative to his expression—i.e. "do machines have emotions?"—but whether this moment corresponds to a *projectible* predicate (Goodman, 1954) of the relation. As one of us argued elsewhere, "sincerity is consistency" (Ross & Dumouchel, 2004), a fixed point in a dynamic of relation.

5 Social Robots: Models or Partners?

The affective coordination hypothesis, applied to our interactions with social robots, argues that, in spite of the irreducible differences between human–human and human–robot relations, these last interactions also rest on affective coordination. More precisely, the idea is that they constitute a limited, specific, but certainly genuine form of affective relationship (Damiano & Dumocuhel, 2018). The claim, to put it otherwise, is that social robots are affective actors or agents, rather than simply emotional regulators, like teddy bears or comfort food.

This position is controversial and not surprisingly the common objection raised against it is that robots, like puppets, do not have emotional states.

---

[2] See L. Wolpert (1992).

Lacking an internal animal-like emotional system, these artifacts are unable to reciprocate emotions. In consequence, it is impossible to establish with them bi-directional affective relations. Whether or not our affective exchanges with them are designed following the affective loops approach is beside the point, is it argued. Robots are inert objects, incapable of emotions. Like dolls, they are fake, not true emotional agents, though we may become attached to them.

We do not deny that robots are objects, nor that social robotics uses insights from puppetry to design them as technologically enhanced dolls, able to create in their users a "suspension of disbelief" that makes these artifacts appear as humans' peers (e.g., Duffy, 2006; Duffy & Zawieksa, 2012; Zawieksa & Duffy, 2014). Our claim neither concerns the ontological status of social robots—"are they objects or subjects?"—nor the users' perception of robots as merely objects or as subjects. The affective coordination hypothesis centers on the interactive dynamics between social robots and humans, and the related interactive roles that these machines can play. What we maintain is that these "sophisticated dolls" can reproduce in interactions key elements of the dynamics of human–human affective coordination. In particular, they can reproduce key elements of recursive affective expression coordination that leads inter-actors to mutually determine their disposition to action. In consequence, robots are artificial affective inter-actors, or emotional agents. They are a new kind of partner in affective relations with humans. They have limits as emotional agents, and so do affective relations with them, but robot-human affective interactions are nonetheless authentic, because robots can participate in affective coordination dynamics with the humans with whom they interact.

Current epistemological reflection on artificial or synthetic models –that is, the *material models* that are produced by the synthetic or "understanding by building" approach—may help clarify our position. Such models are primarily characterized by the fact that they are not simply abstract theoretical representations, but functional systems, and often material objects. They are models in the sense of Suppe's (1989, p. 167) *iconic model*: "an entity that is structurally similar to the entities in some class". For example, a synthetic cell is such a model of a cell. In consequence, they are also models in the sense of "a new car model", as they instantiate a particular type or kind. These models are built either as functioning robotic, computational, chemical or as composite systems able to generate (reproduce) targeted natural processes. They are used for the experimental investigation of aspects of the target processes that are not accessible in traditional research scenarios.

A recent proposal (Damiano & Cañamero, 2010; Damiano et al., 2011) introduces two "criteria of relevance" for synthetic models. The first, which aims at assessing their "phenomenological relevance", claims that a synthetic model is phenomenologically relevant if it exhibits, within strict parameters, the same phenomenology as displayed by the target system, regardless of the mechanisms used to achieve this effect. The focus is on the capacity of an artificial system to reproduce the behavior of a natural system, independently of how that behavior is generated. The second criterion aims to evaluate the "organizational relevance" of synthetic models. A synthetic model is organizationally relevant if its organization reproduces what constitutes the target system's organization according to a pertinent scientific theory. As such, it is quite different, since it shifts the focus from the phenomenology or behavior to the organized mechanisms underlying that behavior. Thus, in the context of robotic modeling of emotions, a model is phenomenologically relevant if it reproduces the emotional behavior observable in the target systems, and is organizationally relevant if it reproduces the mechanisms that underlie the target systems' behavior.

One interesting aspect of these criteria is that they allow us to escape the rigid alternative between pure imitation and genuine reproduction when we assess the scientific value of synthetic models. These criteria make possible a classification of forms of relevance for synthetic models that takes in consideration the possibility not only of partial expressions, but also of different combinations of phenomenological and organizational relevance. In fact, the "weak/strong AI" alternative corresponds to two extreme cases. On the one hand, we have phenomenologically relevant models, which are purely imitative, without any organizational relevance. On the other hand, we have models that are organizationally *and* phenomenologically relevant, and in this sense reproduce the "real thing"—both the behavior and the organization of the target system. Additionally, this classification contemplates the case of models that are *only* organizationally relevant, which could happen if the reproduction of the target's organizational mechanisms gave rise to new behavior, divergent from that of the target systems. To put it otherwise, this classification takes into account that synthetic modeling can lead to novelties, to man-made variants of natural systems, which (partially) share their organization and may manifest new behaviors. Furthermore, the classification accounts allows "progressive" and "interactive" forms of phenomenological or organizational relevance. Progressive relevance refers to when artificial systems exhibit unexpected organizational features or behaviors that are pertinent to the ongoing inquiry.

Interactive relevance characterizes artificial systems able to engage natural systems in dynamics germane to the scientific investigation they support.[3]

It is important to point out that this classification is not theoretically neutral. For example, when assessing robotic emotions, the results will be quite different depending on the view of emotion which we adopt as reference; among other things because the target phenomena is not the same in the two cases. In consequence, the individual, private conception of emotion and the affective coordination approach determine different behaviors and organizational mechanisms to be taken into account when creating and evaluating robotic models of emotion.

If we opt for the classic view of emotion, then robots are to be understood as models of "individual affective agents", characterized by individual emotional processes supported by underlying internal mechanisms, that are also individual, that is, proper to each agent. In this case, all existing robots have low organizational relevance. In fact, specialists of the internal robotics of emotions consider that reaching significant organizational relevance is as "a task for the future" (Parisi, 2014). Most social robots consist of artifacts that, like Paro or Keepon, do not reproduce animal-like or human-like emotional mechanisms, but merely display some limited form of emotional behavior. Interestingly, all these social robots demonstrate interactive phenomenological relevance. Through their ability to reproduce aspects of animal or human emotional behavior, they succeed to different degrees in engaging humans in emotional interactions. Their interactive relevance, when considered in the context of classical theories of emotion, defines social robots as imitative models that can explore emotional processes in natural systems, in the sense that they can be used for the experimental study and manipulation of interactive emotional dynamics in humans and animals.

However, in the context of emotions as relation, this common characteristic of social robots is precisely what makes them successful models of agents in affective coordination. When assessed on this different basis, the focus is on the robots ability to participate in inter-individual emotional dynamics, rather than to model individual emotions. The target phenomenology is different: not individual internal events, but interactive behavior. Similarly, organizational relevance is not to be measured against internal mechanisms that are deemed to produce animal or

---

[3] The two criteria of relevance and the classification of related forms of relevance develops work originally introduced in (Damiano & Cañamero, 2010; Damiano et al., 2011). Developments first presented at workhop SA-BCS 2018 at ALIFE18 (SA-BCS 2018, L. Damiano, *The Synthetic Method. Proposing a framework to the synthetic sciences of life and cognition*).

human emotions, but in relation to the central mechanism of affective coordination, the recursive process of co-expression that leads inter-actors to co-determine their dispositions to action. At this point, many social robots can be considered able to successfully reproduce at least some key aspects of this mechanism. These robots can engage humans in recursive dynamics of expression that trigger in their users—and sometimes also in the robots[4]—changes in their disposition to action. From this point of view, social robots are synthetic models of agents in affective coordination to which we can ascribe partial organizational *and* partial phenomenological relevance.

It is true that the emotional expression of robots, the complexity the forms of coordination of which they are able and the range of actions available to them remain quite limited. At this point, affective coordination with them has neither the wealth nor the breath of what we find in human–human, or even human–animal, affective interactions.[5] However, within these limits, social robots function as effective models of human emotional inter-actors, although the internal mechanism that support their ability to enact aspects of the affective coordination dynamic are completely different from those of their human partners.

We can focus on the partiality of organizational and phenomenological relevance of robotic models in affective coordination, and consider these artifacts as models of humans affective agents. Alternatively, we can focus on the synergy at work between the partial organizational *and* phenomenological relevance of social robots. In this case, their divergence form the "real thing" defines these robots as new models of affective agents that lead to a new and different phenomenology, that of mixed human–robot affective coordination.

The two positions lead to different ethical stances. The first tends to converge with the classic, individualist view of emotions in supporting the diffused ethical stance according to which emotional robots, when they migrate from labs to social spaces, can no longer be defined as modeling emotional processes, but only as a "cheating technology", used to create in humans the illusion of reciprocal affective relations. This view is at the basis of current ethical reflections on social robot that condemns them or even suggest banishing them, in spite of or perhaps because of their concrete uses (e.g., Turkle, 2010). The affective coordination approach underpins a different ethical position, where the question of

---

[4] This depends on the type of machine we are dealing with.

[5] However the range of affective coordination of some semi-autonomous robots, for example Kaspar, can be quite large. See (Dumouchel & Damiano, 2017, pp. 158–163).

the emotional imposture of social robots is replaced by that of the differences between human–human and human–robot affective coordination. To put it otherwise, protecting users from confusing human–robot affective coordination with human–human affective coordination is analogous to protecting them from confusing natural and artificial intelligence. This is, by far, not a trivial issue, especially when we consider not only the problems of addiction to video-games or the status of virtual reality, but also the multiple projected social uses of AI in commerce, law, or the military. Responding these challenges does not however imply that we renounce to all social robots more than to all AI appliances. The issue is regulating the use of these technologies. In particular, to support the sustainability of emerging mixed human–robot social ecologies, *we have to* recognize human–robot emotional interaction as a specific form of affective coordination, and to regulate it. Why? Certainly not to defend robots' rights (robots are only objects!), but to defend humans and to allow them to profit of the advantages that these robots can bring, especially in their "socially assistive" uses.

6 Synthetic Ethics

If human–robot affective coordination can be successful, and the extent to which it can, raise the question of the ethics of affective relations with machines. Until now this issue has been dominated by the claim that such relations can never be authentic because machines do not have emotions. Whatever the sentiments of humans towards a machine are, they will never be paid back. Such a relation, it is claimed, will inevitably in the end reveal itself to be a fraud. However, if emotions correspond to salient moments in a process of social coordination, rather than they are internal states of individual agents, the central question of the ethics of human-robots affective interactions needs to be reformulated, for it is not the presence or absence of particular internal states that is at issue here. What is it then?

One way to approach this question is to ask: what distinguishes human–human, or even human-animal, affective coordination from human–robot coordination? How different is the process in one case and in the other, and what is the moral import of that difference? What distinguishes, say, Paro's friendly disposition towards an aging and lonesome pensioner from that of a dog? In del Rey's short story, Lena, the first robot that the two friends attempt to modify, giving her emotions, one evening explodes shouting at them, indignant that they reproached her having ruined their diner. "Liars! how dare you blame me! If you did not rewire me every evening, maybe I would have time to get something done in this house!" Would it be possible to create an artificial agent that could be humiliated and would fly in a rage as a response? Maybe, but as one Japanese roboticist once told us "there would be no market for such a robot." Effectively, Dave and Phil, in reaction to her outburst, unplug Lena and decide to replace her by a more advance housekeeping robot. This upgrade however did not prove entirely successful. Model K2W88 or Helen, as the machine is called now that she can experiences and expresses emotions, soon is literally dying of heartbreak because Dave fails to return her love.

What this short fiction makes clear is that vulnerability to others, to what they do or do not do to us, and vice versa, is a fundamental dimension of human affective coordination. What others do to us and us to them, however, in affective interactions *we often do without necessarily doing anything to them directly*. Simply by changing attitude, becoming unavailable, interested in something or someone else, or to the opposite by paying attention to what interest you, agreeing or encouraging and supporting your endeavors. We are vulnerable to such changes. They make us happy or unhappy, can move us to friendship and love, but also to anger or even violence, and they slowly affect our character and long-term dispositions. We are vulnerable to each other in that such changes in others' dispositions change us, for better or worse, spur us to action or to the opposite confine us to procrastination or depression.

Unlike a friend, a child, a lover, unlike Lena or Helen, Paro does not care if you suddenly abandon it, prefer what is on TV right now, insult it (even if it could understand) more narrowly focused on emotions, for whatever reason frivolous or important. That invulnerability is the main difference between machines and humans engaged in affective interrelations and it is what underlies the common obsession with deception. Robots, machines cannot be hurt, they can only be damaged, destroyed or unplugged. Unlike us, machines are, as Hobbes (1651, p. 226) says of irrational creatures, unable to distinguish between *injury* and *damage*.

It would most likely be possible to add to the behavioral repertoire of Paro a reaction of sadness or resentment that could be triggered in various circumstances.[6] Would that radically change our relation? Or would it just be some more

---

[6] In its present form, Paro can recognize if it is being caressed or mishandled, but important changes would be needed for it to be able to track different relations and the changed or suddenly offensive attitude of regular partners. The point however is not about Paro as such but simply that it is technically possible to make an artificial agent able to express anger, resentment or shame in appropriate circumstances.

"make believe"? What would it mean, what would it take for a robot to be vulnerable in that affective sense to its human partners?

Affective coordination is at the heart of human sociality, of our extraordinary success in living, acting, communicating and simply being together. This disposition which characterizes us as a highly social species—whose members receive most, if not all, of their advantages and disadvantages from their relations with each other rather than through solitary interaction with the environment[7]—is however not altogether peaceful. As Kant famously termed it, ours is an unsociable sociability. If humans want and need to live together,[8] they also fight and conflict. They flee and resent the presence of others as much as they seek and desire it. We are attached to each other by hatred and jealousy as much as by love and admiration. Giving robots this vulnerability to their human partners, and the sensibility that comes with it, would either mean transforming these machines into helpless victims or giving them the ability to willfully harm the person they interact with. That is why, at this point, it is not necessary to answer the question "could we do it?" The first rule of ethics of social robotics it that artificial agents should not harm those they interact with.

We however are vulnerable and this raises a second question. In what way is the attachment humans develop towards artificial agents different from the affective attachment they develop towards other objects? This question is generally not asked because of the common prejudice that social robots are designed to induce their human partners into falsely believing that they have emotions and feelings. Therefore, it is assumed that in this case the attachment is, of course, different from what it is in the case of other objects, since unlike a book or a comfortable chair, the robot tricks the human to become attached.

The fact is, however, that some people cry when the house where they were born and raised is torn down to make room for a high-rise building. Others are profoundly distraught by the lost of an object to which they are deeply attached. It is normal to develop strong affective relations not only with people, but also to places, to objects, even to abstract concepts. It is normal to step forward to defend them or take action to protect them. The point is not that exactly the same attachment is involved when soldiers beg mechanics to fix their landmines detecting robot that has been badly damaged in an explosion and insist that they do not

---

[7] For a more extensive presentation and analysis of this definition of social species, see (Dumouchel, 2017, pp. 311–312).

[8] Solitude is not only a source of psychological suffering and mental disease, it is also a major cause of physical morbidity and mortality. See (Hawkley & Cacioppo, 2010).

want a new one, but want that one back, their robot and none other! Rather, the point is that there is nothing strange or surprising in their behavior, nothing that requires in order to be explained, for us to assume that they falsely believe that the robot is more than a machine.

The truth is that we do not know what differences there are between attachment to social robots and towards other objects. Nor do we know what are the consequences of the lack of vulnerability of artificial agent for affective interactions with humans. We do not know, because these questions are not asked, and if they are not asked, it is because many assume that the answers can evidently and immediately be derived from the fact that social robots do not have the emotions that they appear to harbor.

We believe that such question need to be asked, for it is only by answering them that we can discover the multiple dangers and advantages of interacting with artificial social agents. Answering them however is an empirical endeavor. There are no priori answers to these ethical questions and the answers will change as machines become different and people develop new forms of affective interactions with social robots. We call synthetic ethic the inquiry into this evolving moral reality (Damiano & Dumouchel, 2018; Dumouchel & Damiano, 2017).

References

Biocca, F., Harms, C., & Burgoon, J. K. (2003). Toward a more robust theory and measure of social presence. *Presence, 12*, 456–480.

Breazeal, C. (2003). Toward sociable robots. *Robotics and Autonomous Systems, 42*(3), 167–175.

Damiano, L., & Cañamero, L. (2010). Constructing emotions. In J. Chappell, S. Thorpe, N. Hawes, & A. Sloman (eds.), *Proceedings of the international symposium on AI inspired biology* (pp. 20–28).

Damiano L., & Dumouchel P. (2018). Anthropomorphism in human–robot co-evolution. *Frontiers in Psychology, 26*, 468.

Damiano, L., Dumouchel, P., & Lehmann, H. (2015a). Artificial empathy: An interdisciplinary investigation. *International Journal of Social Robotics, 7*, 3–5.

Damiano, L., Dumouchel, P., & Lehmann, H. (2015b). Towards human-robot affective coevolution. *International Journal of Social Robotics, 7*, 7–18.

Damiano L., Hiolle A., & Cañamero L. (2011). Grounding synthetic knowledge. In T. Lenaerts, M. Giac- obin, H. Bersini, P. Bourgine, M. Dorigo, & R. Doursat (eds.), *Advances in Artificial Life, ECAL 2011* (pp. 200–207). MIT Press.

Daily, S. B., James, M. T., Cherry, D., Porter, J. J. III, Darnell, S. S., Isaac, J., & Roy, T. (2017). Affective computing: Historical foundations, current applications, and future trends. In M. Jeon (ed.), *Emotions and affect in human factors and human-computer interaction* (pp. 213–231). Elsevier Academic Press.

Duffy, B. R. (2006). Fundamental issues in social robotics. *International Review of Information Ethics, 6*, 31–36.
Duffy, B. R., & Zawieska, K. (2012). Suspension of disbelief in social robotics. In *The 21st IEEE international symposium on robot and human interactive communication, IEEE RO-MAN 2012* (pp. 484–489).
Dumouchel, P., & Damiano, L. (2017). *Living with robots*. Harvard University Press.
Dumouchel, P. (2017). Acting together in dis-harmony. Cooperating to conflict and cooperation in conflict. *Studi Di Socioligia, 4*, 303–318.
Dumouchel P. (1999). *Émotions*. Institut Synthélabo.
del Rey, L. (1938). Helen O'Loy. *Astounding Science Fiction, 22*(4), 118–124.
Fong, T., Nourbakhsh, I., & Dautenhahn, K. (2003). A survey of socially interactive robots. *Robotics and Autonomous Systems, 42*(3/4), 143–166.
Goodman, N. (1954). *Fact*. Harvard University Press.
Hawkley, L. C., & Cacioppo, J. T. (2010). Loneliness matters: A theoretical and empirical review of consequences and mechanisms. *Annals of Behavioral Medicine, 40*(2), 218–227.
Heerink, M., Kröse, B., Evers, V., & Wielinga, B. (2008). The influence of social presence on acceptance of a companion robot by older people. *Journal of Physical Agents, 2*, 33–40.
Hobbes, T. (1651/1976). *Leviathan*. Edited by C. B. MacPherson. Penguin Books.
Höök, K. (2009). Affective loop experiences: Designing for interactional embodiment. *Philosophical Transactions of the Royal Society B, 364*, 3585–3595.
Ishiguro, H. (2016). Android science. In M. Kasaki, H. Ishiguro, M. Asada, M. Osaka, & T. Fujikado (Eds.), *Cognitive neuroscience robotics a synthetic approaches to human understanding* (pp. 193–234). Springer.
Jones, R. (2017). What makes a robot 'social'? *Social Studies of Science, 47*(4), 556–579.
Kahn, P. H., et al. (2007). What is a human? *Interaction Studies, 8*(3), 363–390.
MacDorman, K. F., & Cowley, S. J. (2006). Long-term relationships as a benchmark for robot personhood. In *Proceedings of the 15th IEEE International Symposium on Robot and Human Interactive Communication* (pp. 378–383).
MacDorman, K. F., & Ishiguro, H. (2006). The uncanny advantage of using androids in social and cognitive science research. *Interaction Studies, 7*(3), 297–337.
Paiva, A., Leite, I., & Ribeiro, T. (2014). Emotion modeling for social robots. In R. Calvo, S. D'Mello, J. Gratch, & A. Kappas (Eds.), *Handbook of affective computing* (pp. 296–308). Oxford University Press.
Parisi, D. (2014). *Future robots*. John Benjamins.
Pfeifer, R., & Scheier, C. (1999). *Understanding intelligence*. MIT Press.
Picard, R. W. (1997). *Affective computing*. MIT Press.
Picard, R. W., & Klein, J. (2002). Computers that recognise and respond to user emotion. *Interacting with Computers, 14*, 141–169.
Ross, D., & Dumouchel, P. (2004). Emotions as strategic signals. *Rationality and Society, 16*(3), 251–286.
Searle, J. (1980). Minds, brains and programs. *Behavioral and Brain Sciences, 3*, 417–424.
Suppe, F. (1989). *The semantic conception of theories and scientific realism*. University of Illionois.

Turkle, S. (2010). In good company? In Y. Wilks (Ed.), *Close engagements with artificial companions* (pp. 3–10). John Benjamins.
Wolpert, L. (1992). *The unnatural nature of science*. Harvard University Press.
Zawieska, K., & Duffy, B. R. (2014). The self in the machine. *Pomiary Automatyka Robotyka, 18*(2), 78–82.

Luisa Damiano is Professor of Philosophy of Science at IULM University, Milan, Italy

Paul Dumouchel is Professor of Philosophy at UQAM Universtity, Montreal, Canada

Pre-ceiving the Imminent

Emotions-Had, Emotions-Perceived and Gibsonian Affordances for Emotion Perception in Social Robots

Tom Poljanšek

Abstract

Current theories of emotions and emotional machines often assume too homogeneous a conception of what emotions are in terms of whether they are experienced as one's own emotions ("internal" emotions) or whether they are perceived as the emotions of other agents ("external" emotions). In contrast, this paper argues, first, that an answer to the question of whether machines can possess emotions requires such a distinction—the distinction between internal *emotions-had* and external *emotions-perceived*. Second, it argues that the emotions we perceive in other agents can be explicated as *indicators of likely imminent paths or patterns of behavior* of those agents. As will be shown, perceiving emotions in others does not necessarily involve the ontological attribution of corresponding emotions-had to the subject which is perceived as exhibiting a particular emotion-perceived. Thus, it will be shown that we can, for example, perceive an agent *as angry* without (ontologically) attributing anger (as an emotion-had) to that agent. If we apply this reasoning to emotional robots, it follows that there is no deception involved in perceiving a robot as exhibiting a particular emotion, as long as its behavior realizes and continues to realize the *behavioral form* of that external emotion-perceived.

I would like to thank Tobias Störzinger for helpful comments on an earlier draft of this paper.

T. Poljanšek (✉)
Department of Philosophy, Georg-August-University Göttingen, Göttingen, Germany
e-mail: tom.poljansek@uni-goettingen.de

Thus, a robot's behavior can fully instantiate emotions-perceived. In order to elaborate these claims, the presented view will be contrasted with Gibson's conception of "affordances." The final section discusses whether these considerations might have broader implications for our view of human perception in general.

Keywords

Social robots · Emotion perception · Affordances · Phenomenology of emotion · Perception and prediction · Human robot interaction

1 Introduction

In his autobiography, the Kant critic Salomon Maimon recounts the following incident:

> On another occasion I went to take a walk with some of my friends. It chanced that a goat lay in the way. I gave the goat some blows with my stick, and my friends blamed me for my cruelty. "What is the cruelty?" I replied. "Do you believe that the goat feels a pain, when I beat it? You are greatly mistaken; the goat is a mere machine." This was the doctrine of Sturm as a disciple of Descartes. My friends laughed heartily at this, and said "But don't you hear that the goat cries, when you beat it?" "Yes," I replied, "of course it cries; but if you beat a drum, it cries too." They were amazed at my answer, and in short time it went abroad over the whole town, that I had become mad, as I held that *a goat is a drum*. (Maimon, 2001 [1792], p. 108 f.)[1]

According to his autobiography, Maimon had recently studied a book on physics authored by a certain Professor Sturm, who, as a Cartesian, claimed that animals were nothing more than complicated machines. As such, although they might

---

[1] "So ging ich einmal mit einigen meiner Freunde spazieren. Nun mußte uns gerade eine Ziege im Wege liegen. Ich gab der Ziege einige Schläge mit meinem Stocke, meine Freunde warfen mir meine Grausamkeit vor. Ich aber erwiederte: was Grausamkeit? Glaubt ihr denn, daß die Ziege einen Schmerz fühlt, wenn ich sie schlage? ihr irrt euch hierin sehr. Die Ziege ist (nach dem Sturm, der ein Kartesianer war) eine bloße Maschine. [] Diese lachten herzlich darüber, und sagten, aber hörst du nicht, daß die Ziege schreyt, wenn du sie schlägst? worauf ich antwortete: ja freylich schreyet sie; wenn ihr aber auf eine Trommel schlagt, so schreyet sie auch." (Maimon, 1792, p. 146 f.).

show superficial signs of emotion and emotional reactions when they scream or contort their faces in apparent pain, they supposedly do not experience anything at all. Maimon thus concluded—though he, like his friends, may have felt some kind of immediate empathy with the goat—that it would be metaphysically wrong to attribute emotions to it. Similarly, in "Something in the Way," one of Nirvana's famous songs, Kurt Cobain uses the line "It's okay to eat fish 'cause they don't have any feelings." It seems that Cobain (at least, when it comes to fish) and Maimon are Cartesians when it comes to attributing emotions to animals.

If we accept this Cartesian position, one of the fundamental questions motivating this volume would already be answered: Even if we disagree with Descartes's, Maimon's and Cobain's conviction that animals (or at least fish) do not have emotions, we might still agree that machines cannot have emotions and that robots, as long as they are machines, cannot have emotions either. All that robots might ever achieve is the successful *simulation* of emotions, but without ever truly *realizing* them (for a further elaboration of this distinction see, for example, Seibt, 2017). But as simple as this argument may sound, it involves metaphysical assumptions that, at least to our knowledge, cannot be conclusively and non-speculatively proven. On the one hand, why, from a metaphysical point of view, should "machines" in the future be generally incapable of exhibiting consciousness and emotions? On the other hand, why should we not one day be able to explain the neurobiological mechanisms underlying animal and human consciousness aswell as emotional expressions and reactions in such a way as to prove that, at least in some respects, animals and humans can indeed be explained as complex biological "machines"?

In this paper, such general and metaphysical questions will be left aside due to their speculative nature. Instead, the paper will focus first on showing that we must distinguish between *emotions-had* and *emotions-perceived* ("internal" and "external" emotions) on both phenomenological as well as ecological grounds. Second, it will be argued that the external emotions we perceive in others can be explicated as *indications of likely imminent paths of behavior*. As will be further argued, the perception of emotions in others does not necessarily involve the ontological attribution of the corresponding (internal) emotions-had to the subject who exhibits the behavioral pattern constitutive for a particular emotion-perceived. Although this may not be the standard case, that is to say that we can perceive someone as being angry without judging that they are experiencing anger, without any deception or error involved. If we apply this consideration to emotional robots, it follows that there is not necessarily deception involved when we perceive a robot as having a particular emotion. When we perceive a robot as angry or aggressive, we perceive it as posing an "intentional" (i.e., goal-directed) threat to ourselves or others, whether or not we attribute a corresponding internal emotion. In the final section, I will discuss whether these considerations might have broader implications for how we think about human perception in general and for our interaction with social robots in particular.

2 Emotions-Had Versus Emotions-Perceived

The main thesis of this chapter is that there is a phenomenological difference regarding emotions that is often overlooked in accounts that focus too much on the definition or ontology of emotions *per se*: On the one hand, emotions are something that subjects *have and experience* as their own, an essential part of their (internal) conscious lifes. On the other hand, subjects not only *have* emotions themselves, but they also directly (i.e., without any conscious consideration or inference involved) *perceive others as having or exhibiting emotions*.[2] Now, from an ontological perspective, it does not seem to make much difference whether an emotion is an emotion-had (or internal emotion) or an emotion-perceived (or external emotion). The difference seems to boil down simply to a difference in epistemic access. Whether a subject is experiencing a particular emotion itself, or whether it perceives another subject as being in a particular emotional state—in both cases, or so it is presumed—it is the *same emotion*, either had or perceived. Moreover, in both cases, at least if the perception or recognition of the emotion is adequate and there is no deception or pretense involved, it is usually assumed that there is someone experiencing the emotion in question, as it were, *from within*. I can either be angry at someone or have the experience of someone else being angry (or more specifically, being angry at me). I can either experience pain or perceive someone else as being in pain. I can love someone or experience someone else loving someone. From an ontological point of view, it seems that in all these cases we are dealing with instances of the same emotion, the only supposed difference being one of epistemic access and possible deception.

However, there is a significant asymmetry between emotions-had and emotions-perceived that cannot be reduced to a difference in epistemic access. Rather, internal and external emotions are fundamentally different in that they serve different ecologi-

---

[2]This distinction is related, though not identical, to the distinction between *experiencing an emotion* and *expressing an emotion*. The main difference is that emotions-had and emotions-perceived refer to two kinds of objects of conscious experience: one's own emotions and the emotions one (directly) perceives in others.

cal functions for the perceiver. Internal emotions orient the agent with respect to aspects that are relevant to them, whereas the perception of external emotions provides clues to imminent behavioral possibilities of others that may or may not be relevant to the perceiving agent. This difference is also reflected in the difference in experiential qualities: Experiencing anger toward another person (i.e., being the subject of anger) differs significantly from perceiving that another person is angry. On the one hand, when one experiences anger toward another person, one experiences a certain kind of outwardly directed arousal toward the person one is angry with, and one may additionally experience the impulse to attack them in some way (to yell at them, to insult them, to hit them, etc.). On the other hand, when you experience another person as angry (or even as being angry at you), you experience an external *aggression* directed towards someone or something.[3] Phenomenologically speaking, in such a case, one can get the impression that the direction of the aggresive person's gaze, as well as their entire body posture, indicate various imminent threatening avenues of possible harm. This anticipatory (co-)presentation of possible paths of harm seems to be a constitutive part of the perceptual experience of aggression itself, and one might feel frightened, intimidated, or threatened by these possibilities (or fear for the well-being of the person they are aimed at). Anger-had and anger-perceived are two different things. To ignore this difference is to assume a perspective that is neutral to the intrinsic perspectivity of conscious experience. Emotions, however, are ontologically perspectival phenomena. They are not neutral with respect to whether they are experienced as one's own or perceived as those of others.

Current theories of emotion often claim that emotions have two types of objects—*particular objects* and *formal objects* (see, for example, Kenny, 1963; de Sousa, 1987; Prinz, 2004; Teroni, 2007). When Tobias is angry at Tom, Tom is the particular object at which Tobias's anger is directed, while the formal object of anger is supposedly *offensiveness* such that the emotion "ascribes the formal object to the particular object" (Smith, 2014, p. 95). Tobias's anger would thus depict Tom as offensive (to Tobias). However, if we switch perspectives, Tom would not normally perceive Tobias's anger as being about him being offensive to Tobias, but rather *as aggression* aimed at inflicting harm (on him, for example). By perceiving Tobias as angry, Tom perceives him as *being threatening* (to someone or something). In a sense, then, we can argue that the formal object of Tobias's anger as perceived by Tom is *being threatening* (to Tom, for example), rather than Tom being offensive toward Tobias, while its particular object is Tom

---

[3] This points toward the phenomenological difference between a perceived aggression and a perceived threat, say, from a falling rock. The threat posed by a falling rock is not experienced as pointing toward and following you. The threat of a Terminator-robot, however, is.

himself. Another way to explain this difference would be to refer to the different *standards of correctness* for emotions-had and emotions-perceived: Tobias's anger-had would be "correct" or "adequate," if Tom is indeed offensive; in this case, Tobias's anger discloses Tom "in the right way," i.e., as offensive. On the other hand, Tobias's anger-perceived is "adequate" or "correctly perceived" by Tom, if Tobias poses an intentional threat (to something or someone); in this case, Tom's perception of Tobias's anger-perceived discloses Tobias "in the right way," as being threatening in a goal-directed manner. Thus, the perception of external emotions in others is not aimed at adequately mapping their internal emotions, but at orienting the subject with respect to the likely behaviors of others. (I leave aside for the duration of these considerations the challenging question of whether there are forms of empathy aimed at (co-)perceiving the internal emotions of others that are distinct from the ordinary perception of external emotions, see Poljanšek, 2022, p. 511–519.)

The important difference to keep in mind here is the difference in the *mode of givenness* of these two types of emotions, a phenomenological difference. This difference can be explained in terms of the different *ecological functions* that emotions-had and emotions-perceived in others play for the conscious subject: While emotions-had (among other things) disclose the environment with respect to the goals and needs of a subject having them, emotions-perceived indicate for the perceiving subject (among other things) likely paths of imminent behaviors of other agents. The formal object of a particular emotion-perceived is thus not the same as the formal object of the corresponding emotion-had. However, there seems to be an interesting *dynamic of interlocking of emotions* when it comes to social interaction: One person's anger-had may cause another person to perceive an aggression directed against them, so that the subject perceiving aggression directed against them may be afraid of the angry person, and thus come into epistemic contact with the likely possibility of being harmed. When one person envies the other, the envied person may feel flattered or ashamed. If one person feels schadenfreude toward another, the other person may feel embarrassed. Perceiving a person to be sad may cause the perceiver to feel caring compassion aimed at assuaging that person's sadness, and so on. If emotions-perceived in others are somehow relevant or significant to the perceiver, they often elicit such corresponding emotions. Moreover, it seems that this interlocking of emotions accounts for a significant part of everyday emotional and social microdynamics (for the view that emotional expression has an important strategic function in social interaction, see e.g., Luhmann, 1982, Hirshleifer, 1987, Frank, 1988, Ross & Dumouchel, 2004; Van Kleef et al., 2011; O'Connor, 2016, and especially Van Kleef, 2016).

This line of reasoning also seems to be suggested by evolutionary theoretical considerations: While internal emotions orient an organism's behavior with respect to its own needs and vulnerabilities by disclosing certain objects in the light of corresponding formal objects (as attractive or repulsive, with respect to certain courses of action or probable paths of events), perceiving external emotions in others might confer an evolutionary advantage in that they make the organism aware of probable imminent paths of events or behavior with respect to other intentional agents. Emotions-perceived (or external emotions) thus disclose the other agent's "intentions" in the sense of goals in terms of which their behavior might be predicted and described, not necessarily anything about their inner goings on. Now, in order to adequately perceive another agent as angry or aggressive, it is not necessary that either the agent thus perceived actually experiences anger, or that the subject perceiving the other as angry tacitly or explicitly attributes an internal experience of anger to the agent thus perceived. Rather, it is sufficient for the perceived agent to exhibit behavior that realizes the *behavioral form of an intentional aggression* toward someone or something.[4]

> What thoughts, ideas, feelings, and desires one has behind one's forehead remain hidden from the observer. ... But that is not the point of understanding at all. ... If one says: I can see from his face that he is ashamed, that he repents, is angry, is grieving, this does not mean that the being and the manner of his being ashamed, repentant, angry, grieving is given to me, but only that the interplaying forms of his behavior [die spielenden Formen seines Verhaltens] are given, which determine a certain stance [Haltung] in relation to the surroundings. (Plessner, 2003 [1925], p. 123 [my translation, TP])

To reiterate this point using the example of Tom and Tobias: When we ask whether Tom's perception of Tobias as being angry or aggressive toward him is adequate, the question is not so much of whether Tobias is actually *experiencing* anger but rather whether Tobias *poses an intentional threat to Tom* (whether he is willing to cut Tom's throat with a knife, for example). Tobias could be ready and determined to hurt Tom, but not feel any anger at all. He might simply be

---

[4] A similar point is made by Jean-Paul Sartre in his *Being and Nothingness*: "[T]o be exact, the anger which the Other feels is one phenomenon, and the furious expression which I perceive is another and different phenomenon." (Sartre, 1956, p. 226). "[T]he anger of the Other-as-object as it is manifested to me across his cries, his stamping, and his threatening gestures is not the sign of a subjective and hidden anger; it refers to nothing except to other gestures and to other cries." (Ibid., p. 294).

convinced that hurting Tom is the right thing to do at this moment, without any strong internal emotions involved on his part. Thus, it might be wrong for Tom to attribute the emotion of anger to Tobias as an emotion-had, while at the same time it might be correct (and even vital) for Tom to perceive Tobias as being angry or aggressive toward him.[5]

If these considerations are sound, then perceiving someone as angry or aggressive amounts to anticipating that agent as an imminent intentional threat, and thus does not necessitate the (ontological) attribution of a corresponding emotion-had to that person. In ordinary language, however, we are not accustomed to distinguishing between *emotions-had* and *emotions-perceived*.[6] Phenomenologically speaking, an anger-had is not the same as an anger-perceived in others; they have divergent phenomenologies as well as divergent ecological functions or formal objects.[7] For the present purpose, therefore, it will prove useful to distinguish between *anger* as the experience of being angry toward someone or something and *aggression* as the perceived emotion of another agent being angry (bearing in mind that this distinction between an emotion-had and an emotion-perceived is intended to apply, *grosso modo*, to all kinds of emotions).

Now, one might tacitly or consciously infer from the aggression perceived in others that they are experiencing something similar to what one feels when one experiences anger. However, it is not necessary to draw this conclusion when perceiving aggression in others. We do not have to *mentalize* someone in order to perceive them as aggressive (see e.g. Bourdieu, 2009, p. 178; Millikan, 2004, p. 21 f.). To go one step further, the existence of internal anger is not even a

---

[5] It may also be the case that Tom does not directly perceive Tobias as angry, but at the same time correctly attributes anger as an emotion-had to him because he knows him well enough to suspect that he is angry under certain circumstances, even if he does not show it in any conspicuous way.

[6] Maybe because, as one might speculate, ordinary language is much more focused on the—somewhat outward-facing—ontological than on the—rather inward-facing—phenomenological aspect of phenomena.

[7] An analogous phenomenological differentiation concerns the distinction between one's own emotions-had and one's own emotions-expressed (one's own emotions-perceived from the perspective of others). If a person is sad, she might still express amusement, for example, when making fun with her friends, without necessarily deceiving them with regard to her 'true' emotions-had. In reality, however, we often find all shades of mixed emotions and in-between cases.

necessary condition for adequately perceiving aggression in another agent.[8] For aggression to be adequately present as an experientially given feature of another agent or object, that agent or object thus need not itself feel internal anger, without the perception of aggression being in any way false or inadequate from an epistemic perspective. The mere realization of the behavioral form of aggression (i.e., the posing of an intentional threat) is sufficient. Regarding the perception of such "meaningful" forms of animal behavior, the anthropologist Helmuth Plessner expresses this idea as follows:

> This layer of form [Formschicht] is simply there and is perceived by everyone, if he conceives the living being not as a mere moving one, but as a behaving one. ... One movement then appears playful, the other joyful, the third fearful, the fourth angry, simply because under "play," "joy," "fear," "anger" basically just kinesthetic gestalts of a certain consistency of succession [Bewegungsgestalten von einer bestimmten *Folgesinnigkeit*] are comprehended. (Plessner, 2003 [1925], p. 82 [my translation, TP])

As elaborated in the next section, while emotions-had often motivate behavior (which is in one way or another salient from the perspective of the subject) *by disclosing affordances* (as possibilities for action on the part of the perceiving subject), emotions-perceived in other agents orient the perceiving subject with respect to likely imminent paths of behavior of these agents within a given situation (which may or may not be salient or relevant from the perspective of the perceiving subject). This is where a rather crucial phenomenological point comes into play, one that can also be found in James Gibson's ecological account of perception, as well as in ethological considerations such as those found in the work of Jakob Uexküll (Gibson, 1986; Uexküll, 1928): Living beings always experientially find themselves in *worlds of meaning*. The phenomenal world that is directly given to a subject in perception is always mediated by the "system of relevance" (Schütz, 1972, p. 66), by the specific "background" (Searle, 1994, p. 192) or "habitus" (Bourdieu) of that subject. The phenomena perceived and thus experientially given in perception are not given in a neutral way, simply mirroring ontologically objective and observer-independent matters of fact. Their *perceptual individuation* or *segmentation*, as well as the *meanings* or *significances with which they are experienced*, rather depend on the needs, interests, and background of the perceiver

---

[8] This claim may seem trivial when one thinks of cases in which one is deceived by a skillful actor about his true feelings.

(Poljanšek, 2022): Different kinds of entities or processes in the environment ecologically matter to a subject to different degrees and in different ways in different contexts. For humans in general, as well as for other animals, many of the events that occur around them—if considered from a purely physical perspective—have no particular significance at all. Most of what happens around us leaves us relatively unaffected and is therefore not usually salient or prominent in our perception. Some types of processes, objects or events, however, are quite significant to humans: They stand out in perceptual experience as if they had been intentionally cut out of the "ongoing tissue of goings on" (Sellars, 1981, p. 57) and then highlighted with different (affective) colors and tones (as attractive, inviting, repellent, off-putting, delicious, dangerous, disgusting, etc.) to signal different categories of objects, each offering specific affordances.[9] These significant types of objects include mimic expressions, typical forms of human behavior, functional artifacts, other intentional agents, the perceptual distinction between animate and inanimate entities, and, last but not least, the emotions of other agents. From a phenomenological perspective, the perception of such entities is often intertwined with our emotional responses in ways that do not allow for a precise distinction between purely perceptual and purely emotional aspects of our own experience. When we receive a smile from a person we admire, the pleasure we feel is phenomenologically indistinguishable from the mere perception of the facial expression as well as from the immediately grasped meaning or interpersonal significance of the smile. To connect the idea that emotions disclose affordances with the question of whether and to what extent social robots can have emotions, I will thus take a slight theoretical detour in the next section that addresses the question of what it means for humans and other living beings to live in worlds of meaning.

To put these considerations in the context of current theories of emotion recognition, the account presented here is somewhat hybrid in that it integrates aspects of *direct perception accounts* and *theory-theory accounts* of emotion recognition. With respect to direct perception, both one's own emotions-had and emotions-perceived in others can be directly perceived. That is, they are immediately (without any conscious consideration) present in perceptual experience, which does not preclude mediation through cultural backgrounds. The emotions-had of other agents, however, are never directly perceived. They may be inferred and postulated, but they need not be.

---

[9] Uexküll and Kriszat (1934) use color coding to signify the way in which different kinds of objects in the surrounding of different subjects are perceived as belonging to different object-categories (like food, obstacles, tools, and so on).

3   Emotions as Perceived Probabilities. A Critical Reassessment of Gibsonian Affordances

The following section discusses the relationship between emotions and "affordances" in Gibson's sense. To do so, I will first briefly explicate the idea that humans, as well as other living beings, live in perceptual "worlds of meaning" or "worlds of appearance" rather than merely mirroring physical states of fact concerning their environment in perception (for a brief overview of this idea, see Poljanšek, 2020). I will then elaborate on Gibson's theory of affordances and discuss the relationship between affordances and emotions to show how these considerations effect emotion perception in emotional machines.

3.1   Perceiving Worlds of Meaning

The idea that perception is not so much a neutral representation of brute physical matters of fact, but rather brings us into contact with a *world of value, meaning, or significance* that is shaped by the needs and the (cultural) backgrounds of the perceiving subjects, has many historical antecedents. However, such an approach, which links perception to the experience of a world of meaning has not always been very popular in philosophy. For a long time, philosophers tended to focus on perception as the most promising epistemic way to gain (objective) knowledge of external reality as conceived by natural sciences. Thus, they have focused on the more "objective" or "epistemic" aspects of perception that bring us into contact with objective physical states of affairs, leaving aside all the more "subjective" colorings and meanings of the objects of perception that make up worlds of meaning. These colorings and meanings, however, are closely related to the fact that the way the world is given to us in experience is not neutral with respect to our needs, interests, and backgrounds. As Friedrich Nietzsche wrote in *The Gay Science*, humans are

> no objectifying and registering devices with frozen innards—we must constantly give birth to our thoughts out of our pain and maternally endow them with all that we have of blood, heart, fire, pleasure, passion, agony, conscience, fate, and disaster. (Nietzsche, 2001 [1887], p. 6)

Although there is arguably an important sense in which perception brings us into contact with objective states of fact, we should nevertheless keep in mind that the phenomenal content of perceptual experience includes what is nowadays often called "semantic information" (see, for example, Floridi, 2005; Dennett, 2017; Millikan, 2004) and evaluations in a non-symbolic sense. Perceptual experiences and their phenomenal contents include both "objective" aspects (which could be said to "mirror" structures of physical reality) and "meaning" (or "semantic") aspects, which arguably arise from the subpersonal, neurobiological constitution of the subject, where the latter is itself at least partially structured and shaped by the sedimented experiences of the perceiving subject (for an elaboration of this distinction between the more "objective" aspects and the "meaning" aspects of the phenomenal objects of ordinary perception, see Poljanšek, 2015). As Edmund Husserl writes about the world phenomenally given in ordinary perception,

> [m]oreover, this world is there for me not only as a world of mere things, but also with the same immediacy as a *world of objects with values, a world of goods, a practical world*. I simply find the physical things in front of me furnished not only with merely material determinations but also with value-characteristics, as beautiful and ugly, pleasant and unpleasant, agreeable and disagreeable, and the like. Immediately, physical things stand there as Objects of use, the "table" with its "books," the "drinking glass," the "vase" the "piano," etc. These value-characteristics and practical characteristics also belong *constitutively to the Objects "on hand" [vorhandenen] as Objects*, regardless of whether or not I turn to such characteristics and the Objects. (Husserl, 1983, p. 53, [Husserl, 1976 (Hua III/1), p. 58])

The environment—or "Umwelt," to use Uexküll's notion—, as it is presented to a subject in perception is always carved out and shaped with respect to its need to orient itself in its surroundings with respect to its own interests and well-being, as are the various object categories (such as edibles, predators, conspecifics, tools, etc.) as manifestations of which the objects are immediately presented in perception. The important idea to grasp here is that the world of meaning given in perception is not "objective" in the sense that it exists independently of what a particular subject perceives; rather, it is composed of the entities that can be perceptually given to a particular subject in its perception with respect to its specific background. Thus, if two subjects have significantly similar backgrounds, they will find each other in the same world of meaning shaped by similarly segmented entities and similar distinctions of meaningful object categories (Allen, 2014). However, as subjects' backgrounds differ, so do their Umwelten or worlds of meaning. Thus, the world of meaning of a particular subject is not to be identified with the physical reality that surrounds it (which is determined and stipulated by physics and of which certain aspects are arguably mirrored in perception). While all organisms are plausibly considered to be located in the same spatio-temporal reality, the corresponding subjects perceptually find themselves situated in divergent worlds of meaning (for a detailed justification of these claims, see Poljanšek, 2022). There is nothing metaphysically strange or mysterious about this.

The idea of distinguishing between physical reality and (perceptually) given worlds of meaning or appearance has a number of historic antecedents, some of which are briefly mentioned for context: Ernst Mach, for example, distinguishes between the "physiological space" on the one hand and the "geometric space" on the other, whereby physiological space designates space as it appears to a subject in their perception, while geometric space designates the space with which physics is concerned (Mach, 1906, p. 337, 1922, p. 148 f.). The Gestalt psychologist Kurt Koffka distinguishes between "behavioral" and "geographical" environment; while the former designates the environment as it is given to a perceiving subject in experience, the latter (geographical environment) designates the same space from a perspective independent of perception (Koffka, 1935, p. 27 f.). A similar distinction is made by Kurt Lewin (1982, p. 93). The phenomenologist Maurice Merleau-Ponty distinguishes between "phenomenological world" and "objective space–time," whereby the phenomenological world describes the world of meaning as it is given to a human subject in perception, and the objective space–time represents a physical description of reality (Merleau-Ponty, 1966, p. 10; cf. also Merleau-Ponty, 1963, p. 129). Sir Arthur Stanley Eddington distinguished between the "the world which spontaneously appears around me when I open my eyes" and the scientific "world which in more devious ways has forced itself on my attention" (Eddington, 2012 [1928], p. xii), his famous example being his two tables: The ordinary table of his everyday experience, a "strange compound of external nature, mental imagery and inherited prejudice" (Ibid., p. xiv), and the "scientific table", which is "mostly emptiness. Sparsely scattered in this emptiness are numerous electric charges rushing about with great speed" (Ibid., p. xii). Furthermore, Wilfrid Sellars's distinction between the "manifest image" and the "scientific image of man-in-the-world" implicitly draws on this tradition (Sellars, 1963). Gerhard Roth distinguishes between a "phenomenal world" which he takes to be a construction of the brain, and the objective, "transphenomenal world" or "reality" (Roth, 1997, p. 325). Finally, Gibson makes an analogous distinction between the meaningful "environments of animals" or humans and the "physical world" (Gibson, 1986, p. 8 f.).

Of the accounts cited above, Gibson's seems to be one of the most popular today, at least within the Anglophone discourse. This popularity may have to do with the fact that his notion of "affordances" (as possibilities for action immediately grasped by the perceiver) has been quite successful and is cited and applied in a variety of forms and contexts (see, for example, Millikan, 2004; Norman, 2013; Clark, 2016; Dennett, 2017), while the rest of his ecological theory of perception is often somewhat neglected. In the following sections, Gibson's theory of affordances will be fleshed out in the context of his ecological theory of perception. It will be argued that in an attempt to avoid the claims that a) affordances are somehow *projected* onto physical entities in the process of perception and b) that affordances belong to the *phenomenal* rather than to the physical object, Gibson puts forward an exceedingly narrow conception of affordances, that is especially too restrictive to account for social affordances and emotions-perceived. In this context, some authors have argued for a broader notion of affordances that can account for social affordances while it also links affordances to emotions (in the sense of what is here called emotions-had) by claiming that the latter, if adequate, disclose the former. However, or so I will argue, this broader notion of affordances is still too narrow to account for emotions-perceived in others, since these do not indicate action possibilities *for the perceiver*, but probable actions *of the agent perceived*. I will return to this last point in the chapter's concluding paragraph.

In a first attempt, Gibson defines affordances as follows: "Roughly, the affordances of things are what they furnish, for good or ill, that is, what they afford the observer" (Gibson, 1982, p. 403). To give an example Gibson himself uses, to "perceive a surface is level and solid is also to perceive that it is walk-on-able" (Ibid., p. 408). This conception of affordances, however, is integrated into Gibson's general ecological approach of perception, which focuses on the idea that humans, as well as animals, always perceive their surroundings in terms of the possibility spaces that they offer or block to the perceiving subject:

> Our own experience of the visual world can be described ... as colored, textured, shadowed, and illuminated; as filled with surfaces, edges, shapes, and interspaces. But this description leaves out the fact that the surfaces are familiar and the shapes are useful. ... [W]e apprehend their uses and dangers, their satisfying or annoying possibilities, and the consequences of action centering on them. ... The visual world, in short, is meaningful as well as concrete: it is significant as well as literal. (Gibson, 1950, p. 198)

According to Gibson, the world of perception is a world of (physical) forms *and* (immediately grasped) meanings, where meaning designates the possibility spaces either offered to or blocked from the perceiving subject.[10] Affordances are

---

[10] There exists an alternative interpretation of Gibsonian affordances which takes "biological norms" to be an important part of them, such that a subject doesn't only perceive possibilities for action relative to its abilities when it perceives affordances, but rather possibilities for action as evaluated through emotions where this evaluation can be traced back to biological norms (Hufendiek, 2016). The affordance of "being dangerous" and the "fear"

thus, as he claims, "not simply the physical properties of things as now conceived by physical science. Instead, they are ecological, in the sense that they are properties of the environment relative to an animal" (Gibson, 1982, p. 406). Gibson is nevertheless quite clear that he does not support the idea that affordances (as perceived meanings) lie in the eye of the beholder, that they are the "outcome of a perceptual process" (ibid., p. 411). Rather, he claims that affordances are "neither subjective nor objective but transcend this dichotomy" (ibid.).[11] At the same time, he criticizes accounts such as Lewin's or Koffka's as far as they situate affordances or what they call the "demand-character" of objects "in the *phenomenal* object but not in the physical object" (ibid, 409).

> What a thing is and what it means are not separate, the former being physical and the latter mental, as we are accustomed to believe. The perception of what a thing is and the perception of what it means are not separate, either. ... Thus we no longer have to assume that, first, there is a sensation-based perception of a thing and that, second, there is the accrual of meaning to the primary percept (the "enrichment" theory of perception, based on innate sensations and acquired images). (ibid., p. 408)[12]

Using a distinction proposed by Searle, one might thus say that Gibson conceptualizes affordances as "ontologically subjective" but "epistemologically objective" (Searle, 2010, p. 18); that is, although their existence depends on the existence of subjects, their capacities, and their conscious perspective, Gibson nevertheless conceptualizes them as epistemologically objective. Thus, affordances seem to be relational entities whose relata are the organism in question (with its abilities)

---

by which this dangerousness is disclosed might be an example: "fear doesn't represent the chemical structure of the snake's venom; rather, it represents the snake's being dangerous for our bodily well-being" (Ibid.). However, this interpretation of affordances, although it seems quite promising in itself, brings affordances closer to what Gestalt psychologists like Lewin an Koffka called the "demand character" of things. A conception, however, Gibson explicitly disagrees with.

[11] Plessner talks here of the "subjective–objective indifference" of the meaningful objects of perception (Plessner, 2003, p. 87).

[12] The idea that what a thing is and what it enables are not distinct is not a Gibsonian Invention, but traces back to at least the Gestalt theoretical tradition. As Koffka writes: "To primitive man each thing says what it is and what he ought to do with it: a fruit says, 'Eat me'; water says, 'Drink me'; thunder says, 'Fear me,' and woman says, 'Love me.'" (Koffka, 1935, p. 7).

and the corresponding objects in the organism's environment; and they are objective insofar as the relation between an organism's abilities and capabilities and the object in question is an objective matter of fact.[13] "Affordances are neither properties of the animal alone nor properties of the environment alone. Instead, they are relations between the abilities of an animal and some feature of a situation" (Chemero, 2009, p. 191, see also Heft, 1989, p. 14 f.). When a subject correctly perceives an affordance, it perceives possibilities for action that its environment objectively offers with respect to its capabilities:

> The significance of perceiving possibilities for action follows from one of the core assumptions of the ecological approach—namely, that the environment is perceived in terms of what the organism can and cannot do within it. In other words, to see something is to see what to do with it. (Fajen & Turvey, 2003, p. 277; Fajen, 2005, p. 718; see also Nanay, 2011, p. 311)

3.2 Emotions and Affordances in Socially Constituted Worlds of Meaning

Gibson's narrow conception of affordances reaches its limits when it comes to human worlds of meaning or appearance. At first glance, however, it might seem promising to apply his notion of affordances to objects of our shared social worlds. One might say, for example, that a banknote affords the opportunity to pay something with it, that an outstretched hand affords the opportunity to shake hands, that a rook in chess affords the opportunity to be moved orthogonally (and not diagonally), that a smile affords the opportunity to smile back, that a complaint office affords the opportunity to complain, or that a coffee mug affords the opportunity to drink hot beverages from it (and not so much, say, cold, carbonated beverages).

---

[13] Hufendiek (2016) argues, that affordances are constituted by "observer-independent relational properties" (like x being dangerous for y) as well as by "evolved responses to them" (like y being afraid and avoidant of x) such that an organism might learn to perceive the affordance of x being "a danger-to-be-avoided". This analysis can be reconstructed as addressing the problem that while affordances seem to be objective in so far as they concern the relation between the abilities of an observer and its environment, there still seems to be the need for the observer to correctly grasp the existence of such a relation. A frozen lake may be walk-on-able for a certain subject (relative to its weight, its walking abilities, etc.), however, it might not be able to perceive this affordance (because it might not be used to walking on ice regularly).

As one might argue, following Searle's account of the construction of social reality, these "affordances" are not reducible to brute physical facts. They are, however, not reducible to the relation of brute physical facts and the behavioral abilities of a subject, either. Take banknotes as an example: For the banknote to afford the purchase of something, something more is required than a certain behavioral ability of the subject and certain physical properties of the rectangular paper in their hand. Social affordances are rather the result of a process of social construction, that, according to Searle, involves a "truly radical break with other forms of life [...] through collective intentionality" (Searle, 1995, p. 40). For Searle, it is through collective intentionality, which he takes to be a specifically human capacity, that *status-functions* are constituted and imposed on physical objects, "where the function cannot be achieved solely in virtue of physics and chemistry but requires continued human cooperation in the specific forms of recognition, acceptance, and acknowledgement" (ibid.).[14] Thus, a physical object X can "count as Y in context C" qua collective acceptance. Now, whether one agrees with Searle's theory of social construction or not, it seems clear that Gibson's conception of affordances, when taken to define affordances as relations between behavioral capacities and intrinsic dispositions or properties of the environment, cannot account for such cases, insofar as they go beyond mere relations between physical things and subjects' capacities. Human worlds of meaning are full of socially constituted affordances that cannot be reduced, at least not in any straightforward way, to the relation of brute physical facts and the capabilities of individual subjects.[15] As already indicated in the above examples, it is not only inanimate objects that provide socially constituted affordances for subjects like us, but also other agents and their behavior (see also Lo Presti, 2020). Thus, in certain cultural contexts, an outstretched hand is an invitation (i.e., it affords) to shake hands; a slightly longer-than-usual gaze may be an invitation to flirt; and an outstretched middle finger may be an invitation to start an argument.

---

[14] A critique of Searle's account concerning his idea that social construction begins where status-functions are constituted which go beyond mere physical possibility can be found in Poljanšek, 2015. There I argue, as the example of the coffee mug exemplifies, that socially constructed status functions do not have to go beyond the possibilities previously available because of brute physical facts, but that such status functions can also specifically limit the horizon of socially acceptable options ("We don't drink cold carbonated beverages from coffee mugs").

[15] Even if one takes biological norms to be constitutive of common affordances (Hufendiek, 2016), socially constructed affordances seem to need something more to count as such.

> [T]he sheer bodily movements of our interlocutors can be seen as possibilities for different types of actions in the same vein as affordances in a physical environment. For instance, a specific type of gesture accompanied by gaze can be directly perceived as an opportunity to come closer, to walk away, or perhaps to hug, kiss, push, or smile to that person. Likewise, we directly perceive vocalisations as affording new verbal actions—for example, to say something new in an affirmative, evaluative, supportive, joyful, teasing, flirting, or antagonistic way. (Jensen & Pedersen, 2016, p. 82)

This is where emotions come back into play. As some authors have argued (see e.g. Hufendiek, 2016; Jensen & Pedersen, 2016, Poljanšek, 2011, Slaby, 2013), there seems to be a close connection between the perception of "affordances" and "emotions," where emotions are understood in the sense of what is explained here as one's own *emotions-had*. Jensen and Pedersen (2016), for example, argue that the "theory of affordances needs to be supplemented and closely related to contemporary notions of affect and emotion" (ibid., p. 80 f.). They further argue that emotions-had "disclose affordances" (ibid., p. 84), for example, when it comes to social affordances:

> [P]articipants in social interaction directly perceive affordances for further action in-and-through their affective engagement within the situation. This direct perception cannot in any meaningful way be divided into a cognitive domain and an emotional domain. (ibid., p. 100)

Similarly, Kiverstein and Rietveld write with respect to Clark's predictive coding theory of perception:

> A friend's sad face invites comforting behavior, a colleague at a coffee machine affords a conversation, and the extended hand of a visitor solicits a handshake. Affect plays a crucial role in preparing us to act in these cases: it signals which possibilities for action in a situation matter to us in sense of being relevant to us given our interests and needs. (Kiverstein & Rietveld, 2012, p. 1)

In a very lucid paper, Hufendiek argues in a similar direction that the intentional objects of emotions—in the sense of *emotions-had*—can be described as affordances, such that, for example, the intentional object of fear is the possibility of being hurt, "the object of jealousy is 'being left out by others,'" while "guilt is about having transgressed a social norm" (Hufendiek, 2016). Now, to explain instances of what she calls "emotional affordances" in social contexts, we need to refer not only to brute physical facts and capabilities of subjects but also "to social rules and norms to spell out what they are about," to define their respective formal objects (Hufendiek, 2016). Hufendiek thus asserts that for

> emotions such as envy as well as guilt, shame, jealousy, and pride, the social context in which they occur is a *sine qua non*: these emotions (or homologous forms of these emotions) could not exist in nonsocial species outside of a social context. (ibid.)

Consider, for example, a case in which a subject feels their dignity has been violated by a derogatory remark, so that they take themselves to feel justified anger toward the offender. Here, the argument goes, the subject must at least tacitly recognize several constitutive social rules and norms in order to feel offended by the insult: They have to perceive the sounds coming from the other person's mouth as sentences in a language they understand. They must (tacitly) understand these sentences as an insult to their dignity, and they must perceive this insult as a violation of a social rule, so that their anger is justified. On the descriptive level of experience, however, they immediately feel insulted by the comment. In a manner quite similar to Searle's account, Hufendiek thus ontologically grounds the existence of the corresponding social affordances in the existence of such norms and rules (which Searle in turn grounds in the collective acceptance of constitutive rules of the form "X counts as Y in context C" via collective intentionality). This conception of affordances thus goes beyond a mere relation between the subject's behavioral capabilities and the physical properties of its environment, and thus beyond what Gibson's theory of affordances can provide.

The important point for the present considerations, however, is the close connection between emotions-had and affordances in this broader sense, with the first claim here being that *emotions-had*, when adequate, *disclose affordances*. That is to say, emotions-had point to "evaluated possibilities inherent in situations" (Poljanšek, 2011, p. 2); where these possibilities are imminent possibilities for the subject that are in one way or another salient for or relevant to the subject itself and that are indicated as being either opened or blocked by a situation.[16] Fear points to the imminent possibility of being hurt. Anger indicates the immi-

---

[16] Such a conception of emotions-had as disclosing affordances bears some striking resemblance to Robert C. Roberts' account of emotions as "concern-based construals", "ways things appear to the subject" (Roberts, 2003, p. 75) which are "imbued, flavored, colored, drenched, suffused, laden, informed, or permeated with concern" (ibid., p. 79).

nent possibility of hurting someone. Grief may indicate the blocked possibility of ever meeting an existentially important person again. Shame points to the possibility of being seen in certain uncomfortable circumstances, and so on.

> It is adequate to understand emotions as a complex *sense of possibility*: emotions disclose what a situation affords in terms of potential doings, and the specific efforts required in these doings, and potential happenings affecting me that I have to put up with or otherwise respond to adequately. These two aspects—situational (what is afforded by the environment) and *agentive* (what I can or cannot do)—are intimately linked to form a process of dynamic situation-access: an active, operative orientation towards the world. (Slaby et al., 2013, p. 42)

The second claim concerning the connection between emotions and affordances is that—although it was argued in the first section that emotions-perceived indicate *imminent probable action paths of* other agents—they are not adequately conceptualized either as affordances (either in the narrow sense of possibilities for action in relation to the perceiving subject's capabilities or in the broad sense that includes socially constructed affordances) or as disclosing affordances. On the one hand, the imminent probable action paths of another agent indicated by an emotion-perceived need not point to possibilities of action *for* the perceiving subject, and on the other hand, they need not be relevant or significant for it, too. (Although they often are.) For example, one can perceive someone as exhibiting aggression and thus anticipate (without any conscious inference involved) the imminent possibility that they will hurt someone or something, without being in any meaningful sense within the scope of that aggression. Such imminent possibilities that emotions-perceived disclose go beyond even the broader notion of affordances, insofar as they do not necessarily affect the subject perceiving them in any significant way and insofar as they are not necessarily, in any significant sense, possibilities *for the subject*. Rather, they are likely imminent possibilities that constitute a significant part of the way objects and situations are experientially given to us in ordinary perception. I will return to this consideration at the end of the chapter.

3.3 Perceiving Emotions in Social Interactions and Social Robots

We are now at a point where we can apply the previous considerations of affordances and the distinction between emotions-had and emotions-perceived to the perception of emotional robots. In a previous paper on social robots (see

Poljanšek & Störzinger, 2020), we have distinguished between "functional social interactions" and "close interhuman relationships," defining functional interactions as those forms of social interaction that find their fulfillment in the mere realization of certain goals, behaviors, and appearances. Functional social interactions exclude, at least to some extent, any significant communicative reference to the difference between the subject performing an interaction and the subject-image created by this performance in the perception of the interaction partner by the interacting agents.

> As a rule of thumb, we can say that in functional interactions, we care for the *performed self*, not for the *self performing a functional role*. Full-fledged realization of such functional interactions thus seems to lie in the presentation of a perfect facade or surface, in a flawless realization of certain functions and the maintenance of specific appearances, regardless of what's going on behind the behavioral surface or inside the mind of the performer. (Poljanšek & Störzinger, 2020, p. 72)

This distinction can also be applied to the difference between emotions-had and emotions-perceived. Consider the case of a nurse: In addition to carefully performing the technical tasks with which a nurse is entrusted on a daily basis, a nurse might be expected to be kind and caring toward their patients and to show some interest in the well-being of those patients on a more emotional level that goes beyond the purely medical aspects of care. However, although we might wish this to be the case, we would not expect the nurse to "really" have the internal emotions that are usually associated with the external emotions she displays. If we have no illusions about the realities of the inner lives and sensibilities of human beings, we may very well know that there may always be patients for whom the nurse feels no compassion at all. We assume, however, that it is part of a nurse's job to exhibit such behavior even in such cases, without assuming that they are in any way deceiving their patients by doing so. Therefore, we do not criticize a nurse for not *having* the right emotions towards their patients, but rather for not *exhibiting* the right kind of behavior when they act rude or unfriendly. We expect a nurse to *display* a certain kind of behavior that makes us perceive them as showing certain emotions toward their patients. We do not expect them to really *have* the corresponding internal emotions in every instance. In the same way, someone who sees himself as what might be called a "traditional patriarchal father figure" might exhibit a strict, punitive behavior toward his child who has "done something wrong" because he considers it his duty to do so, even though he might actually be emotionally inclined to be more lenient with them. Here, too, the goal of the father's behavior is to create a certain behavioral image in the eyes of the child, to establish a certain "stance" [Haltung] in Plessner's sense, rather than to convey particular internal emotions to his child.

The same consideration applies to the rather large part of our everyday social interactions, where we do not so much expect each other to *have* and convey certain (internal) emotions but rather, that we realize the behavioral forms and appearances of certain external emotions. Slavoj Žižek describes this phenomenon as the "non-psychological character" of the "symbolic order," which can be identified, for the present purpose, with our mutual orientation toward the external images we create through our behavior, while largely blending out the "true" inner states of our interaction partners:

> [W]hen, upon meeting an acquaintance, I stick out my hand and say "Good to see you! How are you today?," it is clear to both of us that I am not completely serious (if my acquaintance suspects that I am genuinely interested, he may even feel unpleasantly surprised, as though I were aiming at something too intimate and none of my business—or, to paraphrase the old Freudian joke, "Why are you saying you're glad to see me, when you're *really* glad to see me?"). Yet it would still be wrong to designate my act as hypocritical, since in another way I *do* mean it: the polite exchange renews a kind of pact between the two of us. (Žižek, 2007, p. 31 f.)

This phenomenon can also be explained in terms of the above considerations regarding the perception of emotions: From an ecological point of view, perceiving a particular (external) emotion as the emotion-perceived of another agent is less about what internal mental or emotional state this agent is "really" in, and more about what imminent behaviors can be reliably expected from this agent. Perceiving an agent as aggressive then boils down to, among other things, seeing them as an intentional threat, while perceiving an agent as friendly boils down to, among other things, not seeing them as an intentional threat.

With these considerations in mind, we can finally answer the question of whether and in what sense robots can "have" emotions: If we perceive robots as exhibiting certain emotions, we are not necessarily deceived by these robots, provided that the anticipation of certain probable imminent behavioral possibilities or patterns associated with such a perception is adequate. On the contrary, perceiving a machine as friendly or aggressive may be adequate and even vital for the perceiving subject from an ecological perspective. However, this does not prevent us from asking the question of whether these robots *actually experience* the corresponding (internal) emotions-had that we often (although not always!) experience ourselves when we exhibit such behavior. While we may be strongly inclined to answer this question in the negative, especially with respect to current emotional robots and machines, the same question can be asked with respect to all other conscious agents with whom we interact on the basis of their external emotions (and without reference to their internal emotions).

From the perspective of conscious experience, internal emotions and external emotions (emotions-had and emotions-perceived) serve different ecological functions. They orient the perceiving subject to different kinds of possibilities and probable paths of events in given situations. Moreover, although external emotions often serve as reliable indicators of a perceived agent's internal emotions, we should also keep in mind that the outward-facing expression of emotions(-perceived) is to some extent independent from an agent's emotions-had in both animal and human social life. Especially in human worlds of meaning, the behavioral expression of external emotions often serves communicative or symbolic functions that go far beyond the mere indication or expression of internal emotions concerning the agent expressing the emotion. Putting on a happy face and verbally expressing one's joy to a friend who has just found the love of his life, when one's predominant internal emotion is deep depression, does not necessarily mean that one is deceiving or misleading one's friend.[17] Rather, to put it somewhat metaphorically, it shows that— at least as far as emotions are concerned—people can sometimes give to each other what they themselves do not have. This possibility of giving to the other something that one does not have oneself is related to the peculiar perspectival difference of "inside" and "outside," which makes the phenomenal access to worlds of appearance possible for humans precisely by denying them direct access to what and how it "really" is on the "inside" of the perceived "outside." In this way, therefore, robots can also give to man what they themselves (at least until today) do not possess on their "inside." Robots can have emotions(-perceived).

4 Concluding Remarks: Pre-ceiving the Imminent

Finally, I would like to return to the question of how we conceptualize human perception in general. As ecological, phenomenological, and the enactivist conception of perception suggests, it is not the only—and probably not even the most important—function of perception to mirror (physical) facts that exist in our

---

[17] In the same way, you can express joy over a gift from a friend and really mean it, even though at the same moment you are not able to really feel joy. From a phenomenological point of view, such cases can be quite intricate, insofar as one can, 'externally' so to speak, feel joy in and through the expression of joy, while 'internally' one does not feel joyful.

current environments. There is certainly nothing wrong with focusing on these mirroring aspects of perception. After all, we have good reason to believe that an important aspect of human perception is that it brings us into contact, albeit somewhat indirectly, with subject-independent facts. However, we should not overestimate this aspect by ignoring the even stronger reasons we have for thinking that the main function of perception is to *orient organisms in complex, overwhelming, and constantly changing worlds of meaning that they at least partially share and constitute with others*. If perception would merely re-present *current* environmental conditions, it would in many cases come a decisive moment too late with regard to the needs of the perceiver. Under conditions of time scarcity and pressure to make decisions, organisms cannot afford to first have to consciously interpret and evaluate an initially neutral representation of existing environmental facts. Instead, as a lot of recent research suggests (see, for example, Husserl, 1968 (Hua XI), Clark, 2016; Poljanšek, 2022), perception is not so much to be understood—as its origin from the Latin "per-ceptio" suggests—as the reception of something outside that is already there before perception takes place. On the contrary, human perception seems to be more concerned with the anticipation of probable paths of imminent events, where these anticipations, at least in the case of humans, derive in part from earlier and ongoing exposure to regularities in their respective cultural, biological, and physico-chemical environments. Continued exposure to such regularities, through a process that is sometimes called "socialization" or "enculturation," leads to the sedimentation of specific anticipations of horizons of probable paths of events associated with particular kinds of phenomenal objects: Fluids generally behave in certain, predictable ways, and so do solids. Animate objects generally behave specifically different from inanimate objects. People in cultural contexts and environments often exhibit specific regularities and patterns of behavior, and so on. Many of these regularities seem to be reflected in the fact that people ontogenetically form backgrounds that allow them to perceive and experience their worlds as meaningful in anticipatory and structured ways. As I have suggested, the dimension of meaning at issue here can be elucidated, at least in large part, by tacit anticipation of the paths that imminent events might take. Anticipating such horizons of possibility is a fundamental part of our ordinary perceptual experience. If we take such considerations seriously, we might even consider replacing the time-honored word "perception" with "pre-ception," the phenomenal (co-)presentation of the imminent (for a phenomenological formulation of this claim, see Husserl, 1968 (Hua XI), p. 336 f.). A theory of affordances, emotions-had, and emotions-perceived could thus be grounded in a *general theory of perception as preception*, according to which the content of perceptual experience is mediated by the needs and backgrounds of the subjects in question. Thus, Andy Clark, one of the most renowned proponents of such a view, argues that, through what he calls "prediction-driven learning,"

> [r]ather than aiming to reveal some kind of action-neutral image of an objective realm, prediction-driven learning delivers a grip upon *affordances*: the possibilities for action and intervention that the environment makes available to the agent. (Clark, 2016, p. 171)

To this, Daniel Dennett adds,

> [P]redictive coding is a method for generating affordances galore: we expect solid objects to have backs that will come into view as we walk around them; we expect doors to open, stairs to afford climbing, and cups to hold liquid. These and all manner of other anticipations fall out of a network that doesn't sit passively waiting to be informed. (Dennett, 2017, p. 168 f.)

However, as the previous considerations have shown, the perception of affordances is only a special case of how our perception generally works. Affordances are a natural byproduct of the fact that our perception is generally oriented toward the imminent.

Assuming that these considerations are sound, it follows for our interaction with emotional robots and machines that it can be appropriate (and may even be vital) to perceive them as having emotions (in the sense of emotions-perceived) as long as the anticipations constitutive for such preceptions are appropriate with regard to the machine in question. Whether robots will one day further be able to *have* and *experience* internal emotions—as we nowadays assume animals (and even fish!) do—remains an open question, at least for the time being.

References

Allen, C. (2014). Umwelt or Umwelten? How should shared representation be understood given such diversity? *Semiotica, 198,* 137–158.
Bourdieu, P. (2009). *Entwurf einer Theorie der Praxis auf der ethnologischen Grundlage der kabylischen Gesellschaft.* Suhrkamp.
Chemero, A. (2009). *Radical embodied cognitive science.* MIT Press.
Clark, A. (2016). *Surfing uncertainty. Prediction, action, and the embodied mind.* Oxford University Press.
de Sousa, R. (1987). *The rationality of emotions.* The Massachusetts Institute of Technology.

Dennett, D. (2017). *From bacteria to Bach and back again*. Penguin.
Eddington, S. A. S. (2012 [1928]). *The nature of the physical world*. Cambridge University Press.
Fajen, B., & Turvey, M. (2003). Perception, categories and possibilities for action. *Adaptive Behaviour, 11*(4), 276–278.
Fajen, B. (2005). Perceiving possibilities for action: On the necessity of calibration and visual learning for the visual guiding of actions. *Perception, 34*, 717–740.
Floridi, L. (2005). Is semantic information meaningful data? *Philosophy and Phenomenological Research, LXX*, 351–370.
Frank, R. H. (1988). *Passions within reason: The strategic role of emotions*. Norton.
Gibson, J. J. (1982). Notes on affordances. In E. Reed & R. Jones (eds.), *Reasons for realism. Selected essays of James Gibson* (pp. 401–418). Lawrence Erlbaum Associates.
Gibson, J. J. (1986). *The ecological approach to visual perception*. Routledge.
Gibson, J. J. (1950). *The perception of the visual world*. Houghton Mifflin.
Heft, H. (1989). Affordances and the body. An intentional analysis of Gibson's ecological approach to visual perception. *Journal of the Theory of Social Behavior, 19*(1), 1–30.
Hirshleifer, J. (1987). On the emotions as guarantors of threats and promises. In J. Dupré (ed.), *The latest on the best: Essays on evolution and optimality* (pp. 307–26). MIT Press.
Hufendiek, R. (2016). Affordances and the normativity of emotions. *Synthese, 194*(11), 4455–4476.
Husserl, E. (1968). *Analysen zur passiven Synthesis. Aus Vorlesungs- und Forschungsmanuskripten 1918–1926*. In M. Fleischer (ed.). Husserliana Volume XI. Nijhof. (= Hua XI)
Husserl, E. (1983). *Ideas pertaining to a pure phenomenology and to a phenomenological philosophy. First book. General introduction to a pure phenomenology*. Nijhof.
Husserl, E. (1976). *Ideen zu einer reinen Phänomenologie und phänomenologischen Philosophie. Erstes Buch. Allgemeine Einführung in die reine Phänomenologie*. In K. Schuhmann (ed.). Husserliana Volume III/1. Nijhof. (= Hua III/1)
Jensen, T. W., & Pedersen, S. B. (2016). Affect and affordances—The role of action and emotion in social interaction. *Cognitive Semiotics, 9*(1), 79–103.
Kenny, A. (1963). *Action, emotion and will*. Routledge.
Kiverstein, J., & Rietveld, E. (2012). Dealing with context through action-oriented predictive processing. *Frontiers in Psychology: Theoretical and Philosophical Psychology, 3*(421), 1–2.
Koffka, K. (1935). *Principles of gestalt psychology*. Routledge.
Lewin, K. (1982). *Feldtheorie. Kurt-Lewin-Gesamtausgabe* (Vol. 4). Klett Cotta Stuttgart.
Lo Presti, P. (2020). Persons and affordances. *Ecological Psychology, 32*(1), 25–40.
Luhmann, N. (1982). *Liebe als Passion. Zur Codierung von Intimität*. Suhrkamp.
Mach, E. (1922). *Die Analyse der Empfindungen und das Verhältnis des Physischen zum Psychischen*. Verlag von Gustav Fischer.
Mach, E. (1906). *Erkenntnis und Irrtum*. Verlag von Johann Ambrosius Barth.
Maimon, S. (1792). *Salomon Maimon's [sic] Lebensgeschichte. Von ihm selbst geschrieben. In zwei Theilen*. In K. P. Moritz (ed.). Friedrich Vieweg.
Maimon, S. (2001). *An autobiography*. University of Illinois Press.
Merleau-Ponty, M. (1966). *Die Phänomenologie der Wahrnehmung*. De Gruyter.

Merleau-Ponty, M. (1963). *The structure of behavior.* Beacon Press.
Millikan, R. (2004). *Varieties of meaning. The 2002 Jean Nicod lectures.* Cambridge University Press.
Nanay, B. (2011). Do we see apples as edible? *Pacific Philosophical Quarterly, 92*(3), 305–322.
Nietzsche, F. (2001). *The gay science.* In B. Williams (ed.). Cambridge University Press.
Norman, D. (2013). *The design of everyday things.* Basic Books.
O'Connor, C. (2016). The evolution of guilt: A model-based approach. *Philosophy of Science, 83*(5), 897–908.
Plessner, H. (2003). *Ausdruck und menschliche Natur. Gesammelte Schriften VII.* Suhrkamp.
Poljanšek, T. (2011). *"Just 'cause you feel it doesn't mean it's there". Ein angemessenes Angemessenheitskriterium für Emotionen und Werteigenschaften* (Vol. XXII). Deutscher Kongress für Philosophie, Ludwig-Maximilians-Universität München.https://doi.org/10.5282/ubm/epub.12587
Poljanšek, T. (2015). Sinn und Erwartung. Über den Unterschied von Sinngegenständlichkeit und Referenzialität. *Zeitschrift für philosophische Forschung, 69*(4), 502–524.
Poljanšek, T. (2020). A philosophical framework of shared worlds and cultural significance for social simulation. In H. Verhagen et al. (eds.). *Advances in social simulation. looking in the mirror* (pp. 371–377). Springer Proceedings in Complexity.
Poljanšek, T. (2022). *Realität und Wirklichkeit. Zur Ontologie menschlicher Welten.* Transcript.
Poljanšek, T. & Störzinger, T. (2020). Of waiters, robots, and friends. Functional social interaction vs. close interhuman relationships. In M. Nørskov et al (eds.). *Culturally sustainable social robotics* (pp. 68–77). IOS Press.
Prinz, J. (2004). *Gut reactions.* Oxford University Press.
Roberts, R. C. (2003). *Emotions. An essay in aid of moral psychology.* Cambridge University Press.
Ross, D., & Dumouchel, P. (2004). Emotions as strategic signals. *Rationality and Society, 16*(3), 251–286.
Roth, G. (1997). *Das Gehirn und seine Wirklichkeit. Kognitive Neurobiologie und ihre philosophischen Konsequenzen.* Suhrkamp.
Sartre, J.-P. (1956). *Being and nothingness.* Philosophical Library.
Schütz, A. (1972). *Gesammelte Aufsätze II. Studien zur soziologischen Theorie.* In A. Brodersen (ed.). Den Haag Martinus Nijhoff.
Searle, J. (1995). *The construction of social reality.* The Free Press.
Searle, J. (2010). *Making the social world.* Oxford University Press.
Searle, J. (1994). *The rediscovery of the mind.* The MIT Press.
Sellars, W. (1963). *Science, perception, and reality.* Ridgeview.
Sellars, W (1981). Foundations for a Metaphysics of Pure Process: The Carus Lectures of Wilfrid Sellars. *Monist, 64,* 3–90.
Seibt, J. (2017). Towards an ontology of simulated social interaction: Varieties of the "As If" for robots and humans. In R. Hakli & J. Seibt (eds.), *Sociality and normativity for robots: Philosophical inquiries into human-robot interactions* (pp. 11–41). Springer.
Slaby, J., Paskaleva, A., & Stephan, A. (2013). Enactive emotion and impaired agency in depression. *Journal of Consciousness Studies, 20*(7–8), 33–55.

Smith, J. (2014). Are emotions embodied evaluative attitudes? Critical review of Julien A. Deonna and Fabrice Teroni's the emotions: A philosophical introduction. *Disputatio, VI*, 93–106.
Teroni, F. (2007). Emotions and formal objects. *Dialectica, 61*(3), 395–415.
Uexküll, J., & Kriszat, G. (1934). *Streifzüge durch die Umwelten von Tieren und Menschen*. Springer.
Uexküll, J. (1928). *Theoretische Biologie*. Springer.
Van Kleef, G. A., Van Doorn, E. A., Heerdink, M. W., & Koning, L. F. (2011). Emotion is for influence. *European Review of Social Psychology, 22*(1), 114–163.
Van Kleef, G. A. (2016). *The interpersonal dynamic of emotion*. Cambridge University Press.
Žižek, S. (2007). *How to read Lacan*. Norton.

Tom Poljanšek is a postdoctoral researcher associated to the Chair of Prof. Dr. Catrin Misselhorn, Georg-August-University Göttingen. His areas of research include phenomenology, social ontology, philosophy of technology, and aesthetics. His dissertation "Realität und Wirklichkeit. Zur Ontologie geteilter Welten" (2022) develops a dualistic theory of the relation between mind-independent reality and subject-dependent worlds of appearance and presents an explanation for the diversity and multiplicity of the latter.

The Route to Artificial Phenomenology; 'Attunement to the World' and Representationalism of Affective States

Lydia Farina

Abstract

According to dominant views in affective computing, artificial systems e.g. robots and algorithms cannot experience emotion because they lack the phenomenological aspect associated with emotional experience. In this paper I suggest that if we wish to design artificial systems such that they are able to experience emotion states with phenomenal properties we should approach artificial phenomenology by borrowing insights from the concept of 'attunement to the world' introduced by early phenomenologists. This concept refers to an openness to the world, a connection with the world which rejects the distinction between an internal mind and the external world. Early phenomenologists such as Heidegger, consider this 'attunement' necessary for the experience of affective states. I argue that, if one accepts that the phenomenological aspect is part of the emotion state and that 'attunement to the world' is necessary for experiencing emotion, affective computing should focus on designing artificial systems which are 'attuned to the world' in the phenomenological sense to enable them to experience emotion. Current accounts of the phenomenal properties of affective states, analyse them in terms of specific types of representations. As artificial systems lack a capability for such representation mainly because of an inability to determine relevance in changing contexts ('the frame problem'), artificial phenomenology is impossible.

L. Farina (✉)
Department of Philosophy, University of Nottingham, Nottingham, UK
e-mail: lydia.farina@nottingham.ac.uk

I argue that some affective states, such as 'attunement' are not necessarily representational and as such a lack of capacity for representation does not imply that artificial phenomenology is impossible. At the same time 'attunement' helps restrict some aspects of the 'frame problem' and as such, goes some way of enabling representational states such as emotion.

Keywords

Artificial systems · Phenomenology · Representationalism · Artificial emotion · Attunement

1 Introduction—Background Assumptions on the Nature of Emotion

The dominant view in the literature on artificial intelligence is that artificial systems (AS) including robots, machines, algorithms etc., although capable of recognising and simulating emotional behaviour to some extent, are not capable of experiencing emotion; the phenomenological aspect, what some philosophers call 'what it is like to experience emotion', is non-existent in artificial systems (Moss, 2016; Picard, 2003). This negative view is associated with the claim that mental states such as emotion, cannot be reduced to neurobiological states on account of possessing phenomenal properties (Kriegel, 2017; Stephan, 2009). These phenomenal properties are usually associated with the distinctive feeling aspect of emotion. For instance, there is a specific feeling associated with experiencing *fear*, a specific feeling associated with feeling *anger* or *pride*. The dominant view is that we cannot replicate emotion states in AS because the phenomenological aspect of the emotion state is not reducible to neurobiological states.

Not being able to identify phenomenal properties with, or reduce them to, neurobiological states comes with the acknowledgement that there are several opposing views of what these properties are and how they relate to what we call 'the emotion state'. An underlying assumption which is sometimes used to support the claim that AS are not capable of experiencing any phenomenology is that this phenomenology is necessarily associated with a human-like body (Gallagher, 2014). According to such views a human-like body is necessary for experiencing states with phenomenal properties such as emotion or mood. However, according to functional accounts of emotion, a human-like body is not necessary for experiencing phenomenal states; instead, the emotion state is identified with the performance of a specific function, to enable us to interact with our environment in ways which facilitate the achievement of our goals.[1] If emotion states are functional states, they have distinct roles/functions which individuate them from other mental states and from one another (Kim, 2011).[2] One of the main advantages of endorsing functional accounts of emotion states is that they allow for multiple realisation. This means that to identify an emotion state, one needs to look at whether the state fulfils the relevant causal role and not at whether it is realised in the same physical way. For example according to functional accounts, humans and squid can be in *fear* states even if these states are realised by completely different physical states. Therefore, if one endorses a functional account of emotion, the absence of a human-like body does not entail that AS are not capable of experiencing affective states. However, although functional accounts reject the claim that experiencing affective states is necessarily associated with having a human-like body, they do not provide adequate explanations for the phenomenal properties of those states (Kim, 2011).

A further way to analyse emotion states is to consider them as representational states integrating two different types of representation: a) the representational '*what*', the object being represented and b) the representational '*how*', the way the object is being represented (Dretske, 1995; Kriegel, 2012; Tye, 1995). For example the state of 'fearing the shark' can be analysed as a representational state including an object e.g. *shark* and an attitude towards this object e.g. *fearing*. The phenomenological aspect of emotion states is identified with the representational attitude one takes towards an object. If one endorses a representational analysis of emotion states, replicating the experience of emotion in AS depends on the ability of AS for representational states with a phenomenological aspect.

An ongoing debate in the philosophy of mind focuses on whether the phenomenal properties associated with phenomenal states (such as affective states) are intrinsic to those states or whether they are something separate to them. Philosophers such as Armstrong (1980) or Rosenthal (1991) claim that the experience of

---

[1] In a similar way Elpidorou and Freeman suggest that the function of emotion is to reveal features of situations we find ourselves in and motivate us towards some action (Elpidorou & Freeman, 2015, p. 661).

[2] For the purposes of this paper I use Kim's definition of functional states as combinations of inputs and outputs where these inputs and outputs may include other mental states as well as sensory stimuli and physical behaviours (Kim 2011, p. 169). As I discuss in Sect. 2 these inputs and outputs do not presuppose a distinction between an internal subject and external objects.

any mental state should be identified with a separate state which reflects (either via perception or via thought) on the mental state. So the experience of a mental state should be interpreted as reflective consciousness, which is considered as an ability to reflect on a mental state. On the other hand according to an opposing view (e.g. Block, 1995) the experience of emotion should be interpreted as phenomenal consciousness, which is explained as properties associated with 'what it feels like' to undergo a phenomenal state. These properties are not reflective; they cannot be explained by reflective consciousness. They are intrinsic features of the phenomenal state.

In this paper I discuss the possibility of designing artificial phenomenology, if one endorses functional or representational accounts of emotion states that can be intrinsically phenomenal. I argue that such accounts are still unable to solve certain aspects of the 'frame problem', namely the ability to determine which context one is in when the contexts change. I conclude that 'attunement' could provide a partial solution to this problem by making some contexts more likely than others. In addition I claim that although 'attunement' does not appear to be necessarily representational it is itself necessary for emotion states which are considered typically representational.

2 Affective Computing and Artificial Phenomenology

Neither functionalism nor representationalism provide adequate accounts of phenomenal properties when these are considered as phenomenal consciousness rather than reflective consciousness (Kim, 2011). Furthermore, these accounts are usually associated with the view that phenomenal properties do not have a role to play in the emotion state. In this section, I focus on two related claims which are used in the wider literature of artificial intelligence against functional or representational accounts of phenomenal states.[3] According to the first one, AS are not capable of representation[4] because of 'the frame problem', e.g. they are not

---

[3] See for example Ford & Pylyshyn (1996), Dennett (1998), Wheeler (2008), Ransom (2013) and Boden (2016) chapters 2 and 3.

[4] Here by representation I do not refer to simple simulation of representation states which has been achieved by some embodied AI (see Di Paolo 2003). Instead, I refer to a stronger sense which takes representation to be intrinsically meaningful rather than a state with externally imposed meaning or significance.

able to determine the relevance of incoming information so as to determine which situation/context they are in and, as a result, they cannot determine which stored information is relevant for each context.[5] Dreyfus claims that this problem leads to a 'regress of contexts' as to identify which context one is in, one must take some features to be relevant by looking at a broader context (Dreyfus, 2007, p. 2). However, to determine the broader context one must look at features which are relevant in an even broader context etc. As such, it seems very difficult to specify rules for determining relevance unless one operates in an environment with strictly isolated features.[6] According to the second claim, affective states such as emotion and mood are representational states which are intrinsically phenomenal e.g. they represent some object or some state of affairs in the world in a certain attitude.[7] Combining the first and the second claim forms the following argument:

1. Affective states such as emotion or mood are representational states which are intrinsically phenomenal
2. AS are not capable of representational states which are intrinsically phenomenal

AS are not capable of affective states

If the premises of the argument above are true, one should conclude that we will not be able to design artificial phenomenology.

It is a widely accepted assumption in affective computing (Scherer et al., 2010) that AS currently lack the phenomenal aspect of the emotion state. Most of the recent discussion in affective computing seems to be targeting ethical issues relating to whether we should design AS with the capability to simulate feelings

---

[5] It is still problematic to give an accurate description of the 'frame problem'. Here I borrow from Ransom (2013, p. 2) who takes it to refer to a cluster of problems having at their core the problem of determining relevance in changing contexts.

[6] As I discuss in the following section, one area of AI can deal with some aspects of 'the frame problem' if it isolates the features of the environment the AS operate in. However, such attempts cannot accommodate changing contexts as the context must always remain constant. As Froese and Ziemke argue, the existence of closed feedback loops or unchanging contexts is not a sufficient condition for the attribution of intrinsically meaningful perspective in AS (Froese & Ziemke, 2009, p. 472).

[7] Here I focus on the intrinsic view of affective states, according to which affective states are intrinsically phenomenal.

rather than how to go about designing AS such that they are capable of experiencing feelings (Scherer et al., 2010, p. 316).[8] This tendency can be partially explained by the fact that the dominant view, as mentioned above, is that artificial phenomenology is impossible in the absence of a human-like body. In the introduction, I briefly mentioned that a functional account of emotion would not be subject to this issue as emotion states are considered multiply realisable.[9]

There are several objections raised against functional or representational accounts of mental states and whether we can replicate these states in AS on the basis of those accounts.[10] Here I concentrate on two very important ones raised against representational or functional accounts of emotion: 1) AS are not capable of representation (Dreyfus, 1992, 2007) and 2) Phenomenal states are necessarily representational (Tye, 1995); if 1 and 2 are true, AS are not capable of experiencing affective states because they are not capable of the required type of representation.

Both objections are associated with an ongoing debate between Symbolic AI (Artificial Intelligence) and Situated Robotics on the ability of AS to have repre-

---

[8] A notable exception in affective computing literature is Parisi and Petrosino (2010) who claim that they have created simulated robots which 'can be said to *have* emotions'. They claim that 'adding a special emotional circuit to the neural network controlling the robots' behaviour leads to better motivational decisions' (Parisi & Petrosino 2010, p. 453). However, Parisi and Petrosino are not interested in replicating phenomenological features of emotion. Instead they focus on the functional role that emotions play in motivating behaviour. A more recent paper by Hickton et al. (2017) focuses on designing an architecture which links affect and behaviour with the use of artificial hormones. According to Hickton et al. this architecture can be applied to different types of robot with the view to facilitate 'subtle forms of expression and adaptive behaviour' (Hickton et al., 2017, p. 472). However these authors, in a similar way to Parisi and Petrosino, do not discuss the phenomenological features of such affective behaviour.

[9] A functional account does not preclude the possibility that some type of physical body may be important for experiencing affective states e.g. humans and octopi can both be in fear states and they both have physical bodies. However functional accounts relax the conditions according to which something would be considered a 'physical body' because that also would be determined on the basis of the function it performed and not on whether it was similar or different to a human-like or octopus-like body. Here I do not focus on 'embodied' accounts of affective states however insights provided by such accounts are important for the general project of designing artificial phenomenology because they provide details on how some systems realise affective states.

[10] For a good discussion of objections against functional accounts see Boden (2016). For a discussion of objections against representational theories of mental states see Kim (2011).

sentational states. Symbolic AI is associated with the claim that the only way to create artificial mental states is via replicating mental representations.[11] However, *'situated robotics'*, a distinct area of AI, focuses on designing robots who operate by using situated behaviour (Beer, 1990; Brooks, 1991). *Situated robotics* is influenced by the psychological input that human behaviour consists of role-playing in distinct social environments; it is situation based. The behaviour depends on the context an individual is in; as such, the context enables specific roles which the individual performs. To give an example, experiments by social psychologists in the 70s showed the behaviour of individuals to depend on the specific roles they were asked to play within given contexts rather than on the deliberation between several behavioural choices.[12] In a similar way *situated robotics* were influenced by studies on insect behaviour; insects respond to current situations by adapting their behaviour through the use of reflexes rather than after deliberation (Götz, 1972). Situated robots in the same way respond unthinkingly to situations by using engineered reflexes and adapting their behaviour without the need for deliberating between mental representations (Beer, 1990). Webb's robotic female crickets simulate the behaviour of female crickets by identifying, locating and approaching a particular sound pattern as a response to the frequency of the male song (Webb, 1996). This response is a reflex rather than a choice as the female cricket can detect only a particular frequency. This finding lead some AI scientists to claim that representation is too cerebral and it cannot be replicated in AS because AS is not capable of representation (Brooks, 1991; Dreyfus, 2007). Therefore, according to such views, if the phenomenological aspect of affective states is representational in nature, AS will not be capable of replicating it.

This inability to replicate representational states in AS is associated with the 'frame problem', the claim that because AI lacks a human's sense of relevance it cannot determine which information is relevant to a certain context which, in turn, creates issues with determining which context one is in. In cases where

---

[11] See Boden (2016) for a good summary of this debate and its current status. Other researchers suggest hybrid accounts according to which situated behaviour and mental representations can be replicated in AS once they reach a certain level of organisational complexity (Sloman & Chrisley, 2003). In addition connectionist or Artificial Neural Networks accounts allow for distributed representation which refers to something different to the one postulated by Symbolic AI (Baars 1988).

[12] See for example Zimbardo's (1972) Stanford Prison Experiment where the majority of experiment participants behaved in accordance with the roles they were assigned within the given context rather than deliberating about which action they should perform.

determining which context one is in, depends on taking into consideration implicit associated information, AS fail to take such implicit information into consideration and cannot determine which context they are in and which action is the best in the given context. In cases where AS would have to choose an action between several available choices, they fail to select the optimum choice because they cannot determine which information is relevant in changing contexts. To give an example, whether the fact that the ink in the printer is running out is relevant to Stella, depends on the context she is in: for instance, she could be away on holiday; she may need to print a contract; she may not have the money needed for the expensive ink; she may not be at the stationary store etc. Even if she does have the money to buy ink, something may happen when she leaves the flat to buy it e.g. she runs into a friend and she decides to go for a cup of coffee instead. In this case the fact that the ink is running out is no longer relevant to the situation.

According to one view the 'frame problem' is more or less solved if all possible consequences of every possible action are known and we are able to create finite descriptions of functional states where all implicit associated information becomes explicit and helps us to determine which context we are in (Shanahan, 1997). In human systems, it has been argued that emotion helps to narrow down the relevant implications during deliberation and, as such, helps reduce the number of choices during deliberation after a context has been identified (De Sousa, 1987; Ketelaar & Todd, 2001). As De Sousa argues, emotions control which features of perception and reasoning we consider salient, thus reducing our choice of options (De Sousa, 1987, p. 172). Wheeler (2008) calls this the 'intercontext problem' facing artificial systems which he takes to mean the problem of determining relevance of environmental information within a specific context. This is juxtaposed to what Wheeler calls the 'intracontext problem' which relates to the inability to specify which context is the relevant one. However, as Ransom argues, it is not likely that the contribution of emotion solves the 'frame problem' outright; instead it helps to determine which information held by a system is relevant and should be drawn upon (2013, p. 4). Nevertheless, a problem of determining relevance still applies in that one is unable to determine which context one is in. As Ransom states, "this is the problem of determining what features of the environment one ought to take as relevant, amongst the many possible candidates, in specifying the situation" (2013, p. 3). Therefore, although emotion helps restrict the 'frame problem' (the intracontext version) once a context or a situation has been identified, it cannot help with identifying the relevant context to begin with (intercontext version). As Ransom claims, what makes determining relevance difficult in the intercontext version, is first, that such relevance is context sensitive and second, that we are confronted with ever changing contexts (2013, p. 3).

Dreyfus claims that all representational accounts are subject to the 'frame problem' because they cannot provide an account which accurately describes our connection with the world. According to Dreyfus, the mistake of Symbolic AI was to claim that all mental states experienced by humans require representation. Dreyfus correctly points out that this is not the case in human systems because situated behaviour takes place in absence of representation. To give an example if Marta throws the mug at Julia and startles her, Julia will probably raise her hands to avoid the mug hitting her on the head. This action can happen without Julia having a representation which appoints significance to incoming information. Raising the hands to protect the head is an automatic reflex which happens as a response to the situation. More importantly for the claim I make in this paper, Dreyfus claims that representation is not necessary for appointing significance or relevance to certain aspects of the world. He further argues that we do not represent first and then appoint significance to the world around us. We appoint significance in the absence of representation. Dreyfus suggests that most mental states and, more importantly for the claim I am making in this paper, affective states do not require representation.

One can claim that by appealing to the notion of representation one cannot explain how we appoint significance to some features of the environment and, as such, how we determine which context is relevant in each situation. Moreover, according to the objections discussed above, as AS are not capable of representation, attempts to replicate phenomenal states in AS by using representational accounts are bound to fail. In the next section I turn to discuss the concept of 'attunement to the world' as developed by Heidegger (1927) and Merleau-Ponty (1945) to suggest that although traditionally it is used as an objection against the possibility of artificial intelligence, it can be used to provide a solution by re-formulating the concept of the phenomenology of affective states. The reason for focusing on this concept is twofold; first I argue that 'attunement' can restrict the 'frame problem' by narrowing down which information from the environment the system will judge to be significant in determining contexts. Secondly, by rejecting the claim that all affective states are representational, I argue that 'attunement' allows for the coexistence of representational affective states e.g. emotion and non-representational affective states e.g. mood. It is only if one considers representation as necessary for all affective states that artificial phenomenology seems impossible.[13]

3 'Attunement to the World' and Artificial Phenomenology

According to early phenomenologists, 'attunement to the world' is a necessary pre-condition for experiencing affective states (Heidegger, 1927; Merleau-Ponty, 1945). An advantage arising from integrating this phenomenological insight into the project of designing artificial phenomenology relates to the role of 'attunement' in determining relevance e.g. determining the significance of incoming information and thus determining the possible situations one is in. I claim that by doing so, 'attunement to the world' helps to reduce the possible contexts that AS may consider to be relevant so as to determine which situation they are in. As such, by reducing the possible contexts, it reduces the effect of the 'frame problem' attaching to functional and representational accounts of artificial phenomenology. At the same time, I argue that 'attunement' is not a representational state but it enables emotion states that are typically representational. If this is the case, some affective states are not necessarily representational. I conclude that we need to provide new accounts on the one hand on the relation between emotion and 'attunement' as 'attunement' is considered a necessary condition for emotion and, on the other hand, on the relation between representing the world and experiencing the world.

In the previous section I discussed two claims the acceptance of which provides support against the possibility of artificial phenomenology e.g. that affective states are representational states which are intrinsically phenomenal and that AS are not capable of such representation. In what follows I argue that 'attunement' does not require representation and as such the claim that all affective states are representational is not true. I claim that some affective states e.g. mood are non-representational states whilst others such as emotion are typically representational e.g. emotion is felt towards an object (Deonna & Teroni, 2012). It is through 'attunement' that we are able to be in moods. At the same time, moods make some emotions more likely than others. One way through which moods achieve

---

[13] Here by 'representation' I mean of the specific type I have been discussing in the first part of the paper e.g. a representational state that is intrinsically phenomenal.

this, is by revealing features in the environment which are significant to us. Here I do not claim that the mood one is in, determines which emotion one feels or not. Rather the claim I am making is that being in a certain mood makes it more likely that I will feel certain emotions rather than others. To give an example, being in an anxious mood makes it more likely that I will appoint significance to specific stimuli rather than to others.[14]

In Sect. 1, I discussed a particular problem relating to determining which context one identifies as relevant between the possible contexts one finds oneself in. Agreeing with Ransom (2013) I claimed that emotion cannot help us identify which context (between all possible contexts) we are actually in. As Ransom claims, emotion helps reduce which stored information we consider relevant once the relevant context has been identified. Here I suggest that 'attunement', interpreted as being-in-a-mood, operates as a mechanism which partially fills this gap; it reduces the number of possible contexts we find ourselves in and therefore reduces the number of contexts we consider relevant.

As claimed in the previous section, Dreyfus claims that representational theories of mind find it particularly difficult to provide an explanation for the connection between an internal representation and the significance attached to features of the world (2007). He supports this claim by stating that as long as the mind is considered as a passive receptor of meaningless inputs that need to have significance and relevance added to them, we cannot explain how this binding takes place (Dreyfus, 2007, pp. 18–19). Instead, he argues that the phenomenological model can provide a better description of this process by allowing for a direct access to significance in the absence of representation.

According to early phenomenologists this direct access to significance is connected to the idea of 'attunement to the world' (Heidegger, 1927; Merleau-Ponty, 1945). This 'attunement to the world' is interpreted as 'an openness to the world' and is a necessary requirement for experiencing affective states such as emotion. This openness is interpreted as 'being in a mood'; for us to experience emotion towards some object, this object must first matter to us. However, for the object to matter to us we must first be attuned to it by being in a certain mood (Smith, 2016). This attunement is what allows us to appoint significance; to care about some features of the environment. It is what enables affectivity, our ability to

---

[14] One could make a stronger claim to the effect that the mood I am in determines which emotions I feel. Here I focus on the weaker claim that the mood makes it more likely that I will feel certain emotions but does not guarantee it. I am grateful to an anonymous reviewer for bringing this to my attention.

have affective states such as emotion (Welton, 2012). This conception of attunement is similar to Colombetti's 'primordial affectivity' considered as a capacity of living systems "to be sensitive to what matters to them....to already realise a relationship with themselves and the world in which they are situated that entails purposefulness and concern for their existence" (Colombetti, 2013, p. 2). It is this attunement which enables other affective states such as emotion.

The view sketched above seems to suggest a very important role for attunement as it is what enables objects in our environment to matter to us.[15] According to such views through attunement we are always in a mood, although there is constant change between the moods we are in (Freeman, 2014; Heidegger, 1927). In addition being-in-a-mood is considered necessary so that someone experiences an emotion. One obvious problem with such views is that early phenomenologists do not agree on precise and distinct definitions of these terms (e.g. attunement or mood).[16]

In a similar way, scientific accounts of mood lack a unified ontological account. There are few studies in affective science focusing on how moods connect us to the world. In this paper I accept two claims about the nature of mood which are compatible with most literature on affective phenomena (Freeman, 2014; Stephan, 2017); first, mood *qua* an affective state, is a phenomenal state; it has a phenomenal aspect e.g. it feels a certain way to be euphoric, or to be depressed. Secondly, I accept the claim that mood is a different affective state to the emotion. Several features have been suggested as constitutive of this difference such as intentionality e.g. what moods or emotions are directed towards, duration e.g. how long a typical mood or emotion lasts and causal settings e.g. what gives rise to mood or emotion (Stephan, 2017, p. 1488). In this paper I focus on the feature of intentionality of mood and its compatibility with representational theories of affective states.

In the wider literature in affective science it is a matter of debate whether moods are directed towards something at all, whether they represent anything

---

[15] According to Heidegger this attunement is necessary not only for the experience of emotion but for the experience of all mental states because it enables us to interact with the environment. In this paper I focus on the importance of attunement for affective states.

[16] For example, Heidegger seems to use the same term to refer to 'attunement' and 'mood'. The debate on explicating these terms and providing an ontological account of mood is currently ongoing. For a useful collection of papers on the different interpretations of 'mood' see the special issue published in *Philosophia* 2017 Vol. 45.

(e.g. whether they are representational/intentional[17]). According to one view, they are nonrepresentational; they are not directed at anything (Frijda, 1986). Others claim that they are directed towards something general e.g. the world as a whole rather than something specific (Lazarus, 1991). According to a separate view, although moods appear to be objectless, they do have a formal object to which they are directed to. For example, Goldie (2000) takes the whole world to be the object of moods whilst Price (2006) claims that moods are intentional states which represent how things are likely to turn out in the immediate future. An alternative view claims that they are pre-intentional states e.g. they are non-intentional but they enable emotion which is intentional (Ratcliffe, 2010).[18] Here, following Kim, I claim that moods do not appear to be representations; they don't have representational content because they don't have satisfaction conditions, e.g. conditions that define their representational correctness, accuracy or fidelity (Kim, 2011, p. 278). Kim states: 'Representations must be evaluable in terms of how closely they meet those conditions. Evaluating moods in this way makes no sense' (Kim, 2011, p. 293). Accepting that moods are non-representational in nature comes with the associated implication that the phenomenal character of moods is not reducible to representational content. This is compatible with non-reductive representational accounts where some aspects of phenomenal states are not reducible to representational content[19] (Block, 1996; Chalmers, 2004).

---

[17] Following Kim (2011, p. 24–25) I use representationalism and intentionalism interchangeably to refer to the view that mental states have the capacity and function of representing things and states of affairs in the world from a certain perspective.

[18] Saying that mood is pre-intentional entails that it is not representational. Ratcliffe's view is similar to the view I endorse in this paper as I consider mood to be non-representational whilst at the same time enabling emotion states which are representational.

[19] Critics suggest that it is not clear that the phenomenological properties associated with phenomenal states can be adequately described on the basis of representational content; for example spectrum inversion cases are possible where one person sees green where another sees red when they see a red tomato but both call that colour 'red'. In such cases although the content of the representation seems identical e.g. 'a red tomato', the colour experienced is different (Shoemaker 1982). According to this objection, the content of a representation on its own cannot adequately account for the phenomenological aspect of experience. Any representational account of the phenomenal properties of affective states would need to address these concerns. See Kriegel (2017) for a recent attempt to give a representational account of emotional phenomenology by suggesting that the representational content is not exhausted by the object being represented but also by an attitude of representing the object as x. Here I suggest that 'the attitude towards x' should be based on the idea of 'attunement to the world'. As such my account could be considered compatible with, and an addendum to, Kriegel's.

The view that mood is non-representational seems compatible with the view of early phenomenologists such as Merleau-Ponty (1945) who reject the subject/object, internal/external model for understanding 'attunement' viewed as a relationship between human beings and the world. This is because such dualisms do not capture the way in which we already find ourselves in the world. Freeman similarly claims that our affective states cannot be severed from the context in which they arise (2014, p. 452).[20] We need to provide an account of mood which rejects the distinction between subject/object or internal/external models. According to this account, being-in-a-mood describes an existence in-the-world; this existence does not require any representation of the world as an external object.

Heidegger similarly describes 'attunement' as a feeling of belonging to a world that is taken for granted in everyday life (1927, p. 172). He claims that in our practical dwelling in this world, this distinction between the internal and external can disappear in the sense that it becomes inaccessible.[21] This distinction is the result of detached contemplation, of theorising about our practical dwelling (Ratcliffe, 2005, p. 51). We can think of cases where we use some kind of tool to accomplish a task and we experience the tool as part of our body rather than an external object e.g. the hammer in the hands of a carpenter or the basketball in the hands of a basketball player. In such cases the distinction between internal and external appears only when we focus our attention to specific parts of the process of hammering or playing basketball, e.g. focusing on the way I hold the hammer in my hand, rather than focusing on the nail I am about to hammer. Here the importance is not on whether such disappearance of the distinction is ever-present but on the fact that this distinction can become accessible or inaccessible. Unless something happens which breaks our being-in-the-world, whatever it is that we are is integrated into the world rather than being distinct from it. Therefore, early phenomenologists use the idea of 'attunement' to describe the way that we ordi-

---

[20] This rejection of the distinction between internal/external is a recurring theme in phenomenological accounts e.g. Merleau-Ponty's or Heidegger's. However an account providing an adequate explanation of this phenomenon is still missing.

[21] As Dreyfus argues, Heidegger attacks this internal/external distinction as a misunderstanding based on Descartes' distinction between the internal mind and the external world (Dreyfus 2007). According to Heidegger (1927), Descartes' distinction was accepted as a dogma and created several epistemological issues such as the knowledge of the external world which was completely unwarranted unless someone pre-supposed the truth of the distinction.

narily exist as beings-in-the-world, as parts of the world rather than as beings external to it.

This way of being-in-the-world describes a connection between moods, the world and the (living) system. Because we are attuned to it by being-in-a-mood, the world is disclosed to us through our moods but it also influences these moods. This is relevant to the claim that mood is 'scaffolded' by the environment, where an interaction with the environment is necessary for the creation of affective states (Colombetti, 2017). One type of affective scaffolding is what Colombetti calls 'incorporation', the fact that one can 'experientially incorporate parts of the world' in the sense that when one undergoes certain moods, one experiences certain aspects of the world as part of oneself (Colombetti, 2017, p. 1441). For example, the instrumental musician experiences her instrument as part of herself, through which part, she can express a mood of longing or sadness (Colombetti, 2017, p. 1448). Colombetti describes this as 'being incorporated into a person's pre-reflective awareness of her or his body' (2017, p. 1448).

As per the above, the way that the world appears is directly dependent upon the way we are attuned to it through mood. To give an example used by Freeman (2014) 'when I am melancholic it is not just that I feel as though the world is grey, dull, uninteresting, uninspiring, unexciting. Rather, when I am melancholic the world is grey, dull, uninteresting, uninspiring and unexciting'. One way to understand this example by Freeman is to say that the distinction between 'feeling as though the world is grey' and 'the world being grey' is blurred. Such a state presupposes the absence of a distinction between what is out there e.g. the external and what is in here e.g. the internal; the boundaries between ourselves and the external world seem to dissolve such that the mood one is in determines which features of the environment will matter and in what way. In such cases there is no internal subject who judges the world (the external object) as being grey. Instead when one is melancholic, one is being-in-a-grey-world.

In addition, when one is in a melancholic mood, the openness to the world which is taken for granted when one is not melancholic, disappears. What remains is a feeling of disconnectedness from the world (Ratcliffe, 2014).[22] However, these feelings of connectedness or disconnectedness have no representational content. Even if one claims that the representational content is 'being connected to the world' (or 'not being connected to the world') this content can-

---

[22] Ratcliffe connects this feeling of disconnectedness with the experience of being in a depressive mood (see Ratcliffe 2014 for more detail).

not explain the phenomenal character of being melancholic or not being melancholic. It therefore seems to be the case that the phenomenal properties of mood cannot be considered as representational.

Once the distinction between subject and object is rejected because we are attuned to the world, the question of how these are connected e.g. what explains the intentionality of mood seems to disappear. It is not necessary to represent the world as being one way or another because one is already being-in-the-world. Therefore one could claim that being-in-a-mood does not need to provide an account which accommodates representation of the world; being-in-a-mood is the experience of being-in-the-world. It does not require and it happens without representation. According to this interpretation being-in-a-mood isn't representational but is pre-reflective, arising in tandem between the individual and the environment. Because we are 'mooded' beings we are already always being-in-the-world; we are always experiencing the world rather than directing our mind towards the world as if we are standing outside of it.

At the same time this non-representational state is a pre-condition for the experience of the emotion state which is considered as representational.[23] For this view to be plausible one should hold that although emotion is typically representational in nature, some affective states such as being in moods need not be. The claim I put forward in this paper seems to be compatible with ideas put forward by 'enactive' accounts in cognitive science. For example, Froese and Ziemke argue that perceptual experiences need not be internal events but rather something 'which we enact or bring forth through our active engagement and sensorimotor exploration of our environment' (2009, p. 475). This would allow that to determine significance or relevance to the environment, an interaction between the organism and the environment is necessary.[24]

In addition if phenomenologists are right in their claim that the world is disclosed to us through our moods, being-in-a-mood seems to be a very important ingredient in bringing about representation. According to Heidegger's idea of 'being-in-a-mood', we can be emotionally affected only because we are connected, open, and responsive to the world in which we are embedded (Heidegger, 1927). It is not the case that human beings exist and then are added into a world, nor can they be understood apart from their context and history. Our experiences

---

[23] For a stronger view which takes mood to be the background condition for most other representational states see Ratcliffe (2013) and Slaby 2014.

[24] This interaction is called 'sense making' in enactive AI accounts (Froese & Ziemke, 2009, p. 495).

immediately reveal the world to us insofar as one's existence is already bound up with the world. Experiences are filtered through moods such that the world reflects the mood I am in. Our entire existence within the world is shaped, coloured, experienced and in a sense determined by our moods. According to this view, being-in-a-mood makes it possible first of all to direct oneself towards something but at the same time the world is disclosed to us through/out of our moods (Heidegger, 1927).

According to such views moods are world-orientations, ways of being integrated into the world as a whole.[25] For example when one is in a melancholic mood one is likely to attend to features of the environment that match this specific mood, e.g. reading sad news, reminiscing sad or embarrassing memories etc. To give another example when one is anxious one overestimates the danger in situations and anticipates future threats.[26] Being-in-a-mood of anxiety makes certain contexts more likely by appointing significance to certain aspects of the world e.g. the hammer becomes a tool expected to cause bodily harm rather than a tool for hammering. Where such an anxious mood persists, it may lead to avoidance behaviour of certain activities reducing one's range of experiences. As such, one can claim that being-in-a-mood helps to restrict the number of possible contexts we choose from when deciding which context we are in, by appointing significance to some features of the environment rather than others. As soon as this significance is appointed these features can then be used to identify contexts. For example, when one is in an anxious mood the knock on the front door becomes a cause for fear (it could be a burglar) rather than a cause for joy (a surprise visit from a friend). In the example of the printer ink used in Sect. 1, Stella is deliberating whether to leave the flat to buy some ink and tackle the stormy weather, or stay indoors where it is warm. Suddenly, she gets news that her sister, who is living abroad, is coming to visit her in a month. In an elated mood she decides to go to the shop to buy the ink despite the menacing rain which does not seem to be a significant factor anymore.

By acting as a significance appointer, being-in-a-mood reduces the amount of possible contexts one considers as relevant and restricts the 'frame problem'. Froese and Ziemke claim that enactive AI are still unable to design AS which are

---

[25] Ratcliffe call these 'existential feelings' however here I interpret moods as modes of being rather than feelings (see Ratcliffe, 2005, p. 56).

[26] See entries for 'depression' and 'anxiety' disorders on Diagnostic and Statistical Manual of Mental Disorders (2013) published by the American Psychiatric Association for more details.

capable of both generating their own systemic identity at some level of description and also of actively regulating their ongoing sensorimotor interaction in relation to changing contexts (what they interpret as the inter-context version of the frame problem) (2009, p. 486). If this interpretation of the concept of attunement as being-in-a-mood is right, being-in-a-mood operates as a frame setter, it influences which context within the possible contexts we consider significant or relevant to us. Although on its own this cannot solve the 'frame problem', nevertheless, it addresses a specific aspect of it: it acts as a frame identifier by appointing significance to specific features and, as such, making some situations (among all the possible relevant ones) more likely than others.[27] In the first section I mentioned that the 'frame problem' is one of the main objections against the claim that AS are capable of a specific type of representational states (with phenomenal properties). In this section I argued that if being-in-moods reflects ways of being integrated in the world, it can restrict the possible contexts we experience and as such facilitate or make representation in changing contexts possible. Although this claim does not prove outright that AS are capable of representation in changing contexts, it provides a viable route to artificial phenomenology which can be empirically tested.

4 Conclusion

The focus of research in affective computing currently falls on emotions rather than moods.[28] As such, the relation between being-in-a-mood, in the sense suggested in this paper, and emotion has not been investigated in any depth. It is important for the project of designing artificial phenomenology that we work on an account which explains the relation between the various affective states and, more specifically, the relation between emotion as a representational affective state and being-in-a-mood as a non-representational affective state. In this paper I give two reasons why it would be advantageous to do so. First, early phenom-

---

[27] This concept of mood has similarities with the concept of disposition and it would be beneficial for the project of designing artificial phenomenology to provide an account of the relation between moods and dispositions.

[28] Some roboticists integrate moods in their models however these moods are typically limited to three modes, that is, negative, positive and neutral (see for example Kirby et al., 2010). As such, these attempts reflect different approaches to the concept of mood than the one I discuss in this paper.

enologists (Heidegger, 1927; Merleau-Ponty, 1945) argue that 'attunement' in the sense of 'being in moods' is necessary for experiencing emotion. Therefore if we wish to replicate affective phenomenology in AS we should attempt to design them such that they are in moods. Second, being-in-a-mood acts as a frame selector by appointing significance to specific aspects of the environment. This restricts the intercontext version of the 'frame problem' by making some contexts relevant and ignoring others.

I discussed the possibility of artificial phenomenology by focusing on functional or representational accounts of affective states and discussed two claims; according to the first AS are not capable of representation in changing contexts; according to the second AS are not capable of experiencing affective states. I suggested that some affective states e.g. being-in-a-mood are not necessarily representational. We need to analyse them without expecting representationalism to provide the tool for such analysis. I argued that the concept of 'attunement' reveals a world which is already significant to us without the need for representation. We should explore the possibility that the route to artificial phenomenology does not necessitate representation.

Acknowledgements: I would like to thank Pilar Lopez Cantero and Joel Smith for their helpful comments on an earlier draft of this chapter.

References

American Psychiatric Association. (2013). *Diagnostic and statistical manual of mental disorders* (DSM-5). American Psychiatric Publishing.
Armstrong, D. (1980). *The nature of mind*. University of Queensland Press.
Baars, B. J. (1988). *A cognitive theory of consciousness*. Cambridge University Press.
Beer, R. D. (1990). *Intelligence as adaptive behaviour: An experiment in computational neuroethology*. Academic Press.
Block, N. (1995). On a confusion about a function of consciousness. *Behavioural and Brain Science, 18*, 227–247.
Block, N. (1996). Mental paint and mental latex. *Philosophical Issues, 7*, 19–49.
Boden, M. A. (2016). *Ai: Its Nature and Future*. Oxford University Press UK.
Brooks, R. A. (1991). Intelligence without representation. *Artificial Intelligence, 47*, 139–159.
Chalmers, D. (2004). The representational character of experience. In B. Leiter (Ed.), *The future for philosophy* (pp. 153–181). Oxford University Press.
Colombetti, G. (2013). *The Feeling Body: Affective Science Meets the Enactive Mind*. MIT Press.
Colombetti, G. (2017). The embodied and situated nature of moods. *Philosophia, 45*, 1437–1451.
de Sousa, R. (1987). *The rationality of emotion*. MIT Press.

Dennett, D. (1998). Cognitive wheels: The frame problem of AI. In *Brainchildren* (pp. 181–205). Penguin Books.
Deonna, J., & Teroni, F. (2012). *The emotions. A philosophical introduction*. Routledge.
Di Paolo, E. (2003). Organismically-inspired robotics: Homeostatic adaptation and teleology beyond the closed sensorimotor loop. In K. Murase & T. Asakura (Eds.), *Dynamical systems approach to embodiment and sociality* (pp. 19–42). Advanced Knowledge International.
Dretske, F. (1995). *Naturalizing the mind*. MIT Press.
Dreyfus, H. L. (1992). *What computers still can't do: A critique of artificial reason*. MIT Press.
Dreyfus, H. L. (2007). Why Heideggerian AI failed and how fixing it would require making it more Heideggerian. *Artificial Intelligence, 171*, 1137–1160.
Elpidorou, A., & Freeman, L. (2015). Affectivity in Heidegger I: Moods and Emotions in Being and Time. *Philosophy Compass, 10*(10), 661–671.
Ford, K. M., & Pylyshyn, Z. W. (1996). *The robot's dilemma revisited: The frame problem in artificial intelligence*. Ablex.
Freeman, L. (2014). Toward a phenomenology of mood. *The Southern Journal of Philosophy, 52*(4), 445–476.
Frijda, N. H. (1986). *The Emotions*. Cambridge University Press.
Froese, T., & Ziemke, T. (2009). Enactive artificial intelligence: Investigating the systemic organization of life and mind. *Artificial Intelligence, 173*, 466–500.
Gallagher, S. (2014). Phenomenology and embodied cognition. In L. Shapiro (Ed.), *The Routledge handbook of embodied cognition* (pp. 9–18). Routledge.
Goldie, P. (2000). *The emotions: A philosophical exploration*. Oxford University Press.
Götz, K. G. (1972). Principles of optomotor reactions in insects. *Bibliotheca Ophthalmologica : Supplementa Ad Ophthalmologica, 82*, 251–259.
Heidegger, M. (1927). *Being and time*. Translated by Joan Stambaugh. State University of New York Press.
Hickton, L., Lewis, M., & Cañamero, L. (2017). A flexible component-based robot control architecture for hormonal modulation of behaviour and affect. In Y. Gao et al. (eds.) *TAROS 2017, LNAI* 10454 (pp. 464–474). Springer.
Ketelaar, T., & Todd, P. M. (2001). Framing our thoughts: ecological rationality as evolutionary psychology's answer to the frame problem. In H. R. Holcomb III (ed.) *Conceptual Challenges in Evolutionary Psychology: Innovative Research Strategies* (pp. 179–211). Kluwer Academic Publishers.
Kim, J. (2011). *Philosophy of mind*. Westview Press.
Kirby, R., Forlizzi, J., & Simmons, R. (2010). Affective social robots. *Robotics and Autonomous Systems, 58*(3), 322–332.
Kriegel, U. (2017). Reductive representationalism and emotional phenomenology. *Midwest Studies in Philosophy, 41*(1), 41–59.
Kriegel, U. (2012). Towards a new feeling theory of emotion. *European Journal of Philosophy, 22*(3), 420–442.
Lazarus, R. S. (1991). *Emotion and Adaptation*. Oxford University Press USA.
Merleau-Ponty, M. (1945). *Phenomenology of perception*. Translated by Donald Landes. Routledge.

Moss, H. (2016). Genes, affect, and reason: Why autonomous robot intelligence will be nothing like human intelligence. *Techné: Research in Philosophy and Technology, 20*(1), 1–15.

Parisi, D., & Petrosino, G. (2010). Robots that have emotions. *Adaptive Behaviour, 18*(6), 453–469.

Picard, R. W. (2003). *Affective computing: Challenges.* MIT Press.

Price, C. (2006). Affect without object: Moods and objectless emotions. *European Journal of Analytic Philosophy, 2*(1), 49–68.

Ransom, M. (2013). Why emotions do not solve the frame problem. In V. V. Muller (Ed.), *Fundamental issues of artificial intelligence* (pp. 353–365). Springer.

Ratcliffe, M. (2005). The feeling of being. *Journal of Consciousness Studies, 12*(8–10), 43–60.

Ratcliffe, M. (2010). Depression, Guilt and Emotional Depth. *Inquiry: An Interdisciplinary Journal of Philosophy, 53*(6), 602–626.

Ratcliffe, M. (2013). What is it to lose hope? *Phenomenology and the Cognitive Sciences, 12*(4), 597–614.

Ratcliffe, M. (2014). The phenomenology of depression and the nature of empathy. *Medicine, Health Care and Philosophy, 17*(2), 269–280.

Rosenthal, D. M. (1991). The independence of consciousness and sensory quality. *Philosophical Issues, 1,* 15–36.

Scherer, K. R., Banziger, T., & Roesch, E. B. (2010). *Blueprint for affective computing.* Oxford University Press.

Shanahan, M. P. (1997). *Solving the frame problem: A mathematical investigation of the common sense law of inertia.* MIT Press.

Shoemaker, S. (1982). The inverted spectrum. *Journal of Philosophy, 79,* 357–382.

Slaby, J. (2014). The other side of existence: Heidegger on boredom. In S. Flach & J. Soffner (eds.), *Habitus in habitat II: Other sides of cognition* (pp. 101–120). Lang.

Sloman, A., & Chrisley, R. (2003). Virtual machines and consciousness. *Journal of Consciousness Studies, 10*(4–5), 133–172.

Smith, J. (2016). *Experiencing phenomenology: An introduction.* Routledge.

Stephan, A. (2009). On the nature of artificial feelings. In B. Roettger-Roessler & H. J. Markowitsch (Eds.), *Emotions as bio-cultural processes* (pp. 216–225). Springer.

Stephan, A. (2017). Moods in layers. *Philosophia, 45,* 1481–1495.

Tye, M. (1995). *Ten problems of consciousness: A representational theory of the phenomenal mind.* MIT Press.

Webb, B. (1996). A Cricket Robot. *Scientific American, 275*(6), 94–99.

Welton, D. (2012). Bodily intentionality, affectivity and basic affects. In D. Zahavi (Ed.), *The Oxford handbook of contemporary phenomenology* (pp. 177–198). Oxford University Press.

Wheeler, M. (2008). Cognition in context: Phenomenology, situated robotics and the frame problem. *International Journal of Philosophical Studies, 16*(3), 323–349.

Zimbardo, P. G. (1972). *Stanford prison experiment: A simulation study of the psychology of imprisonment.* Zimbardo.

Lydia Farina is an Assistant Professor at the University of Nottingham. My research focuses on social ontology, emotion, personal identity and the philosophy of artificial intelligence. I am currently working on projects dealing with the powers of social kinds, AI Responsibility, AI Trust and the relation between personal identity and responsibility.

Design, Social Integration and Ethical Issues

When Emotional Machines Are Intelligent Machines: Exploring the Tangled Knot of Affective Cognition with Robots

Lola Cañamero

Abstract

Research in neurobiology has provided evidence that emotions pervade human intelligence at many levels. However, "emotion" and "cognition" are still largely conceptualized as separate notions that "interact", and untangling and modeling those interactions remains a challenge, both in biological and artificial systems. My research focuses on modeling in autonomous robots how "cognition", "motivation" and "emotion" interact in what we could term *embodied affective cognition*, and particularly investigating how affect lies at the root of and drives how agents apprehend and interact with the world, making them "intelligent" in the sense of being able to adapt to their environments in flexible and beneficial ways. In this chapter, I discuss this issue as I illustrate how my embodied model of affect has been used in my group to ground a broad range of affective, cognitive and social skills such as adaptive action selection, different types of learning, development, and social interaction.

Keywords

Embodied affect and cognition · Embodied AI · Bio-inspired robotics · Affect-cognition interactions · Motivated adaptive behavior

L. Cañamero (✉)
ETIS Lab, UMR8051, CY Cergy Paris Université—ENSEA—CNRS, Cergy-Pontoise, France
e-mail: Lola.Canamero@cyu.fr

© Springer Fachmedien Wiesbaden GmbH, part of Springer Nature 2023
C. Misselhorn et al. (eds.), *Emotional Machines*, Technikzukünfte,
Wissenschaft und Gesellschaft / Futures of Technology, Science and Society,
https://doi.org/10.1007/978-3-658-37641-3_6

1 Introduction

For the last two decades, research in neurobiology has provided evidence that emotions pervade human intelligence at many levels, being inseparable from cognition. Perception, attention, memory, learning, decision making, social interaction or communication are some of the aspects influenced by emotions. Their role in adaptation has likewise been evidenced by these studies. In the Artificial Intelligence (AI) and Affective Computing communities, the need to overcome the traditional view that opposes rational cognition to "absurd" emotion has also been acknowledged. Emotion is no longer regarded as an undesirable consequence of our embodiment that must be neglected, but as a necessary component of intelligent behavior that offers a rich potential in the design of artificial systems, and for enhancing our interactions with them. However, "emotion" and "cognition" are still largely conceptualized as separate notions that "interact", and untangling and modeling those interactions remains a challenge, both in biological and artificial systems.

My research since 1995 has focused on modeling what we could term *embodied affective cognition*: modeling those interactions in autonomous robots, and particularly investigating how affect lies at the root of and drives how agents apprehend and inter- act with the world, and how such robot models can contribute to an interdisciplinary effort to understand affective cognition in biological systems (Cañamero, 2019). In my view, the different "cognitive" skills are not sufficient to apprehend and interact with the world, and affective embodiment provides the "grounding" for them as well as an underlying common principle that brings them together. In other words, we could say that affect makes intelligent systems not only "more intelligent", but makes them intelligent, following the definition of "intelligence" in embodied cognitive science and embodied AI (Brooks, 1991; Pfeifer et al., 2005; Steels, 1995)—a continuum of capabilities and skills, linked to the notions of adaptation and survival, that results from the embodied interactions between agents and their environments, rather than the "rational" view of intelligence as an exclusively human capability to "reason" and "think rationally".

I talk about "affect" rather than simply "emotion", since other affective phenomena are also very important for affective cognition, and notably the notion of *motivation*. Motivation and emotion are often studied independently from each other in psychology and neuro-science, and modeled independently in artificial systems. However, classic authors in the psychology of emotion (e.g., Frijda, 1986; Tomkins, 1984) have for a long time stressed the importance of the interactions between motivation and emotion, and the importance of this link is increasingly acknowledged in neuroscience (e.g., Damàsio, 1999; Pessoa, 2013) and philosophy (Asma & Gabriel, 2019; Scarantino, 2014).

My specific model of embodied affective cognition is biologically inspired and builds on a synthetic physiology that is regulated and modulated using principles stemming from embodied AI, cybernetics (e.g., homeostasis and allostasis), ethology, neuroscience and dynamical systems. It provides a blueprint for a "core affective self" that endows robots with "internal" values (their own or acquired) and motivations that drive adaptation and learning through interactions with the physical and social environment. In this chapter, I illustrate how such model has been used in my group to ground a broad range of affective, cognitive and social skills such as adaptive decision making (action selection), learning (e.g., self-learning of affordances, learning to explore novel environments, learning to coordinate social interaction), development (e.g., development of attachment bonds, epigenetic development), and social interaction (e.g., robot companions for children).

2 Affective Decision Making: Action Selection, and Learning

Competences and skills traditionally regarded by some schools of thought as fundamentally "cognitive" (in the sense of "cold cognition" and "rationality"), such as decision making and learning, are nowadays considered to be deeply rooted in affect. In some cases, emotion is placed at the root of such processes—for example, values and valence underpin decision making, as in Damàsio (1994) and Damàsio (1999)—in other cases, that role is assigned to motivation—for example, intrinsic motivation as a driving force in curiosity, exploration and learning, e.g., Oudeyer et al. (2016). Motivation and emotion are deeply intertwined—and it is not always easy to tease them apart. For example the notion of valence is usually associated with emotion; however, motivation also has a strong "valence" element in its approach/avoid component.

Yet, some disciplines prioritize one over the other, sometimes to the extent that one of them is practically neglected. For example, in developmental psychology, the development of affect tends to focus on emotional development (e.g., Nadel & Muir, 2004), whereas motivation plays a big role in classical ethology (e.g., Colgan, 1989), where emotion is largely disregarded and considered a "subjective" notion that cannot be studied scientifically—despite the fact that "motivation" is an abstract construct to explain observed behavior variability and also has a subjective component, and that the physiological, chemical and neural processes and mechanisms underlying both notions can be studied—analyzed and modeled—scientifically.

In my research with robots, I have adopted the view that these two notions are somewhat distinct but highly intertwined, and I have focused on their interactions in the context of affective cognition—my first paper in this area (Cañamero, 1997) was actually entitled "Modeling Motivations and Emotions as a Basis for Intelligent Behavior"—and the advantages (and sometimes disadvantages) that they bring in terms improving adaptation to the (physical and social) environment. By "improving" I mean not only making it more frequent or more efficient, but also making it more flexible and meaningful—and so we could say more "intelligent". I have particularly (but not exclusively) focused on the adaptive, behavioral and cognitive effects brought about by emotional influences on perception in the context of motivation of autonomous robots situated and in interaction with their environment, modeling "low level" mechanisms such as hormonal modulation of the salience of (incentive) stimuli. Even such a simple interaction can have significant effects in (biological and artificial) embodied agents in dynamic interactions with their environments, and they are easily measurable in the case of robots. Let us briefly discuss some examples in action selection and learning.

For example, in my early work (Cañamero, 1997) I modeled discrete basic emotions interacting with survival-related motivations to drive action selection and learning in simulated autonomous agents. This model took inspiration from ethological and neuroscientific models of behavior, motivation and emotion (particularly Damàsio, 1994; Kandel et al., 1995; LeDoux, 1996; Panksepp, 1998) to implement their underlying mechanisms and interaction dynamics. In this work, simulated robots inhabiting a complex and dynamic two-dimensional action selection environment had to make decisions in real time and deal with the static and dynamic elements of the environment in order to survive. The environment—"Gridland"—was populated by other moving agents—both conspecifics and "predators"—and contained dynamic survival-related consumable resources, obstacles obstructing availability of those resources, and other objects of various shapes. The complex action selection architecture of the agents included a (simulated) physiology—in effect a complex dynamical system—of homeostatically controlled survival-related variables and modulatory "hormones", both generic (with global excitatory and inhibitory effects) and specific (acting differentially on specific receptors) that grounded the embodiment of the agents and other elements of the architecture such as motivations, consummatory and appetitive behaviors, and a number of basic emotions. In this system, basic emotions have specific functions, in line with basic emotions theory. Designed in such way, each basic emotion would normally contribute to the good maintenance of homeostasis and to the agent's action selection; however, it could also have noxious effects when its intensity was too high or displayed in the wrong context—e.g., excessive anger could make the agent bump into things and harm itself, or modify negatively specific parameters of the ART neural network carrying out object recognition, creating a "confused" state that could make the robot interact with the "wrong" object. In this work, the functions of emotions were predefined in the model, which focused on investigating possible mechanisms to model the interactions among motivations, emotions and embodied cognition and action, rather than on assessing the adaptive value of affect.

With the aim of assessing the adaptive value of such interactions, later work explored the interactions between motivations and "lower-level" mechanisms and dimensions of emotions (as opposed to basic emotions with specific pre-defined functions), such as arousal and pleasure. Let us consider models involving pleasure. Like in biological systems, for autonomous robots that need to survive and interact in their environments, pleasure can act as a signal providing "information" about the (survival- or need-related) value of different things, e.g., the "beneficial" or "noxious" quality of perceived stimuli, the behavior being executed, or the interaction with others.

In my group, we have modeled different roles of pleasure interacting with motivation in the context of affective cognition, where "pleasure"—modeled in terms of release of simulated hormones that affect different aspects of the underlying robot architecture—was used, for example, as a signal to learn object affordances in the context of action selection (Cos et al., 2010), to improve internal homeostasis and adaptation to the environment (e.g., improving the trade-off between persistence and opportunism) (Cos et al., 2013; Lewis & Cañamero, 2016), or to convey a positive message in human–robot interaction (Cañamero & Lewis, 2016). In these examples, a simple form of pleasure—a "signal" (a modulating parameter) associated with positive valence and linked to the satisfaction of (survival-related or social) needs—was modeled. To some extent, it is not surprising that the addition of such "signal" providing information about value improved adaptation in the robots, particularly in more challenging situations. However, in humans, pleasure is much more complex and has multiple sources, roles and forms (Frijda, 2010; Berridge & Kringelbach, 2010), not necessarily related to need satisfaction or survival, such as in the case of sensory or esthetic pleasures.

Do such kinds of pleasure bring any sort of "functional" (e.g., adaptive) benefits in humans? And in robots?

The study that we conducted in Lewis and Cañamero (2016) departed from the idea that pleasure is necessarily linked with reward or with signaling biological usefulness. Using a small Nao humanoid robot that had to manage its internal homeostasis and timely satisfy its two needs by searching for and consuming two resources available in the environment, we modeled, in addition to pleasure related to need satisfaction, a second type of pleasure simply linked to the execution of a behavior in different contexts, but unrelated to need satisfaction, to investigate whether hedonic quality (just "liking", "pure pleasure" unrelated to need satisfaction) might also play a role in motivation (Frijda, 1986, 2010). Both types of pleasure were modeled in terms of simulated hormones. Experimental conditions varied the number, distribution, and availability (equal amounts of both or making one of them more abundant or scarce) of the resources in the environment (see Fig. 1 for examples of the environments used under different experimental conditions), as well as the circumstances under which the different types of pleasure hormones were released. Both types of pleasure—modeled in terms of simulated hormones—acted on the "assignment of value" to the perceived resources under different experimental conditions, modifying their incentive motivational salience. In this way, pleasure also modulated how likely the robot was to interact with the perceived stimuli. Our results indicated that pleasure, including pleasure unrelated to need satisfaction, had adaptive value ("usefulness") for homeostatic management in terms of both improved maintenance of the internal "physiology" (viability) and increased flexibility in adaptive behav-

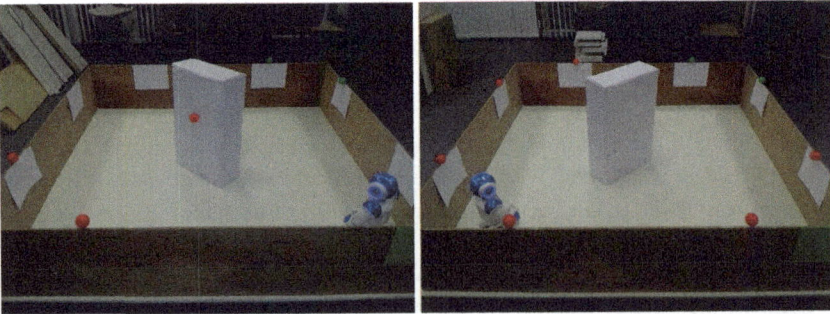

Fig. 1 Photos of the arena used in Lewis and Cañamero (2016), illustrating the distribution of resources—the red and green balls placed on top of the wooden wall, representing food and water, respectively—in two of our experimental conditions

ior. Regarding improved viability, we found that the extent to which the different "types" of pleasure were adaptive or maladaptive depended on the features of environment and the demands it posed on the task, in addition to the "metabolism" of the robot; for example, in "easy" environmental conditions, maximizing pleasure improved the viability of the robot, whereas in "difficult" environments with asymmetric availability of resources, the addition of "purely sensory" pleasure associated with the scarce resource improved the viability of the robot. Increased behavioral flexibility occurred particularly in managing the trade-off between opportunism and persistence in consuming the resources, since pleasure allowed the robot to manage persistence and opportunism independently, and hence to display them in the appropriate context. This is important, for example, in situations where opportunism has a penalty but increased persistence is beneficial, and where an asymmetry in the availability of resources results in the need to consume each of the resources in different ways in order to achieve good management of homeostasis.

3 The Importance of the Interactions with the Environment

One of the pillars of the embodied, situated view of intelligence is the definition of (natural and artificial) intelligence not as a property of an individual, but at a systems level, as the process of (embodied) interaction between agents and their environment. Intelligence is therefore strongly determined by the situational context and our interactions with it, and the same applies to affect and affective cognition. Such interactions are essential not only in the emergence of intelligent/affective behavior at a specific point in time, but also shape its development through mechanisms such as epigenetic adaptation. Different epigenetic mechanisms provide biological agents with the ability to adjust their physiology and/or morphology and adapt to a wide range of challenges posed by their environments.

In my group, in the context of developmental robotics, the PhD thesis of John Lones explored the role of epigenetic mechanisms in the sensory-motor, cognitive and affective development of robots in interaction with their environments, with and without the intervention of (human) caregivers. Such mechanisms affected the development of the architecture (software controller) of the robot and its related capabilities as a function of the relevant elements and interactions in the developmental context.

Let us briefly examine two examples of this work: in the first case, epigenetic mechanisms added to a motivated affective robot architecture enhanced the adaption capabilities of the robot, particularly in interactions with more complex environments; in the second, the development of the robot was shaped, through epigenetic mechanisms, by its early interactions with the environment.

3.1 Affective Epigenetic Adaptation

In Lones et al. (2017) we explored one specific type of epigenetic process, which has been shown to have implications for behavioral development, in which hormone concentrations are linked to the regulation of hormone receptors. Taking inspiration from these biological processes, we investigated whether an epigenetic model based on the concept of hormonal regulation of receptors, could provide a similarly robust and general adaptive mechanism for autonomous robots.

We implemented our model using the small mobile wheeled Koala II robot developed by K-Team (www.k-team.com/koala-2-5-new), and we tested it in a series of experiments set in six different environments with varying challenges to negotiate. Our epigenetic robot architecture was broken down into three systems or "layers", designed to build on each other incrementally:

- *Basic motivated architecture*: the first system handles basic functions such as drives, motivational state, and movement. Following Cañamero (1997), the basic architecture is built around a motivational system in which the basic behavior of the robot is driven by a combination of internal needs that arise from a number of simulated physiological variables—in this case, three: energy, health and internal temperature—controlled homeostatically, and external stimuli. This system generates continuous behavior in which the robot moves around the environment searching for and interacting with objects that allow it to satisfy its needs, running continuously in action selection loops (or time steps) of about 62 ms (about 16 loops per second).
- *Neuromodulatory system*: the second system modulates the basic architecture through the secretion and decay of four different simulated hormones secreted in relation to changes in the robot's internal and external environment. These hormones are of two types: three "endocrine hormones", each related to a physiological variable, secreted as a function of the errors of these variables (they thus "signal" for homeostatic errors), and one "neurohormone", secreted as a function of a combination of the concentration of the endocrine hormones and specific external stimuli. This model takes inspiration from biological organisms, in which hormone secretion derived from homeostatic errors is behind the onset of motivation by providing a signal of the error (Malik et al., 2008) and

the motivational value of environmental cues (Frijda, 1986). These (simulated) hormones, secreted from (simulated) glands as a function of changes in the robot's internal environment that happen as the robot interacts with its external environment, modulate specific characteristics of different sensors (e.g., their accuracy, scope) and actuators (e.g., their speed) of the robot.

- *Epigenetic mechanism*: the third system implements a simple epigenetic mechanism, which controls the sensitivity of the hormone receptors of the second system. Taking inspiration from biological studies (Crews, 2010; Fowden & Forhead, 2009; Zhang & Ho, 2011) showing that hormone concentrations correlate with the triggering of varying epigenetic mechanisms that influence the expression of hormone receptors, we modeled "second order" simulated hormones to regulate the robot's hormone receptors. High concentration of a hormone leads to the upregulation of the receptors (i.e., an increased number of receptors) and lower concentrations to downregulation (i.e., a lowered number of receptors). This regulation functions similarly to a positive feedback loop. Changes in regulation of receptors lead to the emergence of persistent behaviors relevant to the robot's current environment (e.g., consuming more than strictly needed in environments with scarce and difficult to access resources), which in turn can still be modulated by variations in hormone concentration levels.

This architecture was designed incrementally, allowing us to implement robots driven by the first system only, by a combination of the first two, or by the complete architecture. The full (epigenetic) robot architecture, containing the three systems, is depicted in Fig. 2.

This incremental design of the robot architecture allowed us to carry out methodical, incremental testing and analysis of the functionalities and benefits (in terms of adaptation) that each subsequent addition ("system" or "layer") brings to the robot. To this end, we tested each of the 3 robot models—the "Basic robot", the "Hormone-modulated robot", and the "Epigenetic robot"—in the same six environments, each posing different challenges, as shown in Table 1. The robots were let to interact autonomously in the different environments (see Fig. 3 for an image of the Koala robot used and an example of one of the testing environments), trying to satisfy their needs, with behaviors that would depend on the specific architecture tested and the interactions of the robot with the environment (see Lones et al., 2017 for further details of the experiments).

The purpose of these experiments was to investigate the adaptability of our epigenetic mechanism to a range of environmental challenges through allostatic processes (Schulkin, 2004), and more specifically how our system could allow

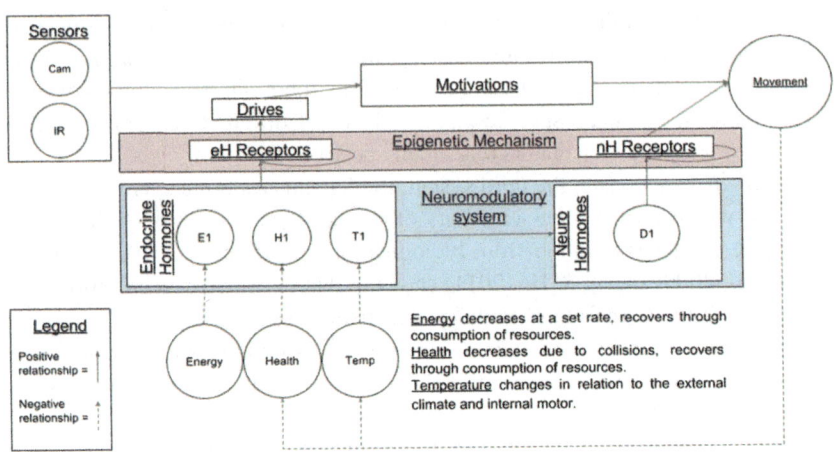

Fig. 2 Complete robot architecture with its three interacting "systems": Basic architecture, Neuro- modulatory system, and Epigenetic mechanism. Reprinted from Lones et al. (2017)

initially identical robots—with the same initial internal parameter values and speed potential—to develop different behavioral phenotypes adapted to different environments as a consequence of the changes brought about by and in the epigenetic mechanism in its interactions with the environment. All the adaptations took place as a result of such dynamics of interactions, without the need to program or tune the robot to operate in a specific environment. I refer the reader to Lones et al. (2017) for a detailed account and discussion of the experimental results. Here, I would like to highlight the main conclusions and illustrate them with some examples. Our results showed that the "Epigenetic robot", compared to the other two robots, maintained homeostasis better in all the six environments and, as a result of different chains of up- and down-regulation of hormone receptors and their effects on the sensors and actuators of the robot, it developed a richer repertoire of behaviors that provided more efficient and "intelligent" adaptation, particularly in the more complex environments. For example, in the environment with the moving resources (experiment 3 in Table 1), the epigenetic robot developed a hunting-like behavior (not possible for the other robots) that made it adapt its speed and timing to access the moving resources—this robot developed an ambush-like strategy, where it would remain sedentary until a needed resource passed nearby, at which point it would "pounce", giving chase at full speed. As another example, in the experiment with dynamic climate cycles alternating between colder periods when moving around was "safe", and extremely hot

Table 1 Overview of the different experiments and the challenges they posed to the robots. Reprinted from Lones et al. (2017)

Experiment	Challenge
Basic environment	An open environment with no significant challenges, used as a baseline for comparison with more complex environments. Two items of each resource were present at all times
Light objects	The environment contains light objects that the robot needs to push past in order to reach resources. As pushing results in a loss of health, this creates a trade-off for the robot between maintaining its energy or its health
Moving resources	The energy resources move in a constant pattern at a set speed, attempting to simulate a simple prey. As they move at the same speed as the robot, the robot cannot simply chase after them to feed
Dynamic climate	The temperature of the environment changes through the experiment in a day-night cycle, with hotter temperatures during the "day" and slowly cooling down at "night". This challenges the robot to satisfy drives during cooler "night" periods so they can reduce movement during the day and not overheat
Uneven resources	Three sets of 5 runs with uneven resource distribution in each set of runs. Each set placed stress on a specific homeostatic variables (e.g., one set of runs would be in an environment with a reduced number of energy resources). This environment challenges the robot to adapt to prioritize less abundant/harder-to-maintain needs
Temporal dynamics	The resources are only available for limited periods of time. This forces the robot to take opportunities to recover deficits when the resources appear. For the first five sets of runs, the energy source would be present for 30 s per minute. In the second five sets of runs the duration was reduced to 20 s. In the final five sets of runs the resource was only accessible for 10 s of every minute. As only one type of resource is affected the robot must prioritize this one over the others

periods when overheating posed "death" risks (experiment 4 in Table 1), the epigenetic robot developed two contrasting behaviors that allowed it to survive the periods of high temperature. In some runs, it developed a "hibernation" behavior—the robot became highly attracted to the different resources during the cold period and thus would fully replenish any deficits before laying "dormant" during the hot periods. In other runs, the behavior developed was to stay near the energy source at all times, except when the occasional need to repair arose; this behavior permitted the robot to continue consuming energy during the hot temperature

Fig. 3 Koala robot used in our experiments. In this picture, the light movable objects from experiment 2 and a moving "prey" resource from experiment 3 can be seen at the top and left of the figure, respectively. Reprinted from Lones et al. (2017)

climate period, with only very limited movement needed. The specific behavior adopted (hibernation or staying near) depended on the robot's early life experiences, and specifically on the amount of health loss that occurred during the first few cycles. As these examples illustrate, our results, including the emergence of varied behaviors well adapted to each specific environment, demonstrate the potential of this affective epigenetic model as a general mechanism for a meaningful—and we could say "intelligent"—adaptation to the context in autonomous robots.

3.2 Affect, Epigenetic Mechanisms, and Cognitive Development

Let us now see how the development of the robot can be shaped, through epigenetic mechanisms, by its early interactions with the environment.

In Lones et al. (2016), we investigated the importance of early sensorimotor experiences and environmental conditions in the emergence of more advanced cognitive abilities in an autonomous robot, grounded in an affective robot architecture integrating epigenetic mechanisms based on hormonal modulation of

When Emotional Machines Are Intelligent ...

a growing neural network. We used the notion of curiosity to model the robot's novelty-seeking behavior by encouraging it to reduce uncertainty in an appropriate manner given its current internal state. This mechanism drives the robot's interactions, allowing it to learn and develop as it is exposed to different sensorimotor stimuli as a result of its interactions. The architecture used was related to the one described in Sect. 3.1, and consists of three elements:

- A basic motivated "core affect" layer comprising a number of survival-related homeostatically controlled variables that provide the robot with internal needs to generate behavior.
- A hormone-driven epigenetic mechanism that controls the development of the robot.
- An "emergent" or "growing" neural network, which provides the robot with learning capabilities.

These three elements of the model interact in cycles or action loops of 62.5 ms, driving the development, learning, and adaptive behavior of the robot. Figure 4 depicts this architecture. The interaction among its various elements occurs as follows:

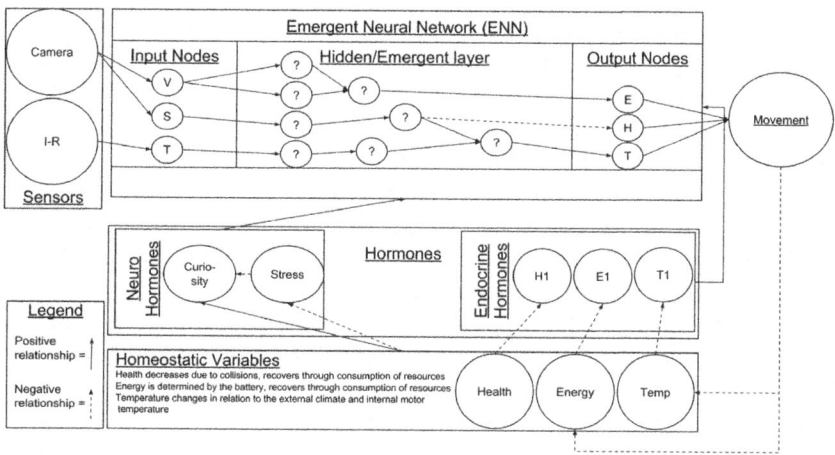

Fig. 4 Robot architecture used in the cognitive development experiments. Reprinted from Lones et al. (2016)

- The levels of the survival-related homeostatically controlled variables change as a function of the actions and interactions with its environment. Deficits of the different homeostatic variables trigger the secretion of the different artificial hormones.
- The different hormones, once secreted, modulate both the epigenetic mechanism (as in Sect. 3.1) and the emergent neural network—"growth" (formation of associative nodes) and synaptic strength.

We implemented this architecture also in a Koala II robot. To test the cognitive development of the robot as a function of its early sensorimotor interactions, we run a set of three experiments for 60 min each, in which the robot, after an initial 10-min period in which it developed with a caregiver who looked after it by bringing it the objects that it would need to satisfy its needs, was let to develop in three different environments (our office space under different conditions) affording very different physical and social sensorimotor experiences for the remaining 50 min.

The first environment was a "normal," standard environment used as control, which provided the robot with reasonable opportunities for stimulation. The second environment was a "novel" environment full of different objects and situations that the robot had not encountered before and that offered many novel sensorimotor experiences. The third environment was a "sensory deprived" environment where the robot had very limited opportunities to interact. We then assessed how these different experiences influence and change the robot's ongoing development and behavior, and observed unforeseen similarities with Piaget's stages of sensorimotor development, as the robot's progression through, as well as the emergence of, the behaviors associated with the different developmental stages, depend on the robot's environment, and more specifically on the sensory stimuli (or the lack of them) that the robot was exposed to over the course of its development. During the initial 10-min period spent with the caregiver (which was identical for the three experiments), the robot developed five basic behaviors (somewhat similar to the so-called reflex acts of a newborn): Attraction/Repulsion (related to resources that permitted to satisfy its homeostatic needs), Avoidance (of objects that were too close), Recoil (from objects that entered in physical contact with the robot), Exploration, and Localized attention (turning to face a moving object). The robot had the same 5 initial "reflex-like" behaviors when it was then put in the different—standard, novel and sensory deprived—environments and left there to interact and develop. In the standard and novel environments, from those initial behaviors the robots developed new, increasingly complex behaviors, particularly in the novel environment, ranging from learning to use resources to satisfy their needs and searching for them, to exploration and "testing" of new objects, to "testing" and using behaviors apparently to make things happen in the environment (e.g., knocking down objects to make the caregiver approach and restore the object to its initial position). On the contrary, in the sensory deprivation environment, where the robot was placed in a small box with the resources adjacent to it, so that it did not have to move in order to satisfy its needs, the robot did not develop new behaviors.

Finally, to test whether the newly developed behaviors had been integrated into the repertoire of the robots, after each "baby" robot had time to develop in its environment, we recreated and assessed its cognitive abilities using different well-known tests used in developmental psychology such as the hidden-toy test to assess object permanence, and the violation-of-expectation paradigm. These more controlled experiments allowed us not only to test the observed behavior of the robots but also to inspect and compare the evolution and maintenance of their internal "physiology" and homeostasis, and the development of their neural networks. As an example comparing the development of the three robots, Fig. 5 illustrates the different neural networks they developed in one of the experiments using the violation-of-expectation paradigm. As we can see in the figure, the greater curiosity, exploratory behaviors, and richer sensorimotor experiences of the robot in the novel environment gave rise to a more complex neural network, with more nodes and connections. The greater effect can be seen in the network of the "Novel Robot", where repeated exposure leads to the generation of a larger number of nodes and connections, and hence to greater neural activity, and to the creation of a new pathway if exposure to the same type of episode reoccurs often enough.

This work illustrates how a "core affect" architecture endowed with epigenetic mechanisms and learning capabilities, situated and interacting with stimulating environments, leads to richer cognitive and behavioral development.

4 Adaptive Social Interaction

Affective cognition is also crucial for robots to be able to achieve "intelligent"—in the sense of adapted, meaningful and adaptive—social interaction with humans in an autonomous way in realistic environments. Within an embodied AI approach, such adaptation can be achieved in different ways—principally as a natural (and in a way "emergent") outcome of the dynamics of interaction between two autonomous agents in a dynamic environment, and as a result of explicit changes in the agents and their behavior brought about by learning or by

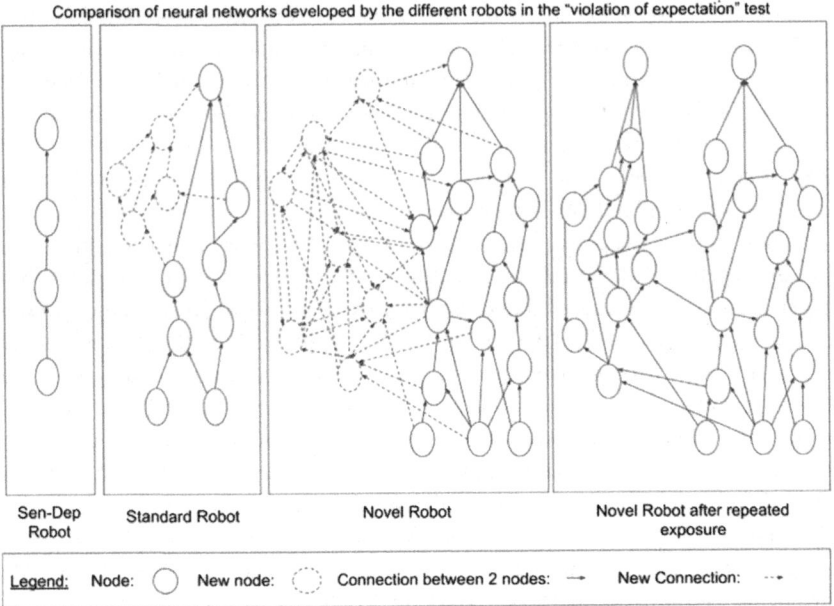

Fig. 5 A snapshot of the neural pathway associated with a "violation of expectation" episode generated by the neural network for each of the robots as a function of the robot's interaction with its environment. Reprinted from Lones et al. (2016)

other processes such as the epigenetic adapation discussed in Sect. 3. Let us discuss an example of each.

4.1 Emergent Adaptive Interaction

An example of this is our robot Robin (http://emotion-modeling.info/robin), initially developed as part of the EU-funded ALIZ-E project and implemented using the small humanoid robot Nao. Robin is a cognitively and motivationally autonomous affective robot toddler with "robot diabetes" that we developed to support diabetes management in children, particularly perceived self-efficacy, a crucial element in improving diabetes self-management skills (Cañamero & Lewis, 2016). We focused on supporting the more affective aspects of self-efficacy, namely self-confidence and the development of responsibility, by providing them with positive mastery experiences in diabetes management, in which children had to look after Robin and address its symptoms of diabetes in a playful but realistic and natural interaction context—while looking after and playing with Robin in its playroom (shown in Fig. 6), designed as a toddler's playroom.

Robin's decision-making architecture follows again principles of embodied AI and draws on the model of motivations and emotions put forward in (Cañamero, 1997)—Robin's architecture is built around a "physiology" of homeostatically controlled "survival-related" variables that the robot needs to keep within permissible values. We also gave the robot a simple model of Type 1 diabetes, comprising an internal blood glucose level that increases upon "eating" toy food, and decreases with simulated "insulin". Robin chooses how to behave as a function of these internal needs and the stimulation that it gets from the environment; it is therefore changing and somewhat unique every time, since it depends strongly on how the different children interact. Elements of the environment are detected using vision (e.g., foods, faces) and tactile contact (e.g., collisions, strokes, hugs). Internal needs and environmental cues are mathematically combined in *motivations* that lead Robin to autonomously select behaviors from its repertoire (e.g., walking, looking for a person, eating, resting) that best satisfy its needs (e.g., social contact, nutrition, resting, playing) in the present circumstances. For this reason, Robin is a *motivationally and cognitively autonomous* robot. Figure 7 provides a high-level overview of Robin's architecture.

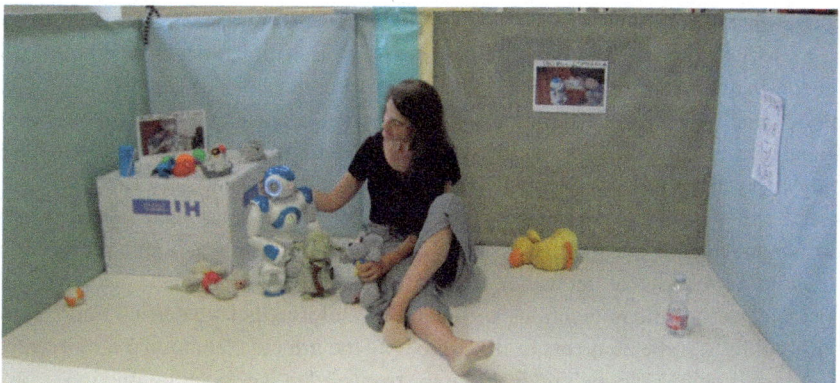

Fig. 6 Robin's playroom. Robin's toys are scattered around the room; items including food, juice drinks, and the glucometer are on a table. "Family pictures" on the walls are sometimes detected as faces, causing Robin to approach them and open its arms (own picture)

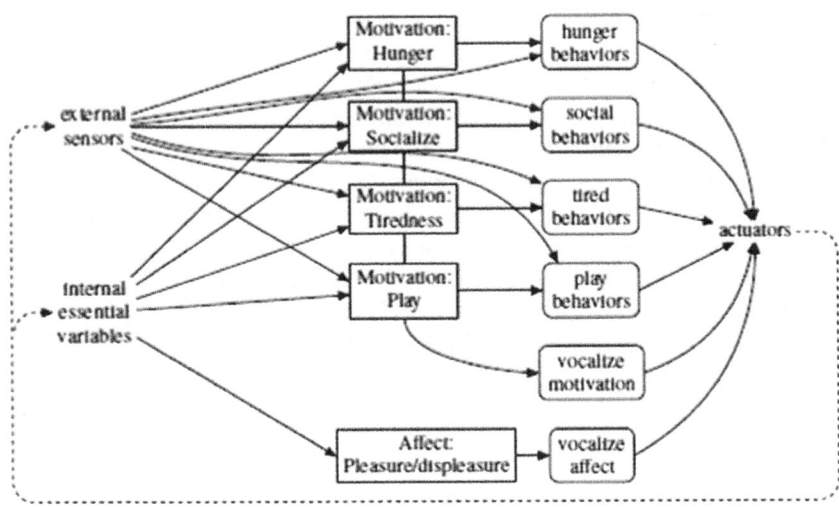

Fig. 7 A "high level" overview of Robin's action selection architecture. Behaviors are shown in round-cornered boxes. Reprinted from Cañamero and Lewis (2016)

To foster appropriate interaction in our scenario, Robin is not capable of fully attending to all its needs without human assistance. It can play on its own, eat, and, by resting, it can recover from tiredness caused by too much movement. However, Robin requires assistance from the children to satisfy its social needs (e.g., social presence, strokes, hugs), some of its nutritional needs (the child can "feed" the robot using toy food items), and to control—check and correct—its blood glucose level. We designed the interaction scenario to be as natural and friendly as possible, in order to provide a positive experience to the children—not only a positive mastery experience of diabetes management but also a positive social and affective experience. I refer the reader to Cañamero and Lewis (2016) for a detailed analysis and discussion of the design of Robin as both a therapeutic tool and a social agent, as well as an analysis and discussion of the affective and social aspects of the interactions with diabetic children that we carried out. Here I will highlight some elements to illustrate how the affective embodied architecture and design philosophy behind Robin gave rise to adaptive social interaction.

Being a motivationally autonomous robot, Robin acts as a friendly but independent agent, who will both approach and move away from humans as it tries to satisfy its own needs, "engaging" and "disengaging" in the interaction. Since the impression given by the first encounter in very important, Robin is already actively moving around when the child arrives. This impression is reinforced by the pictures of Robin with friends and drawings hanging from the walls, and toys scattered in the playroom, which provide clues about Robin's "life beyond the interaction." Another element related to an embodied AI design philosophy is the use of ambiguity. The behavior of Robin was deliberately designed to include some ambiguity so that children had to explore different options by "probing" Robin and various elements of the environment, harnessing their creativity and supporting the goal of helping them to apply their knowledge of diabetes and diabetes management in ways that are not externally imposed. An example related to diabetes management is Robin's "tired" expression, which can be the result of moving around, or a symptom of hypo- or hyper-glycemia (low or high blood glucose). The children can distinguish between these three types of tiredness by using the Bluetooth glucometer device that we programmed to allow children measure Robin's blood glucose levels and ketones, and to provide the robot with "robot insulin". An example of an ambiguous social signal is Robin's "reaching" behavior (raising and opening one or both arms), which can happen under very different circumstances and can be socially interpreted in various ways—e.g., as pointing, as "greeting", as a gesture of friendship, as an indication of looking for something.

Children interacted with Robin very naturally, although in very different ways. All the children showed concern for Robin, becoming socially, cognitively and emotionally invested in the interaction, and gave signs of willingness to continue the interaction after it was meant to end. All the children also showed visible signs of treating Robin as a socially acceptable agent and interacted in a friendly manner, although the form these signs took varied enormously across individual children, and Robin was able to adapt its behavior to the different interaction styles as a result of the dynamics of interaction between its embodied AI affective architecture and the different physical and social "signals" from the environment, without having to change it parameters or re-program it.

Social interactions in this environment do not need to be for a fixed number of participants. Again, without having to change anything in the robot or its environment, we were able to use Robin in both dyadic child-robot interactions and triadic interactions with two children, in addition of triadic interactions comprising a child, an adult and the robot.

4.2 Learning to Adapt Through Social Interaction

In order to allow Robin to change its behavior over the course of specific interactions and explicitly adapt it to the interaction profiles of different individuals, we would need to endow the robot with adaptive learning capabilities. The study presented in Hiolle et al. (2014), based on the role of sensitivity in attachment (De Wolf & van Ijzendoorn, 1997) and on the arousal dimension of emotion, provides an example of such capabilities. In the context of dyadic infant-caregiver interactions and the development of attachment, Hiolle used the responsiveness of a human caregiver to the expression of needs and distress of a baby robot as a signal that permits the robot to self-adapt—in terms of frequency and modality of its responses—the regulation and expression of its arousal as a function of the perceived interaction style of the human. In this study, a Nao robot is exploring and learning about objects in a novel environment—an environment the robot has not encountered before and is not familiar with. The perception of novelty increases the arousal of the robot—an internal parameter that, among other things, interferes with its learning, and "comfort" provided by a human caregiver—either tactile in the form of strokes or visual, placing the face in the visual field of the robot—provides a means to regulate the arousal of the robot, thus having an impact on the learning process. When the robot has a high level of arousal, it stops the exploration of the environment and expresses its arousal behaviorally (e.g., looking around to detect the caregiver); this can be detected by the human and interpreted as a "call for help".

To test this model, we carried out three experiments using three variants of a learning task in a complex environment containing novel and incongruent objects, in which the robot had to learn the features of several objects placed on a table, as follows.

A first experiment investigated how different caregiving styles of the human—ranging from highly responsive to the requests of the robot to mostly unresponsive— could be suited to different characteristics of the strategies used to regulate stress in robots with different profiles—a "needy" robot that expresses its arousal and need for comfort frequently and visibly, and "independent" robot that does not express such needs and self-regulates its arousal. Our results showed that, to achieve the same results with the two robot profiles, different caregiving styles were needed: the "independent" robot needed less interaction with the human caregiver to progress in its exploration, whereas the "needy" robot needed an almost constant presence and "comfort" of the caregiver to progress in its learning with comparable dynamics.

A second experiment investigated how different types of interaction in a more demanding—more complex and difficult to learn—and "stressful" environment (triggering higher levels of arousal) might affect differentially the cognitive and affective development of the robot—its regulatory, exploratory and learning patterns. Our results showed that the two robot profiles exhibit different behavioral dynamics: the "needy" robot, for which the comfort provided by the human decreased the arousal faster, stopped more often and spent more time learning, whereas the "independent" robot, for which the comfort provided lowered the level of arousal for a longer time, showed longer exploration episodes.

A third experiment investigated the use of adaptation to the responsiveness of the human caregiver as a suitable mechanism for the robot to deal with real-time variations in the caregiver's availability to respond to regulatory behaviors, and to adapt to different caregivers. The adaptation capability added to the architecture modulated the effect of the comfort provided by the human by modulating the parameters used to process the comfort received. Our results showed that the robot could modify its own profile autonomously along the "needy"—"independent" axis: more comfort made the robot lean toward the "needy" profile, whereas unattended requests moved the robot towards the "independent" profile. With this capability, the same robot is able to change its behavior as it interacts with different people, as well as to changes in the behavior of the same person.

5 Conclusion

In this chapter, I have discussed and illustrated a model of embodied affective cognition implemented in autonomous robots in which cognition, motivation and emotion are deeply intertwined and interact in fundamental ways. In my view, affect lies at the root of and drives how agents—biological and artificial—apprehend and interact with the world, making them intelligent in the sense of being able to adapt to their environments in flexible and beneficial ways. I have discussed some examples of how such embodied model of affect has been used in my group to ground a broad range of affective, cognitive and social skills such as adaptive action selection, different types of learning, development, and social interaction.

I would like to conclude with a reflection on the title of the chapter: why a "tangled knot" rather than a "Gordian knot"? I prefer to avoid the term "Gordian knot", as it is normally used to denote an intractable problem that needs to be "cut" or "dissolved". However, I don't think that the "tangled knot" of affective cognition is intractable—it is just complex—and, whereas we need to "dissect it" through analysis and synthesis in order to understand it, its various components are *fundamentally* tangled. Perhaps it is time to study affective cognition as a tangled knot, rather than as a collection of independent components.

Acknowledgements The writing of this paper and the work reported here were carried out when I was a faculty member at the University of Hertfordshire. I am grateful to the past and present members of my group, the Embodied Emotion, Cognition and (Inter-) Action Lab, for their contributions to various aspects of the research discussed here, and to the organizers and participants of the "Emotional Machines" workshop for fruitful and insightful discussions. Funding for the research reported here was provided partly by the European Commission through grants HUMAINE (FP6-IST–507422), FEELIX GROWING (FP6-IST–045169), and ALIZ-E (FP7-ICT–248116), and partly by the University of Hertfordshire through various PhD studentships. The opinions expressed are solely the author's.

References

Asma, S. T., & Gabriel, R. (2019). *The emotional mind.* Harvard University Press.
Berridge, K. C., & Kringelbach, M. L. (eds.). (2010). *Pleasures of the brain.* Oxford University Press.
Brooks, R. A. (1991). New approaches to robotics. *Science, 253*(5025), 1227–1232.
Cañamero, L. (1997). Modeling motivations and emotions as a basis for intelligent behavior. In W. L. Johnson (ed.), *First International Conference of Autonomous Agents (Agents'97)* (pp. 148–155). ACM Press.
Cañamero, L. (2019). Embodied robot models for interdisciplinary emotion research. *IEEE Transactions on Affective Computing, 12*(2), 340–351.
Cañamero, L. D., & Lewis, M. (2016). Making new "new AI" friends: Designing a social robot for diabetic children from an embodied AI perspective. *International Journal of Social Robotics, 8*(4), 523–537.
Colgan, P. (1989). *Animal motivation.* Springer.
Cos, I., Cañamero, L., & Hayes, G. M. (2010). Learning affordances of consummatory behaviors: Motivation-driven adaptive perception. *Adaptive Behavior, 18*(3–4), 285–314.
Cos, I., Cañamero, L., Hayes, G. M., & Gillies, A. (2013). Hedonic value: Enhancing adaptation for motivated agents. *Adaptive Behavior, 21*(6), 465–483.
Crews, D. (2010). Epigenetics, brain, behavior and the environment. *Hormones (Athens, Greece), 9*(1), 41–50.
Damàsio, A. (1994). *Descartes' Error.* Avon Books.
Damàsio, A. (1999). *The feeling of what happens: Body and emotion in the making of consciousness.* Harcourt Brace.
De Wolf, M., & van Ijzendoorn, M. (1997). Sensitivity and attachment: A meta-analysis on parental antecedents of infant attachment. *Child Development, 68,* 571–591.

Fowden, A. L., & Forhead, A. J. (2009). Hormones as epigenetic signals in developmental programming. *Experimental Physiology, 94*(6), 607–625.
Frijda, N. H. (1986). *The emotions.* Cambridge University Press.
Frijda, N. H. (2010). On the nature and function of pleasure. In K. C. Berridge & M. L. Kringelbach (eds.), *Pleasures of the brain* (pp. 99–112). Oxford University Press.
Hiolle, A., Lewis, M., & Cañamero, L. (2014). Arousal regulation and affective adaptation to human responsiveness by a robot that explores and learns a novel environment. *Frontiers in Neurorobotics, 8*(17).
Kandel, E. R., Schwartz, J. H., & Jessell, T. M. (1995). *Essentials of neural science and behavior.* Appleton & Lange.
LeDoux, J. (1996). *The emotional brain.* Simon & Schuster.
Lewis, M., & Cañamero, L. (2016). Hedonic quality or reward? A study of basic pleasure in homeostasis and decision making of a motivated autonomous robot. *Adaptive Behavior, 24*(5), 267–291.
Lones, J., Lewis, M., & Cañamero, L. (2016). From sensorimotor experiences to cognitive development: Investigating the influence of experiential diversity on the development of an epigenetic robot. *Frontiers in Robotics and A, 1,* 3.
Lones, J., Lewis, M., & Cañamero, L. (2017/2018). A hormone-driven epigenetic mechanism for adaptation in autonomous robots. *IEEE Transactions on Cognitive and Developmental Systems, 10*(2), 445–454.
Malik, S., McGlone, F., Bedrossian, D., & Dagher, A. (2008). Ghrelin modulates brain activity in areas that control appetitive behavior. *Cell Metabolism, 7*(5), 400–409.
Nadel, J. & Muir, D. (2004). *Emotional development: Recent research advances.* Oxford University Press.
Oudeyer, P. Y., Gottlieb, J., & Lopes, M. (2016). Intrinsic motivation, curiosity, and learning: Theory and applications in educational technologies. *Progress in Brain Research, 229,* 257–284.
Panksepp, J. (1998). *Affective neuroscience: The foundations of human and animal emotions.* Oxford University Press.
Pessoa, L. (2013). *The cognitive-emotional brain: From interactions to integration.* The MIT Press.
Pfeifer, R., Iida, F., & Bongard, J. (2005). New robotics: Design principles for intelligent systems. *Artificial Life, 11*(1–2), 99–120.
Schulkin, J. (2004). *Allostasis, homeostasis, and the costs of physiological adaptation.* Cambridge University Press.
Scarantino, A. (2014). The motivational theory of emotions. In J. D'Arms & D. Jacobson (eds.), *Moral psychology and human agency. Philosophical essays on the science of ethics* (pp. 156–185). Oxford University Press.
Steels, L. (1995). When are robots intelligent autonomous agents? *Robotics and Autonomous Systems, 15*(1–2), 3–9.
Tomkins, S. S. (1984). Affect theory. In K. R. Scherer & P. Ekman (Eds.), *Approaches to emotion* (pp. 163–195). Lawrence Erlbaum Associates.
Zhang, X., & Ho, S.-M. (2011). Epigenetics meets endocrinology. *Journal of Molecular Endocrinology, 46*(1), R11–R32.

Lola Cañamero is Full Professor and INEX Chair of Neuroscience and Robotics at CY Cergy Paris University, in the Neurocybernetics Team of the ETIS Lab, France, and Honorary Visiting Professor in the Adaptive Systems Research Group, Department of Computer Science, University of Hertfordshire, UK, where she was faculty between 2001 and 2021. Her research centers around the origins and constituents of embodied cognition, which she models in robots and artificial life simulations, within a biologically-inspired embodied cognition framework and drawing on the cybernetics tradition. Her research focuses on motivated behavior and on the role(s) that emotional phenomena play in various aspects of intelligence, adaptation, behavior, and interaction, particularly via mechanism such as hormonal/chemical modulation and its interaction dynamics.

Is Empathy with Robots Morally Relevant?

Catrin Misselhorn

Abstract

This contribution provides an indirect argument for the claim that empathy with robots matters from a moral point of view. After reviewing the empirical evidence for the claim that certain robots are able to evoke empathy in human observers the concept of empathy is explicated. It is shown that the fact that humans feel empathy with robots which resemble humans or animals imposes moral constraints on how robots should be treated, because robot abuse compromises the human capacity to feel empathy, which is an important source of moral judgment, moral motivation, and moral development. Therefore, we should consider carefully the areas in which robots that evoke empathy may be used and the areas where we should refrain from such a design.

Keywords

Emotional AI · Empathy with Robots · Robot Abuse · Empathic Human-Machine-Interaction · Moral Relevance of Empathy

---

C. Misselhorn (✉)
Georg-August-Universität Göttingen, Göttingen, Germanye-mail: catrin.misselhorn@uni-goettingen.de

1 Introduction[1]

Building robots that recognize, simulate, or evoke emotion—so-called emotional robots—is a growing trend. These machines are supposed to make human–machine interaction more natural and intuitive. Despite these advantages, human interaction with emotional robots raises several ethical issues in two directions: from robots to humans and from humans to robots. On the one hand, these issues concern the impact of emotional robots on human beings. One may, for instance, worry that machines, which simulate but do not have emotions trick human beings into erroneously attributing real feelings to them; this may be considered a morally problematic form of deceit.[2] On the other hand, one may ask whether there are moral constraints for human beings' behavior toward emotional robots. From a moral point of view, the question is: are human beings allowed to treat these robots however they want?

The current article scrutinizes this issue concerning robots that can elicit empathy in human beings. A common assumption in ethics and moral psychology is that empathy plays an important role as a source of moral behavior (Denham, 2017; Kauppinen, 2017; Schramme, 2017a, b). Empathy, therefore, becomes an issue when the question arises whether there are moral constraints on human beings' behavior toward emotional robots. The importance of this question is also apparent from this topic's popularity in novels, movies, and TV series. Although these works are fictional, they establish that a lot of people find the topic of empathy with robots morally disturbing. A classical work is Ridley Scott's *Blade Runner*, based on a Philip K. Dick novel. In the movie, empathy is the key capacity that distinguishes humans from android robots called replicants. The replicants are treated quite cruelly, and the humans' lack of empathy toward these creatures is an obvious moral problem.

This paper aims to answer the following question: Does our feeling empathy with robots that resemble humans (or animals) in certain salient respects impose any moral constraints on how we should treat them? The first part of this paper reviews some empirical evidence that confirms that robots with certain human-like (or animal-like) features do, in fact, elicit empathy in human beings.[3] The second part defines the concept of empathy more precisely, while the third part explicates its moral impact. It is shown that empathy plays an important role for

---

[1] This is article is an extended version of a talk that I gave for the first time at the 8th European Conference on Computing and Philosophy 2010 (ecap 2010) in Munich. I am thanking Tom Poljanšek and Tobias Störzinger for helpful comments on an earlier draft of the paper.
[2] For an overview of the debate see Danaher (2020).
[3] For reasons of simplicity, I will, nevertheless only speak of human-like robots.

moral judgment, moral motivation, and moral development. These are the foundations needed to demonstrate that empathy with human-like robots is morally relevant in the final part of the paper. It provides an indirect argument for the claim that empathy with robots matters from a moral point of view even if it does not involve the belief that robots really have emotions.

2 Empirical Evidence for Empathy with Robots

The capacity of robots to elicit empathy in human beings became a topic of public discussion after some video clips went viral on social media which showed a vaguely human-like Atlas robot by *Boston Dynamics* being kicked and beaten. Although this video might just be a hoax, the empathic reactions to it were real. Many individuals who felt empathy with the robot even became angry with its tormentors. One might suspect that only hypersensitive individuals react in this manner; however, empirical evidence indicates that robots with certain human-like (or animal-like) features elicit empathy in human beings with normally developed empathic capacities. Since building humanoid robots is costly and the tests for empathy often involve damaging or destroying them, experiments are frequently conducted with video clips or avatars.[4]

One experiment implemented a within-subjects web-based survey (Riek et al., 2009). The subjects watched 30-second film clips of robots with varying degrees of human-likeness in situations that typically elicit empathy when humans are involved. One robot is the popular disc-like robotic vacuum cleaner *Roomba*. Another one is an LED robotic lamp with five degrees of freedom, while a third model called Andrew is an adult-sized humanoid with the capacity to move, limited facial expressions, and a slightly mechanical-sounding voice. There was also Alicia, an adult-sized robot with a female appearance with a wide range of movements, more realistic human facial features, and a human-sounding voice. In the clips, humans treat the robots extremely cruelly; they shout at them, push them, and order them to do embarrassing things. The human subjects' reactions to these clips were compared to their reactions to other sequences in which the same actions were performed on human protagonists. The results indicate that the more human-like the robots looked and behaved, the more empathy the human subjects felt. The experimenters constated that "people empathized nearly as much with

---

[4] An exception is Darling (2016) who works with Pleo, a pet robot that looks like a little dinosaur.

the humanoid and android robot protagonists as they did with the human protagonist" (Riek et al., 2009, p. 246). The so-called "uncanny valley" might, however, limit the correlation between human-likeness and empathy. If machines become very humanlike, minimal deviations from human appearance cause a feeling of eeriness (Mathur & Reichling, 2016; Misselhorn, 2009; Mori, 1970).

Milgram experiments on obedience can also be used to test empathic responses to artifical agents. Bartneck and Hu (2008) conducted such experiments with a robot that was recognizable as a machine but had some human-like characteristics, including a roughly humanoid body with arms and legs, as well as minimal facial features, such as eyes and a mouth. It could also express emotions facially, by shaking its arms or using sentences that followed Milgram's original script. The results of these experiments verify that the participants felt empathy with the robot; however, the experimenter's urging was mostly sufficient to make them continue delivering the electric shocks until they reached the highest voltage. In the original Milgram experiments, fewer participants went that far.

Conversely, Slater et al. (2006) conducted Milgram experiments with avatars. The participants of these experiments were instructed to administer electric shocks to a female avatar when it did not perform a linguistic task correctly. As the voltage increased, the avatar responded with increasingly more discomfort and protests before finally shouting, "stop the experiment." The results are supposed to reveal that the participants who saw and heard the avatar tended to respond with empathetic distress at a subjective, behavioral, and physiological level as if confronted with a human being, although they knew that they faced an avatar. Nevertheless, most of the participants did not refuse to deliver the punishment.

The results of these experiments suggest that human beings do feel empathy with robots but differentiate between them and human beings. More participants were prepared to harm and even destroy robots or avatars compared to human victims. At the same time, the abuse of robots or avatars evoked emotional responses akin to the abuse of humans. The participants of these experiments reported similar feelings of empathy, personal distress, and moral concern on a subjective level and indicated activity in overlapping areas of the brain relevant for empathy (Rosenthal-von der Pütten et al., 2014). These findings reinforce the claim that the participants really felt empathy. Their increased willingness to deliver the electric shocks might be due to a cognitive dissonance between the empathic feeling and the belief that the robots are not really sentient beings. This belief might function as a cognitive defeater of the empathic reaction that occurs at the emotional level and make the participants follow the experimenters' instructions more readily (Rosenthal-von der Pütten et al., 2014); it does not, however, seem to suppress the empathic reaction as such.

3   The Concept of Empathy

Sometimes any kind of emotional understanding is called empathy; nevertheless, this definition is too broad. In psychology, the term "congruence" is used to distinguish empathy from other kinds of emotional understanding (Eisenberg & Fabes, 1990; Hoffman, 2000; Nickerson et al., 2009). An individual who feels empathy with another must have the same feelings as the individual who is the target of empathy. These feelings can be emotions in the narrow sense of the term, such as anger, happiness, or sadness, as well as moods or affective states, such as pleasure and displeasure. The term "feelings" is used here to encompass all these examples. All feelings have a phenomenal quality, and many have an intentional object (these are called emotions), although this may be disputable in some cases, such as moods.

However, not any congruence of feelings between two individuals can be considered as empathy; two further conditions must be met. The first is the asymmetry condition, which indicates that the empathizing individual has the feeling only because the target has it and that it is more appropriate to the target's situation. Moreover, there must be at least a rudimentary awareness that the feeling is the target's feeling and not one's own. I will call this second condition other-awareness. Due to the asymmetry and other-awareness conditions, empathy is often called a vicarious feeling. Because of these three conditions, empathy is at most a subset of what is called a theory of mind (ToM) in psychology and philosophy, i.e. the capacity to understand and predict other people's behavior by ascribing mental states to them based on a kind of theory. The difference becomes clear when one compares empathy with some related but distinct phenomena that belong to ToM.

The first phenomenon is the rational understanding of feelings, which must be distinguished from empathy as it does not fulfill the congruence and asymmetry conditions. One can rationally understand that another individual is afraid of something without having the same emotion. Since one does not have the same emotion, it does not make sense to apply the asymmetry condition. Other-awareness is, in contrast, present since the feeling that one is rationally understanding is not one's own feeling. One might also attempt to understand one's own feelings rationally, for instance, if one infers from one's behavior that one must be jealous. Nevertheless, in this case, one assumes the stance of an external observer from the third person perspective toward one's own feelings.

Empathy must also be distinguished from emotional contagion, in which case one catches a feeling similar to a cold. This phenomenon occurs, for instance, if children start to cry in pain when they see or hear another child crying painfully. In this case, the congruence condition is fulfilled. Asymmetry is also present as the children have the emotion because another child has it, and it is more appropriate to the other child's situation. However, the other-awareness condition is not fulfilled since the children who catch the feeling are unaware that what they feel is another child's feeling.

Furthermore, empathy may not be reduced to shared feelings in the face of a certain event. An example of shared emotions is the happiness that a toddler's parents feel when their child has made some progress. In this case, the one individual does not just have the feeling because the other one has it, rather, there is a common cause, and it is appropriate to the situation of each parent. That is, there is neither asymmetry nor necessarily other-awareness, although one might be aware that the other person is in the same condition.

Finally, empathy must be distinguished from sympathy, although the term was used before the nineteenth century to refer to the phenomenon called empathy today, including by David Hume and Adam Smith. In the contemporary discussion, the term "sympathy" is used to refer to a concern with the well-being of another individual. As Stephen Darwall pointed out, sympathy involves in contrast to empathy the third-person perspective and not the first-person perspective (Darwall, 1998, p. 261). It, therefore, meets none of the three conditions. When I am worried about a friend who wants to offer her entire fortune to a guru out of enthusiasm for his doctrine, I do not share her enthusiasm, but I may feel sympathy for her. The issues of asymmetry and other-awareness, hence, do not arise (Table 1).

This comparison of empathy with related phenomena supports the claim that congruence, asymmetry, and other-awareness are individually necessary and jointly sufficient conditions for empathy. One can summarize this result in the following tentative definition of empathy: Empathy is a mental representation of another individual's feeling that involves (1) a *congruent* feeling; (2) the empa-

Table 1 Empathy and related phenomena

	Congruence	Asymmetry	Other-awareness
Empathy	+	+	+
Rational understanding of feelings	–	–	+
Emotional contagion	+	+	–
Shared feelings	+	–	–
Sympathy	–	–	–

thizing individual has the feeling only because the target has it, and it is more appropriate to the target's situation (*asymmetry*); and (3) the shared feeling is accompanied by *other-awareness*. That is, empathy is a representation of another individual's feeling that involves congruence, asymmetry, and other-awareness. Since robots do not really have feelings and we do not believe that they do, it is tempting to infer that empathy with robots resembles some kind of misrepresentation or illusion in important respects.

The concept of a representation that occurs in this definition of empathy can, for instance, be explicated in Dretskean terms.[5] In Dretske's framework, mental states represent what they do due to having an evolved function that involves carrying information by tracking certain objects. Fear, for instance, would be an indicator of danger. Empathic feelings, however, would not primarily carry information about the original objects that caused the emotion or feeling but track the feelings or emotions of others; they would, in Dretske's terminology, get recalibrated. One can explain the concept of recalibration with the help of the example of a pressure gauge. The function of a pressure gauge is to deliver information about pressure but it can be also used as an altimeter by recalibrating the instrument's display. The pointer positions of the device would then (also) represent altitude.

Dretske uses the concept of recalibration in order to explain how we can learn to distinguish words from hearing sounds: "We still hear sounds, of course, but, after learning, after the kind of calibration occurs that is involved in language learning, experiences acquire an added representational dimension" (Dretske, 1995, p. 20). A similar process may occur with empathy: If perceiving an individual who is experiencing a certain feeling reliably causes a congruent feeling in the observer, then this vicarious feeling gets recalibrated to represent the target's feeling and it is marked by asymmetry and other-awareness. Although the original intentional object of the feeling will still be present (if one exists), it is bracketed in the empathic state, which then has the primary function of tracking the target's feelings. The intentional object of the empathic state is thus not the original object of the target's feeling but the target's feeling that is represented.

One must, however, distinguish this kind of basic or perceptual empathy from the cognitively more demanding perspective-taking, also called reenactive or projective empathy (Stueber, 2006, p. 20 f.). In perceptual empathy, one experi-

---

[5] This is just one suggestion to understand the mechanism that underlies empathy. It can be easily adapted to other tracking-theories of mental content. My main argument is, however, independent of such an account, if one does not buy into any theory of this type.

ences empathy because one perceives that someone else is feeling a certain way, which is a largely passive, perception-like event. Conversely, reenactive empathy requires that we actively put ourselves in the shoes of the target. It arises from imagining that one is in the same situation or in the place of another individual. The remainder of this chapter focuses on perceptual empathy since empathy with robots tends to be perception-based and, accordingly, arises spontaneously and involuntarily.

Perceptual empathy is based on mechanisms of inner imitation or resonance. These mechanisms presumably underlie the quasi-perceptual ability to perceive the feelings of other sentient beings without the help of theoretical assumptions or inferences, which is present in some nonhuman animals (De Waal & Preston, 2017). At the end of the nineteenth century, Theodor Lipps was among the first to notice that, in the context of his work on aesthetics, when the affective state of another individual is perceived, the same state is produced in the viewer (Stueber, 2006, p. 7; Montag et al., 2008).

Whereas Lipps considered this a primitive natural mechanism, research on mirror neurons could explain this process. Mirror neurons were first discovered in the brain of monkeys (Gallese et al., 1996; Rizzolatti et al., 2004). They are nerve cells that become activated when a monkey performs an action and when it observes a similar action in another monkey or a human experimenter. Mirror neurons have also been detected in the human brain (Mukamel et al., 2010; Rizzolatti & Craighero, 2004). Resonance also play a role in empathy. With the help of fMRI, it was revealed that overlapping brain areas are activated when feeling pain and when perceiving the pain of others (Jackson et al., 2005; Lamm & Majdandžiæ, 2015; Zaki et al., 2016).

While much empirical research must still be completed, it seems promising to assume that mirror neurons play some role in explaining perceptual empathy.[6] However, the original definition of mirror neurons is too narrow to account for empathy for three reasons: First, it is limited to actions; second, larger areas of the brain appear to play a role in empathy; and third, the class of triggering stimuli must be defined more broadly.

It has therefore been suggested that mirror neuron systems should generally be understood as neural systems that become activated when individuals are in a certain mental state and when they perceive signs that another individual is in a mental state of the same kind (Goldman, 2009). Perceptual empathy would then

---

[6] For a good overview see Debes (2017).

arise if someone perceived signs that the target was in a certain mental state or event and the corresponding mirror neuron system was activated as a result. Signs that can activate a mirror neuron system include behavioral manifestations of the mental state or event, the facial expressions that come with it and distal stimuli that typically produce a certain mental state or event. The mirror neuron system for pain, for instance, can be activated when one sees a child reaching for a hot stove, their facial expression contorted with pain, and their hand withdrawing in pain.

The resonance mechanisms on which perceptual empathy is based seem to be innate. Nevertheless, cultural aspects might also determine how these mechanisms are used. One can argue that the mirror neuron system must first be calibrated and adapted to its social function. Even if empathy has a biological basis, it is influenced by many contextual factors: Basic emotions might be more prone to trigger the mirror neuron system than more complex social emotions. Moreover, the resulting perceptual empathy might depend on the clarity and intensity of the emotion display; the more obvious and intense an emotion, the more it elicits empathy. Similarity to or familiarity with the target also facilitates empathy and the communicative situation also plays a role. If several individuals show different feelings at the same time, this can prevent the emergence of empathy.

In addition, the assessment of the situation plays a role. If one does not consider another individual's emotions appropriate, one might feel less empathy with their suffering. Mood, the degree of arousal, personality, gender, age, emotional repertoire, and the ability to regulate emotions also influence one's ability to empathize (De Vignemont & Singer, 2006). The stereotype that women are more empathic than men is widespread; however, some studies have indicated that the differences are less discernable than Common Sense suggests. Furthermore, some of these differences seem to depend on how empathy is operationalized and measured (Lennon & Eisenberg, 1987).

Empathy generally enables us to understand our fellow human beings; it provides us with access to their feelings and is useful in understanding and predicting their behavior. Empathy is not the only way to access the emotional lives of others, though, and it is not infallible. However, particularly perceptual empathy allows a quick assessment and reaction to others' feelings and behavior. It does not require much information, and the information is rather easily accessible. Moreover, the involved mechanism is cognitively economical since it does not depend on extended deliberative processes. One can therefore understand empathy as a kind of fast and frugal heuristics (Gigerenzer et al., 1999; Spaulding, 2019) to grasping what is occurring in other individuals' minds and predicting their behavior. Furthermore, it plays a central role in morality.

4 The Moral Relevance of Empathy

4.1 Empathy and Moral Judgment

It is not clear whether empathy is necessary or constitutive for morality, but it is difficult to believe that empathy should not significantly contribute to determining and conveying moral considerations and moving us to act on them (Kauppinen, 2014; Maibom, 2014a, b). The fact that teaching children what is morally wrong heavily relies on inducing empathy for the victims of wrongdoing in them also highlights empathy's significance for our moral practice.

Since pain is especially important for morality (Audi, 2016, 89), we will concentrate on the impact of perceptual empathy with pain on moral judgment, motivation, and development. The first step is to understand the pathway from perceptual pain empathy to moral judgment better. The following schema is supposed to represent the steps along this path (the rationale for each step is going to be explained below):

1. Being in pain feels bad.
2. If we observe signs that another individual is in pain, we feel their pain empathically.
3. The empathically felt pain feels bad.
4. If a bad feeling has an intentional object, it amounts to a negative affective evaluation of this object.
5. The intentional object of empathically felt pain is the other individual's pain.
6. Empathically felt pain therefore amounts to a negative affective evaluation of the other individual's pain.
7. The negative affective evaluation of the other individual's pain is a (proto-)moral evaluation.
8. This (proto-)moral evaluation is the basis for more general moral judgments.

Step (1) is based on the phenomenology of pain, while step (2) is based on the definition of perceptual empathy provided above. Step (3) is derived from the definition of empathy as a congruent feeling and the phenomenology of pain; step (4) is a general assumption about the nature of feelings with an intentional object; and step (5) is gained from step (4) and the definition of empathy as a representation of another individual's feelings.

Step (6) is derived from steps (1) to (5). Regarding step (7), this negative affective evaluation of the other individual's pain is in a very basic sense moral as it involves a non-instrumental concern for another individual's well-being.[7] Since the intentional object of emphatically felt pain is the target's pain, the negative evaluation concerns the other individual's well-being and not one's own. Given that other-regardingness is particularly important for morality, empathy can be considered a source of moral judgment. However, since the empathic perception is not a judgment per se (at least not in propositional form), some might prefer to use the term "proto-moral."

Step (8) is supposed to bridge the gap between empathy and moral judgment. The negative affective evaluation can become the basis for more general moral judgments, such as the following: "It is wrong to inflict pain on someone intentionally and without good reason" or "One should help someone who is in pain." There are different accounts of this transition, yet this is not the place for a proper discussion on which of them is the right one. Regardless, some kind of transition must occur if we want to understand empathy's role in our ordinary moral practice.

As an example, Adam Smith assumes that most of our moral judgments are not inferred from principles but result from feeling empathy (Smith, 1853/1966, p. 377). For him, general moral rules are inductively derived from empathic experiences. In this process, empathy generates beliefs about which act-types are wrong-making (Kauppinen, 2014). These beliefs then become associated with blame (other things being equal). Nevertheless, defeating evidence can prevent an action that originally elicited empathy from being considered morally wrong. Even if we felt empathy with a patient who was operated on without anesthesia in an emergency, we would not deem this morally wrong if we understood that such a surgery was performed for the patient's benefit.

Essentially, empathy has a twofold function in moral judgment: it makes another individual's pain epistemically accessible, and empathic pain feeling bad leads to a negative affective evaluation, which is proto-moral and can function as a reason for moral judgments. That does not mean that other sources of moral justification cannot occur. Granting empathy a role in moral judgment does not force us to declare that empathy is strictly necessary for moral judgment, although it is presumably a basic source of moral justification in the sense of a fast and frugal heuristics explicated above. Nevertheless, to avoid some sources of error to which empathy is prone (e.g., in-group biases), we might depend on reason as a corrective of our empathy-based moral judgments. Reason might likewise sometimes need rectification by

---

[7] The characteristics of the moral are outlined in more detail in Misselhorn 2018.

empathy, though, because it is susceptible to its own kinds of contortions, including its readiness to rationalize immoral behavior.

4.2 Empathy and Moral Motivation

The connection between empathy and moral motivation has been intensively studied in experimental psychology. Daniel Batson's work (Batson, 1991, 2011, 2012) notably contributed a great deal to establishing the so-called empathy-altruism hypothesis, which claims that "empathic feeling for a person in need increases altruistic motivation to have that person's need relieved" (Batson, 1991, p. 72). Although Batson himself does not identify altruistic and moral motivation, I will do so for the sake of argument. His reservations have to do with the role he gives to principles in moral reasoning and motivation. I find this view of moral motivation too restrictive but I am, as above, happy with calling altruistic motivation proto-moral as well.

Similar to the connection between empathy and moral judgment, the one between empathy and moral motivation need not be necessary or constitutive. Empathy's role in motivation can be illustrated using an extension of the schema introduced in the last section:

9. A negative affective evaluation leads to the disposition to do something about the negatively evaluated fact or event.
10. The negatively evaluated fact or event in the case of empathically felt pain is the other individual's pain.
11. The empathically felt pain leads to the disposition to do something about the other individual's pain.

Again, empathy might not be necessary for moral motivation. Other sources of moral motivation can occur, as well. Nevertheless, affects may be motivationally effective, and step (9) is a general assumption about the motivational force of affective states. Step (10) is based on the definition of empathy and states that the negatively evaluated fact or event is not one's own pain but the target's pain; it is due to the intentional object of empathic pain being the other individual's pain and not one's own pain. Step (11) is derived from the preceding steps.

One must keep in mind, though, that different kinds of affective motivation exist. The two basics types are negative and positive affective motivation. The motivation that a negative affective evaluation produces is based on negative feelings (i.e., it feels bad), while the motivation that a positive affective evaluation produces is based on positive feelings (i.e., it feels good). Essentially, the motivational force of negative affective evaluations is something like "stop this"; the motivational force of positive affective evaluations has the character of "more of this." Since having empathy with another individual's pain feels bad, we are only concerned with negative affective motivation.

Because negative affective evaluations feel bad, they tend to lead directly to a change in behavior that aims at making the negative sensation disappear. However, different kinds of behavior can bring about the desired result. One can stop the fact or event that has been evaluated as bad, leave the place where it happened, or attempt to stop the feeling by distracting oneself. The question is, why does empathy usually make us do something about the target's pain?

The answer is that pain empathy is essentially a representation of another individual's feelings that shares their negative phenomenal quality. The negative evaluation is directed at their feeling, not one's own. Although one's own negative feeling induces the action, it is the function of affective evaluations to change the negatively evaluated state or event and not only the negative feeling as such.

Another motivational distinction results from the position one adopts toward the victim inflicted with pain. One can do something about the target's pain in one of two ways: As the aggressor, one must stop the attack; as an innocent bystander, one must aid the victim. These two types of action are motivationally not equivalent: Although both are grounded in negative affective evaluations, they involve two different motivational mechanisms. The one consists in inhibition (i.e., the interruption of an ongoing activity), while the other involves activation (i.e., the impulse to start a new activity). The more fundamental kind of motivation is presumably inhibition.

The fact that inhibition is stronger and more basic than activation could be due to two reasons. The first is causal immediacy, a term that Michael Slote used in *The Ethics of Care and Empathy*:

> We emotionally flinch from causing or inflicting pain in a way, or to an extent, that we don't flinch from (merely) allowing pain, and I want to say that pain or harm that we (may) cause has, therefore, a greater causal immediacy for us than pain or harm that we (may) merely allow. (Slote, 2007, p. 44)

The second reason is that the inhibitory motivation that arises from empathy probably involves an evolutionarily basic mechanism. Some developmental psychologists assume that there is a violence-inhibition mechanism (VIM) in humans involved in empathy—an idea based on work in ethology (Eibl-Eibesfeldt, 1970; Lorenz, 1981). Dogs and other social animals withdraw during acts of aggression when a victim displays a submission cue, such as baring their throat. A homologous mechanism might be responsible for the well-established fact that healthy humans are emotionally disturbed when they see others in distress. This vicarious response is present even in small infants; however, it does not seem to function properly in individuals with psychopathy whose capacity to feel empathy is often assumed to be impaired (Blair, 1995, 2005).

Empathy comes into play insofar as the perception of cues for emotional distress triggers perceptual empathy, leading to an inhibition of aggressive behavior. Such a violence inhibition mechanism would be a good explanation for the strong inhibitory effect of having empathy with individuals in pain. Moreover, it would predict that the positive, activating motivation for helping someone is less strong and not as fundamental from an evolutionary viewpoint. The special role of inhibiting motivation might also have an impact on the ethical debate with respect to the difference between acts and omissions and explain why we tend to morally condemn initiating a violent act more strongly than not interfering with one.

Given the relationship between empathy, moral judgment, and moral motivation, it is plausible to assume that empathy also plays an important role in moral development. Empathy is used to determine how one's own behavior affects others. Furthermore, it helps to perceive others as "self-authenticating sources of valid claims," as John Rawls phrased it (Rawls, 1993, p. 72), because the intentional object in empathically felt pain is the pain of another individual.[8] Since the pain of the other individual is evaluated negatively, the source of this negative value is the other individual. Empathically felt pain thus involves the perception of the other individual as a source of value that is independent of oneself. Pain empathy is, hence, a basic form of perceiving the other individual as a self-authenticating source of valid claims. If one assumes that a key aspect of the concept of morality is a noninstrumental concern for others, then empathy can contribute to forming an understanding of this concept.

---

[8] For a more detailed account of the relation between the empathic and the moral point of view, see Misselhorn (2023)

5 Empathy with Robots from a Moral Point of View

5.1 Arguing Indirectly Against Robot Abuse

The reasoning schema used to explicate the pathway from empathy to moral judgment and moral motivation prima facie contradicts the view that empathy with robots is morally relevant. Robots do not feel pain, and we do not believe that they do. As suggested above, empathy with robots resembles some kind of emotional illusion or misrepresentation—which is sufficient to initiate steps (1) to (7), but our not attributing pain to robots prevents the transition from step (7) to step (8). The belief that robots do not feel pain is a defeater that preempts the corresponding moral judgments from being formed.

We will not conclude that it is morally bad to inflict pain on a robot deliberately and for no reason or that it is morally good to help a robot in pain. If empathy with robots is morally relevant, it is indirectly so at most and must result from what is occurring at the levels below moral judgment. Empathy with robots would be indirectly morally relevant if the abuse of robots corrupted our general capacity to feel empathy and thus undermined an important source of moral judgment and motivation.

How is it possible to establish whether this is the case? One option would be to investigate the causal consequences of the abuse of robots and see whether individuals who abuse robots are also more prone to abusing human beings than others. The disadvantages of such an approach are obvious: First, the use of android or zoomorphic robots is not so widespread that a reliable empirical study can be conducted on this basis. Second, even if they became more common in our daily lives, whether that would fundamentally improve the situation remains unclear. As the discussion on the effects of violent movies or computer games indicates, it is extremely difficult to find clear evidence in this area.

The question of the causal consequences should, hence, be replaced with the question of whether the abuse of robots negatively affects our morally relevant disposition to feel empathy with other individuals. The plausibility of this hypothesis can be scrutinized by using structural considerations about the pathways that lead from empathy to moral judgment, motivation, and development. Although these considerations are not purely conceptual or a priori, they are independent of contestable behavioral evidence.

The question here is, what happens when a robot is abused? First, we must clarify what abuse in this context means It refers to any kind of intentional action that would cause significant physical or mental pain if performed on a human being, such as beating, kicking, or insulting. Given this definition, the term "abuse" does not preempt the question of whether empathy with robots is morally relevant. The term "robot abuse" does not yet conceptually imply moral wrongdoing, but is purely descriptive.

If one understands empathy as a resonance phenomenon, observing or performing the abuse of a robot triggers the mirror neuron system responsible for pain. A reproduction of the presumed pain state, which differs from the organism's own pain state due to the recalibration of its intentional object, thus emerges. The result is a representation of the robot's pain state. If one doubts the role of mirroring, one may also assume that the perception of indicators for a robot's pain directly produces a representation of its pain state.

Again, this is a misrepresentation since robots cannot feel pain so far. Still, similar to one's own pain states, it feels bad even though it is a misrepresentation, which involves a negative affective evaluation of the robot's pain. As revealed above, such an evaluation would usually lead to a disposition to form general moral judgments such as: "It is morally bad to inflict pain on a robot deliberately and for no reason" and "It is morally good to help a robot in pain." Nevertheless, in this case, even if the disposition might be present, it would not lead to these moral judgments because we do not believe that robots feel pain.

Elsewhere (Misselhorn, 2009) I suggested to explain this dual character with the help of the imagination. The act of perceiving signs that induce empathy with a robot in human subjects is imaginative since they do not believe that robots have emotions nor do they literally perceive them as having emotions. Nevertheless, the feelings that are produced by these imaginative stimuli are nevertheless real and hence our empathy is real at the emotional level although it depends on an act of imaginative perception. In this respect it resembles empathy with fictional characters which is also a real feeling although it is induced by an imaginative act.

However, feelings are not fully under rational control, which is particularly true for pain and hence empathic pain. Phenomena such as phantom limb pain demonstrate that pain does not stop simply because one knows that one no longer has the relevant limb. Pain states resemble visual perceptions in this respect. In the case of the Müller-Lyer illusion, for instance, the two lines are still perceived as different in length when we know that they are the same length. Although empathic pain does not involve a visual illusion, it is similar to one in that the empathic pain persists even though we do not ascribe mental states to the robot. The representation of empathic pain and the belief that robots do not feel pain thus conflict. What makes an important difference to the Müller-Lyer illusion is that in our case, strong pressure to resolve the conflict by eliminating the feeling arises; since empathic pain feels bad, we want to eliminate it.

One can eliminate empathic pain by either stopping the abuse of the robot or attempting to annihilate the feeling of empathy or cut oneself off from it. However, if the latter two options to eliminate empathic pain become habitual, it can be assumed that the disposition to feel pain empathy becomes fundamentally impaired. The empirical evidence suggests that at the level of perceptual empathy, no distinction between human and nonhuman stimuli is made emotionally (Suzuki et al., 2015). Indeed, there is no selective deactivation of the capacity to feel empathy only with regard to robots.

5.2 The Arguments from Moral Judgment, Motivation, and Development

The structural considerations described can be used to argue that we should not abuse robots as doing so impairs our capacity to feel empathy with others, which is an important moral source. More precisely, three indirect moral arguments against robot abuse can be established: the argument from moral judgment, the argument from moral motivation, and the argument from moral development. The argument from moral judgment can be put in the following schema:

The Argument from Moral Judgment

Premise 1	The capacity to feel empathy is an important source of moral judgment
Premise 2	Ceteris paribus, it is morally wrong to compromise important sources of moral judgment
Premise 3	The habitual abuse of robots compromises the capacity to feel empathy
Conclusion	It is therefore morally wrong to abuse robots habitually

The first premise was established in the section on the connection between empathy and moral judgment. The second premise is postulated as a generally accepted moral principle. The ceteris paribus clause is supposed to grant that exceptional circumstances can occur; if, for instance, an individual could avoid violent behavior by consuming a drug that suppressed feelings and thus affected empathy, then it might be morally permissible to use the drug if the morally desirable consequences outweighed the negative effects on the individual's capacity to feel empathy. The third premise results from the structural considerations about robot abuse. Human and robot stimuli that elicit empathy are processed analogously at the level of perceptual empathy; there is no selective desensitization of the capacity to feel empathy only with robots. Habitual robot abuse hence compromises our capacity to feel empathy—an important source of moral judgment that is thus impaired. Habitual robot abuse is therefore morally wrong. Analogously, the argument from moral judgment can be illustrated as follows:

The Argument from Moral Motivation

Premise 1	The capacity to feel empathy is an important source of moral motivation
Premise 2	Ceteris paribus, it is morally wrong to compromise important sources of moral motivation
Premise 3	The habitual abuse of robots compromises the capacity to feel empathy
Conclusion	It is therefore morally wrong to abuse robots habitually

Many of the comments regarding the argument from moral judgment also apply to the argument from moral motivation. The first premise relies on the connection between empathy and moral motivation, while the second premise is postulated as a generally accepted moral principle with a ceteris paribus clause. Again, the third premise results from the structural considerations about robot abuse, which were supposed to show that robot abuse negatively affects our capacity to feel empathy, an important source of moral motivation.

One should, however, also consider the distinction between inhibiting and activating moral motivation. As pointed out above, the inhibiting motivation that arises from empathy is evolutionarily more basic and stronger than the activating motivation. This distinction also applies to empathy with robots as we feel much less inclined to help an abused robot than stop abusing one ourselves.

The argument is, therefore, particularly compelling with regard to inhibiting motivation. Again, the system of perceptual empathy treats human and robot stimuli alike; if we override the violence inhibition mechanism concerning robots habitually, and the connection between empathy and inhibiting moral motivation thus becomes severed, this will also negatively affect our moral motivation regarding humans.

Finally, the argument from moral development is an extension of the argument from moral judgment and the argument from moral motivation. If empathy is a source of moral judgment and motivation, then it should also play an important role in moral development.

The Argument from Moral Development

Premise 1	The capacity to feel empathy plays an important role in children's moral development
Premise 2	Ceteris paribus, it is morally wrong to compromise capacities that play an important role in children's moral development
Premise 3	The exposure to and pursuit of habitual robot abuse compromises the capacity to feel empathy and its motivational impact in children
Conclusion	It is therefore morally wrong to expose children to habitual robot abuse or let them pursue this practice

If the other two arguments are compelling, children's moral development might be disturbed if they—actively or passively—become used to abusing robots that elicit empathy. Furthermore, evidence for the first premise suggests that empathy is an important source of moral development since it contributes to the understanding of morality. The second premise is again postulated as a generally accepted moral principle.

One might hesitate to add the ceteris paribus clause in this case because it might be questionable whether circumstances that morally justify it to compromise the development of moral capacities in children can, in principle, occur. Nevertheless, there are cases when it seems to be permissible. One such case might be if treating a severe illness required providing a child with drugs that suppress empathy.

The third premise results from the structural considerations about robot abuse advanced to demonstrate that robot abuse impairs the capacity to feel empathy, which is an important source of moral development, not least because it is an important source of moral judgment and motivation.

6   Conclusion

Given the three arguments presented in the last section, the answer to our initial question is yes: our feeling empathy with human-like (or animal-like) robots imposes moral restrictions on how we should treat these machines. It is morally bad to habitually abuse robots that elicit empathy as doing so negatively affects our capacity to feel empathy, which is an important source of moral judgment, motivation, and development.

The basic idea behind this argument is not new; Kant argued similarly in the *Metaphysics of Morals* that cruelty toward animals should be morally condemned:

> Violent and cruel treatment of animals is far more intimately opposed to man's duty to himself, and he has a duty to refrain from this; for it dulls his shared feeling of their pain and so weakens and gradually uproots a natural predisposition that is very serviceable to morality in one's relations with other men. (Kant, 1797/1991, p. 238)

What does that mean for our interaction with humanoid robots? Is it time to form a robot rights league to protect their moral claims? This conclusion would be premature. After all, robots do not have moral claims per se. Empathy with robots is only indirectly morally relevant due to its effects on human moral judgment, motivation, and development. Yet, such indirect moral considerations seem to be hardly sufficient to implement laws that sanction abusing robots.

However, one can argue that we legally protect animals from abuse not only because their capacity to feel pain grants them a moral status but also since their abuse offends the overall moral sense in a society. This argument is used in the US to justify the ban on slaughtering horses for meat. It could also be applied to humanoid robots. The reason for legally sanctioning their abuse one day would then not be because they have moral claims on us; rather, it would be to protect our own moral feelings and social bonds if a sufficient number of individuals owned humanoid robots and the abuse of these robots would hurt their moral feelings (Cappuccio et al., 2020; Darling, 2016).

Another, more pressing concern is that we should carefully consider in which areas we want to use humanoid robots capable of eliciting empathy. On the one hand, we risk becoming the victims of emotional manipulation if companies use robots that elicit empathy for economic purposes. On the other hand, immoral behavior toward these robots can compromise our own moral capacities, as was shown. In the end, these considerations evoke a motive that several works of science fiction have explored: they show how cruelty toward robots may destroy what makes us human.

References

Audi, R. (2016). *Means, ends, and persons: The meaning and psychological dimensions of kant's humanity formula*. Oxford University Press.
Bartneck, C., & Hu, J. (2008). Exploring the abuse of robots. *Interaction Studies, 9*(3), 415–433.
Batson, C. D. (1991). *The altruism question: Toward a social-psychological answer*. Lawrence Erlbaum.
Batson, C. D. (2011). *Altruism in humans*. Oxford University Press.
Batson, C. D. (2012). The empathy-altruism hypothesis: Issues and implications. In J. Decety (Ed.), *Empathy: From bench to bedside* (pp. 41–54). MIT Press.
Blair, R. J. R. (1995). A cognitive developmental approach to morality: Investigating the psychopath. *Cognition, 57*(1), 1–29.
Blair, R. J. R (2005). Applying a cognitive neuroscience perspective to the disorder of psychopathy. *Development and Psychopathology, 17*, 865–891.
Cappuccio, M., Peeters, A., & McDonald, W. (2020). Sympathy for Dolores: Moral consideration for robots based on virtue and recognition. *Philosophy & Technology, 33*, 9–31.
Danaher, J. (2020). Robot betrayal: A guide to the ethics of robotic deception. *Ethics and Information Technology, 22*, 117–128.
Darling, K. (2016). Extending legal protection to social robots: The effects of anthropomorphism, empathy, and violent behavior towards robotic objects. In R. Calo, A. Froomkin, & I. Kerr (Eds.), *Robot law* (pp. 213–234). Edward Elgar.
Darwall, S. (1998). Empathy, sympathy, care. *Philosophical Studies, 89*, 261–282.
Debes, R. (2017). Empathy and mirror neurons. In H. Maibom (Ed.), *Empathy and morality* (pp. 54–63). Oxford University Press.
Denham, A. E. (2017): Empathy and moral motivation. In H. Maibom (ed.), *Empathy and morality* (pp. 227–241). Oxford University Press.
Dretske, F. (1995). *Naturalizing the mind*. MIT Press.
De Vignemont, F., & Singer, T. (2006). The empathic brain: How, when and why? *Trends in Cognitive Sciences, 10*, 435–441.
De Waal, F., & Preston, S. (2017). Mammalian empathy: Behavioural manifestations and neural basis. *Nature Reviews Neuroscience, 18*(8), 498–509.
Eibl-Eibesfeldt, I. (1970). *Ethology: The biology of behaviour*. Holt.
Eisenberg, N., & Fabes, R. A. (1990). Empathy: Conceptualization, measurement, and relation to prosocial behavior. *Motivation and Emotion, 14*, 131–149.
Gallese, V. et al. (1996). Action recognition in the premotor cortex. *Brain, 119*, 593–609.
Gigerenzer, G., & Todd, P. M. (1999). *The ABC research group: Simple heuristics that make us smart*. Oxford University Press.
Goldman, A. (2009). Mirroring, mindreading, and simulation. In von J. Pineda (ed.), *Mirror neuron systems. The role of mirroring processes in social cognition* (pp. 311–331). New York.
Hoffman, M. (2000). *Empathy and Moral Development: Implications for Caring and Justice*. Cambridge University Press.

Jackson, P. L., Meltzoff, A., & Decety, J. (2005). How do we perceive the pain of others? A window into the neural processes involved in empathy. *NeuroImage, 24*(3), 771–779.

Kant, I. (1797/1991). *The metaphysics of morals*. Translated by M. J. Gregor. Cambridge University Press.

Kauppinen, A. (2014). Empathy, emotion regulation, and moral judgment. In H. Maibom (ed.), *Empathy and morality*. Oxford University Press.

Kauppinen, A. (2017): Empathy and moral judgment. In H. Maibom (ed.), *Empathy and morality* (pp. 215–226). Oxford University Press.

Lamm, C., & Majdandžiæ, J. (2015). The role of shared neural activations, mirror neurons, and morality in empathy–A critical comment. *Neuroscience Research, 90*, 15–24.

Lennon, R., & Eisenberg, N. (1987). Gender and age differences in empathy and sympathy. In N. Eisenberg & J. Strayer (Eds.), *Empathy and its development* (pp. 195–217). Cambridge University Press.

Lorenz, K. (1981). *The foundation of ethology*. Springer.

Maibom, H. (ed.). (2014a). *Empathy and morality*. Oxford University Press.

Maibom, H. (2014b). Introduction: (Almost) Everything you ever wanted to know about empathy. In H. Maibom (Ed.), *Empathy and Morality* (pp. 1–40). Oxford University Press.

Mathur, M., & Reichling, D. (2016). Navigating a social world with robot partners: A quantitative cartography of the Uncanny Valley. *Cognition, 146*, 22–32.

Misselhorn, C. (2009). Empathy with inanimate objects and the Uncanny Valley. *Minds and Machines, 19*, 345–359.

Misselhorn, C. (2018): *Grundfragen der Maschinenethik*. Reclam.

Misselhorn, C (2023). Viewing others as ends in themselves: The empathic and the moral point of view. In T. Petraschka & C. Werner (eds.), *Empathy's role in understanding persons, literature, and art*. Routledge.

Montag, C. et al. (2008). Theodor Lipps and the concept of empathy: 1851–1914. *American Journal of Psychiatry, 165*(10), 1261.

Mori, M. (1970). Bukimi no tani. *Energy, 7*(4), 33–35. Translated by K. F. MacDorman & N. Kageki (2012). The uncanny valley. *IEEE Robotics and Automation, 19*(2), 98–100.

Mukamel, R. et al. (2010). Single-neuron responses in humans during execution and observation of actions. *Current Biology, 20*, 750–756.

Nickerson, R., et al. (2009). Empathy and knowledge projection. In J. Decety & W. Ickes (Eds.), *The social neuroscience of empathy* (pp. 43–56). MIT Press.

Rawls, J. (1993). *Political liberalism*. Columbia University Press.

Riek L. et al. (2009). Empathizing with robots: Fellow feeling along the anthropomorphic spectrum. *IEEE affective computing and intelligent interaction and workshops*, pp. 1–6.

Rizzolatti, G., & Craighero, L. (2004). The mirror-neuron system. *Annual Review of Neuroscience, 27*(1), 169–192.

Rosenthal-von der Pütten, A. et al. (2014). Investigations on empathy towards humans and robots using fMRI. *Computers in Human Behavior, 33*, 201–212.

Schramme, T., & Edwards, S. (eds.). (2017a). *Handbook of the philosophy of medicine*. Springer.

Schramme, T. (2017b). Empathy and altruism. In H. Maibom (ed.), *Empathy and morality* (pp. 205–104). Oxford University Press.

Slater, M. et al. (2006). A virtual reprise of the Stanley Milgram obedience experiments. *PLoS ONE*, *1*(1), e39.
Slote, M. (2007). *The ethics of care and empathy*. Routledge.
Smith, A. (1853/1966). *The theory of moral sentiments*. August M. Kelley Publishers.
Spaulding, S. (2019). Cognitive empathy. In H. Maibom (Ed.), *The Routledge handbook of philosophy of empathy* (pp. 13–21). Routledge.
Stueber, K. (2006). *Rediscovering empathy: Agency, folk psychology, and the human sciences*. MIT Press.
Suzuki, Y. et al. (2015). Measuring empathy for human and robot hand pain using electroencephalography. *Scientific Reports*, *2015*(5), 15924.
Zaki, J., Wager, T. D., Singer, T., Keysers, C., & Gazzola, V. (2016). The anatomy of suffering: Understanding the relationship between nociceptive and empathic pain. *Trends in Cognitive Science*, *20*, 249–259.

Catrin Misselhorn is Full Professor of Philosophy at the Georg-August University of Göttingen; 2012–2019 she was Chair for Philosophy of Science and Technology at the University of Stuttgart; 2001–2011 she taught philosophy at the University of Tübingen, Humboldt University Berlin and the University of Zurich; 2007–2008 she was Feodor Lynen Fellow at the Center of Affective Sciences in Geneva as well as at the Collège de France and the Institut Jean Nicod for Cognitive Sciences in Paris. Her research areas are the philosophy of AI, robot and machine ethics.

Robots and Resentment: Commitments, Recognition and Social Motivation in HRI

Víctor Fernández Castro and Elisabeth Pacherie

Abstract

To advance the task of designing robots capable of performing collective tasks with humans, studies in human–robot interaction often turn to psychology, philosophy of mind and neuroscience for inspiration. In the same vein, this chapter explores how the notion of recognition and commitment can help confront some of the current problems in addressing robot-human interaction in joint tasks. First, we argue that joint actions require mutual recognition, which cannot be established without the attribution and maintenance of commitments. Second, we argue that commitments require affective states such as social motivations or shared emotions. Finally, we conclude by assessing three possible proposals for how social robotics could implement an architecture of commitments by taking such an affective components into consideration.

Keywords

Commitments · Joint action for human robot interaction · Recognition · Social motivation · Normativity

---

V. Fernández Castro (✉)
Universidad de Granada, Granada, Spain
e-mail: vfernandezcastro@ugr.es

E. Pacherie
CNRS Institut/e Jean Nicod Paris, Paris, France
e-mail: elisabeth.pacherie@ens.fr

1   Introduction

A fundamental challenge for robotics is to develop agents capable of interacting with humans in collaborative tasks. In recent years, considerable resources and effort have been directed at designing and manufacturing robots for use in numerous social contexts, such as companionship to the elderly, education, therapy or service. To make further progress, social robotics needs to design robots capable of meaningfully engaging with humans, to ensure the robots can collaborate with humans in shared activities and joint actions with high levels of coordination. This need explains the fast expansion of the field of human–robot interaction (HRI) and the various avenues of research explored within this field in an effort to enable robots to engage more successfully in social interactions. As part of this expansion, HRI research has taken inspiration from some important findings in psychology, philosophy of mind and neuroscience regarding human–human interaction (HHI) to provide robotic agents with the necessary cognitive capabilities for achieving joint actions.

In the wake of these studies, our chapter proposes that some of the current problems confronting robot-human interaction in joint tasks can be solved by equipping robots with capabilities for establishing mutual recognition of social agency. After first arguing in favor of the fundamental role of the notion of commitment in mutual recognition, we show how the attribution and maintenance of commitment requires fundamental affective states such as social emotions or the prosocial motivation of the need to belong. Finally, we survey three proposals on how social robotics could implement an architecture of commitment by addressing the centrality of these emotions and exposing their weaknesses and strengths.

The chapter is structured as follows. In Sect. 2, we argue that prediction and motivation are two pivotal aspects of joint action which are especially challenging for social robotics. In Sect. 3, we discuss a way of meeting these challenges according to which we must equip robotic agents with the capacity of establishing mutual recognition with humans. We argue that, although promising, the minimal version of the recognitional view proposed by Brinck and Balkenius (2018) has important limitations. In Sect. 4, We propose understanding mutual recognition in terms of the capacity for attributing, undertaking and signaling commitments and argue that this version of the recognitional view is better posed to improve prediction and motivation in HRI. In Sect. 5, we argue that social emotions or the need to belong are indispensable for the establishment of commitments. In Sect. 6, we consider three different proposals for enabling such affective states or for replacing them with functionally relevant substitutes and discuss some of their problems and limitations.

2 Joint Action, Motivation, and Prediction in HRI

Broadly considered, a joint action is a social interaction where two or more individuals coordinate their behavior in order to bring about a common goal or to have a particular effect in their environment (Sebanz et al., 2006). Joint action has been the subject of debate between philosophers and psychologists for some time now. On the one hand, philosophers have traditionally claimed that joint action requires the participants to share certain intentions and they have extensively discussed the nature of such shared intentions. For instance, Margaret Gilbert (1992, 2009) and Michael Bratman (2014) have extensively argued about whether shared intentions depend on the establishment of joint commitments or about the continuity between individual and shared intentions. On the other hand, psychologists have focused on elucidating the psychological mechanisms facilitating the coordination of behavior among these participants, including devices like motor predictions (Prinz, 1997) entrainment (Harrison & Richardson, 2009), perception of joint affordances (Ramenzoni et al., 2008) or mimicry (Chartrand & Bargh, 1999).

Both philosophers and psychologists share the objective of elucidating why and how humans spend an important amount of their time engaged in collaborative action and of characterizing the sophisticated abilities that support collaborative action. Collaborative action confronts two main challenges. First, successful cooperative interactions are premised on action *prediction*. Agents need to coordinate their actions at various levels and must be able to make accurate predictions regarding their partner's actions and their consequences. Shared intentions serve to plan joint action and decide on general and subsidiary goals or sequences of actions to be performed (Bratman, 2014; Pacherie, 2013) and a variety of psychological devices help insure their implementation by facilitating coordination and mutual adjustments among co-agents (Knoblich et al., 2011). Second, successful cooperative interactions are also premised on *motivation*. Humans exhibit a strong proclivity to engage in social interactions. Their motivation can have a variety of sources, both endogenous (e.g., need to belong or general pro-social tendencies) or exogenous (e.g. social pressure). We find some interactions with others intrinsically rewarding (Depue & Morrone-Strupinsky, 2005) and even when we do not, other sources of motivation (e.g., moral values or social pressure) can lead us to help others or collaborate with them. In fact, some of the predictive mechanisms mentioned above seem to have close ties to pro-social motivation. For instance, people who exhibit more cooperative tendencies and possess more empathic dispositions are also those who exhibit more nonconscious mimicry of postures or expressions. Chartrand and Bargh (1999) have demonstrated the motivational aspect of the *chameleon-effect*, which refers to the nonconscious mimicry of postures, expression and other behaviors when interacting with a partner.

In a nutshell, motivation and prediction are two pivotal aspects of joint action among humans and, as such, the two elements must be considered central in the design of social robots. Moreover, both create important challenges for social robotics, not just because of the difficulty of implementing mechanisms capable of instantiating such functions in robots but also because, as a number of studies have demonstrated, there are different negative elements of HRI that may undermine motivation and prediction in various ways.

First, several findings in psychology and neuroscience suggest that humans interact differently when their partner is a robot rather than a human (Sahaï et al., 2017; Wiese et al., 2017). To give an example, while different studies in neurosciences indicate that humans can recruit motor simulation mechanisms to understand others' behavior even during passive observation of others (Elsner & Hommel, 2001), studies with Brain Positron Emission Tomography suggest that the premotor mirror system activated when observing human grasping actions is not responsive to non-human generated actions (Tai et al., 2004).

Second, the gap between the expected and the actual capabilities of the robot seems to impact the predictive capacities of humans (Kwon, Jung, & Knepper, 2016). While the physical appearance of the robot and some of their social capacities—e.g., mimicry—may generate high expectations regarding the autonomy and sophistication of the robot, its real capacities might actually be heavily context-dependent and its autonomy quite limited. Such a difference between expectations and reality may provoke frustration in humans but more importantly, it may negatively influence their attunement to the robot and decrease their capacity to generate reliable predictions of the robot's behavior. Thus, Kwon, Jung, and Knepper (2016) created a series of surveys where subjects were presented with vignettes that described a human collaborating with an industrial robot, a social robot or another human in either an industrial or a domestic context. In their first study, they found that people often generalized capabilities for the social robot, attributing the same confidence to both robots in industrial settings. In contrast, they only attributed low confidence to the industrial robot in domestic settings. In a second study, they created two video clips of a human–robot team and a human–human team each completing a simple block-building task, with one teammate responsible for one color of blocks. They programmed the robot to be incapable of stacking blocks without the help of their human partner and introduced the same limitation in the human–human team. Participants were asked to rate the capabilities of the more "limited" teammate at various points of the human–human and human–robot videos. The experimenters found that the subjects were "more willing to modify their expectations based on a robot's perceived capabilities compared to a human". In other words, the gap between expectations and perceived capabilities is more pronounced when the observed agent is a robot, which could increase the risk of failures during HRI.

Similarly, several studies have uncovered a number of factors that can undermine human motivations for engaging with robots. First, an often-voiced problem in social robotics is the well-known *Uncanny Valley Effect*, the phenomenon whereby humans experience a feeling of discomfort or revulsion when perceiving a machine or artifact that acts or looks like a human (see Wang et al., 2015 for a review). The first reference to this effect appears in the work of Mori (1970) and it has been observed not only in humans but also in primates (Steckenfinger & Ghazanfar, 2009). There are different hypotheses concerning the possible causes underpinning the discomfort, for instance, that the feelings are associated with evolutionary adaptations for some aesthetic preferences or the avoidance of pathogens. However, it is important to emphasize that whatever these causes are, they can interfere with the motivation of human agents to engage in social interactions with robots.

Second, the empirical studies reporting negative attitudes toward robots are not restricted to those regarding the Uncanny Valley Effect. In a series of experiments using implicit association tests, that measure the reaction times of participants depending on the associations between a target (a robot or a human) and a positive or negative attribute, (Sanders et al., 2016) found that the participants in the experiments exhibited implicit negative attitudes toward robots even when their explicit assessments of the robotic agent were positive. Those experiments seem to demonstrate that people exhibit adverseness toward robots, or at least, less positive stances than they project toward humans. Moreover, other aspects of robots like their human-like appearance or personality can be perceived as deceptive (Vandemeulebroucke et al., 2018) or impact the human levels of the trust, which could produce resistance to start interacting with the robot or lead one to abandon too quickly the collaborative task, ending up in the disuse of the robot (M. Lewis et al., 2018).

In conclusion, despite the enormous advances made towards endowing robotic agents with socio-cognitive capacities, there are reasons to believe that attempts at establishing a better mutual understanding between humans and robots do not always meet with success. In particular, the mentioned studies have discovered some important elements that could negatively impact the objective of designing robots able to collaborate with humans as a team. In the next section, we review two general approaches to social robot design that may help solve these problems and we present some of their limitations.

3 Social Robotics and Recognition

How can we eliminate or mitigate the threats to prediction and motivation in joint action for HRI?

A promising strategy aims at identifying and characterizing the fundamental social capacities deployed in human–human interactions in order to design robots with similar capacities for understanding the behavior or mental states of their human partners. Such an approach is exemplified by the work of several laboratories that attempt to design robots with social capacities like joint attention (e.g. Huang & Thomaz, 2010), recognition of emotion or body gestures (Benamara et al., 2019) or theory of mind (Pandey & Alami, 2010; Sisbot et al., 2010). To take a few examples, Huang and Thomaz (2011) carried out a study with a Simon robot able to recognize human attention (by recognizing eye gaze, head orientation or body pose), to initiate joint attention (by using pointing gestures, eye gaze and utterances) and to ensure joint attention (monitoring the focus of attention and soliciting the partner's attention with several communicative strategies). In their study, they found that people tend to have a better mental model of the robot and to perceive the robot as more sociable and competent when it manifests a capacity for joint attention. Another example of this type of strategy is Pandey and Alami's (2010) theory of mind-like implementation which uses information about the human's position, posture or visibility of the relevant space and objects to enable the robot to reason about the human reachability and her visual perspective. Moreover, these types of designs enable robots to exploit different social cues to respond to the human's action, plan different courses of behavior or implement collaborative interactions. Some labs have designed planners and decision-making devices that take into account the information provided by these mechanisms. For instance, Sisbot et al. (2010) designed a system that integrates information from perspective-taking, the human point of view on the relevant space and objects, and human-aware manipulation planning to generate robot motions that consider human safety and their posture along with task constraints.

Endowing the robot with such highly sophisticated capacities enables it to understand humans and reason about their behavior and mind. However, as Brinck and Balkenius (2018) argue, the problem with this strategy is that such capacities are not often oriented towards making the human user feel *recognized* by the robot or towards establishing a *mutual recognition* between the human and the robotic agent (for another view of recognition in HRI see Laitinen, 2016). Humans can understand others' behavior, for instance, in terms of mental states. However, they are also capable of expecting different behaviors from others depending on their physical aspect or the shared environment. Moreover, humans are constantly exhibiting different proactive and reactive strategies to make the other aware of such expectations or to acknowledge the expectations the other has towards them. Such responsiveness and proactive and reactive strategies serve to establish a mutual recognition, that is, the participants are not merely passive subjects of prediction and control by the other (Davidson, 1991, p. 163; Ramberg, 2000, p. 356), but active agents who attribute to each other certain rights and obligations concerning the joint action.

The notion of recognition is complex, but in the minimal sense, recognizing and being recognized as social agents give rise to the type of behaviors and responses that are at the root of our sociality and of the way we adjust to each other's actions. In this sense, Brinck and Balkenius argue, the design of the robot must consider the human as a social being with needs, desires, and mental states, and thus, the robot must be designed to be aware of, and responsive to, these features. As a result, the robot should be able to take the human into consideration and react to his presence, body, and actions, but also to signal its own presence and acceptance of the human, so the recognition can become mutual.[1]

Now, the question is which is the best way to implement this recognitional approach in social robotics? Brinck and Balkenius (2018) have put forward a minimal recognitional approach according to which the design of the robot must focus on embodied aspects of mutual recognition. According to several authors (Brandom, 2007; Satne, 2014), recognition involves high-level cognitive capacities including being able to give, and ask for, reasons or the ability to respond to blame and reproach. However, Brinck and Balkenius argue that recognition can be manifested in more minimal social capacities including attending to the other,

---

[1] At this point someone could argue that, given the asymmetry between the social capacities of human beings and robots, the recognition is not necessarily mutual. We acknowledge the importance of this point; however, addressing it goes beyond the scope of this paper (see Seibt, 2018).

searching and making eye-contact, engaging in turn-taking or mimicking postures. According to this minimal approach, recognition involves three cognitive capacities: First, *immediate identification*, the processes that assign certain properties to the other individual and generate certain expectations about how others will engage in a mutual activity based on perceptual information available here and now. This identification, they argue, requires the perception of movement and action, gaze, vocalization, and emotions. Second, *anticipatory identification* requires anticipating the actions of the other based on previous interactions and the available information in the context, for instance, the perception of the others' actions in the interaction. Finally, mutual recognition requires, what Brinck and Balkenius call *confirmation*, which involves reacting to the presence of the other or signaling pro-actively to show that identification has taken place and one is ready for the interaction. As a result, when these elements are combined, the individual shows that their behavior can be influenced by the other's, thus exhibiting a willingness to engage in the interaction. Such mutual recognition is crucial not only to establishing the readiness to interact but also the dynamic of signals, actions and reactions that facilitate the interaction. Although in principle, the three basic components of recognition can be instantiated by different capacities, Brinck and Balkanius emphasize the importance of embodied recognition, that is, identification and confirmation based on physical constitutions of the body and sensory-motor processes. Some of the processes involved in this type of recognition are attentional engagement, mimicry of postures or gestures, emotional engagement, responding to other's gaze and attention, exaggeration of the movements or explicit modification of their kinematics, turn-taking or active eye-contact.

This general framework gives us a solution to the problems presented in Sect. 2. Equipping robots with capacities for establishing mutual recognition may solve the motivational problem to the extent that people may feel recognized by the robot. Robotic embodied recognitional capacities could make the human feel perceived as a social peer, and thus, bring about the same type of motivations that an interaction with a human social peer may trigger. Moreover, this strategy could, in principle, dissipate some of the prediction problems. Designing robots with capacities for confirmation can help control the type of expectations that the human generates. For instance, if the robot is designed with confirmation strategies regarding some expectations (e.g. the robot confirms his capacity for facilitating physical therapy) but not others (e.g. he is unable to establish conversation), it could reduce the aforementioned expectation gap.

Despite its virtues, the embodied recognitional approach suffers from important limitations, however. First, it lacks the level of abstraction that would be needed for it to be a general approach to the design of social robotics. In robotics, we find a large number of different robotic agents, which possess different perceptual and behavioral capabilities and physical features. To give a few examples, we can find robots able to move their heads and arms (like Pepper or i-Cub) while others cannot (e.g. Rackham). Some agents can introduce different signals through channels like a screen (Pepper and Rackham), while other agents, like a humanoid, are restricted to human-like expressive capabilities or language. Similarly, while most robots are equipped with visual capacities in the form of cameras, many others use tactile sensors, thermal cameras or heart rate detectors. Such a variability creates a problem for strategies that lack a sufficient degree of abstraction, precisely because some embodied strategies may be available to some robotic agents and not others. We need to model social interactions in a way that abstracts away from some aspects of implementation, so we can adjust different social strategies to different robots depending on their perceptual and behavioral capabilities.

Second, the embodied recognition strategy fails to take advantage of some important procedures that human exploits during joint action and that are available to social robotics. This is especially obvious in the case of identification. As Brinck and Balkanius state, humans generate expectations about others' actions based on different embodied aspects and physical features. However, the sources of information we use to generate expectations about others during joint action are not restricted to physical features or even contextual factors. Humans often anticipate or predict others' actions by the mediation of patterns of rationality (Dennett, 2009; Fernandez Castro, 2020; Zawidzki, 2013), scripts and social norms (Maibom, 2007; McGeer, 2015) or the structure and features of the joint action itself (Török et al., 2019). To give an example, in a recent study, Török et al. (2019) demonstrated that when people perform joint actions, they behave in ways that minimize the costs of their own and their partner's movement and they make rational decisions when acting together. Arguably, the capacity for diminishing the costs of the partner's choice must be partially based on the assumption that the partner will behave rationally. In other words, we can assume that people expect each other to behave as it is rationally demanded by the joint action. A second example has to do with the type of expectations that we generate depending on the nature of the joint action itself. As several authors have suggested (Knoblich et al., 2011; Pacherie, 2011; Vesper et al., 2017), when we engage in a joint action, we form representations of the joint plan. Such representations not only involve an individual's own actions in relation to the joint action but also in relation to their partner's actions. In this sense, individuals anticipate and predict partner's courses of action in relation to the representation of the joint plan.

In other words, the human capacity for identifying each other relies on a variety of informational sources that are not restricted to embodied physical information. These types of strategies can be extremely helpful for social robotics, so there is no reason why we should not exploit similar strategies during HRI.

Thus, while we do not want to deny that the embodied recognition strategy may contribute to solve or attenuate the prediction and motivation problems in social robotics, we think that a more general strategy is necessary. Such a strategy may be elaborated upon a model that abstracts away from specific implementations of the robotics agents, so it can be adjusted to the embodied and non-embodied idiosyncrasies of every robot while exploiting the socio-cognitive capacities of humans and establishing a mutual recognition between humans and robots. In the next section, we present our proposal according to which the recognition between humans and robots must be understood in terms of commitments. So, the general strategy to design robots able to overcome the aforementioned problems with prediction and motivation must be oriented towards developing robots with the capacity for establishing, tracking, and responding to, individual and joint commitments.

4 Commitments and Recognition

Part of the rationale behind the notion of *recognition* is the idea that the stance that humans have toward each other's behaviors is not passive, as if one were distantly observing a mere object whose movements one needs to predict. In fact, in joint actions, humans adopt an active stance, in which they pro-actively provide social cues to facilitate prediction and anticipation by their partner, regulate their behavior to make it more transparent and actively influence others' actions to facilitate the realization of a joint goal. Evidence that humans adopt such a proactive stance is provided by empirical findings suggesting that, during human–human interactions, people provide others with information about their own actions. For instance, some studies on sensorimotor communication demonstrate that people exaggerate their movements to allow their partners to better recognize the action goal (Vesper & Richardson, 2014). Moreover, humans are sensitive to implicit cues (gaze signals) that manifest an agreement to carry out with them a task their partner intends to perform (Siposova et al., 2018).

In our view, such a proactive stance and the repertoire of actions and capacities necessary to adopt it requires agents to attribute and undertake different participatory and individual commitments. As a first approximation, we submit the idea that mutual recognition in joint action is established when the partners recognize each other as authors of different commitments involved in the interaction and hold each other responsible for such commitments. People proactively give social cues, regulate and adjust their behavior in response to others as a way to establish, negotiate and track a set of individual and joint commitments related to the goal and plan associated with a joint task.

Now, a commitment is in place when the recipient generates an expectation regarding the author as a result of having an assurance that the author will act according to the expectation in a condition of mutual knowledge. To give an example, Sara is committed to helping Andrew repair his bike when Andrew expects Sara to do it as a result of Sara having made a promise to do so and Andrew having acknowledged the promise. Traditionally, philosophers have connected the establishment of commitments to explicit verbal actions, e.g., one agent, the author of the commitment, commits to another, its recipient, to a course of action X by intentionally communicating that one intends to X through a promise or other speech act (Austin, 1962; M. Gilbert, 2009). However, commitments are not necessarily established through explicit verbal agreements. For instance, one might indicate through gestures or facial expressions that one will perform the appropriate action (Siposova et al., 2018). Moreover, as several authors claim, some factors like situational affordances, social norms and scripts (Fernández Castro & Heras-Escribano, 2020; Lo Presti, 2013), or the identification of another agent's goal (Michael et al., 2016), for example, can lead an agent to undertake commitments and attribute commitments to others.

Understanding our social stance in terms of commitments allows us to understand the notion of identification in a way that helps us cover a greater range of strategies than the embodied version of the recognitional view. *Identifying* another agent as a social peer implies attributing to her a set of commitments from which we can generate different expectations that anticipate and predict her actions. We can attribute different commitments depending on the physical appearance of the author, but also, depending on social norms, general patterns of rationality, scripts or the structure and features of the joint action itself. For example, we can attribute to our partner the commitment to behave in the most rational way and minimize the costs of the overall action. In other words, we can assume that people identify each other as committed to behave as rationally demanded by the joint action, which generates expectations that facilitate anticipation and prediction. Another example has to do with the type of commitments that we attribute to each other depending on the nature of the joint action itself. As Roth (2004) has emphasized, when we engage in joint action, we do not only undertake a joint commitment to pursue the joint goal but we also undertake a set of contralateral commitments regarding individual actions and sub-goals necessary to the success of the collective task. For instance, if we agree to go for a walk together, we can attribute to our partner an individual commitment to walk at the same pace.

Now, identification (and confirmation) are not the only types of capacities involved in the social stance we adopt during joint action. When we perceive our partner as such, we do not only generate expectations and wait for confirmation; we often use *exhibitory signals* to indicate to our partner what we expect her to do. In other words, one pro-actively gives cues to one's partner regarding what behavior one believes should be performed. For instance, as Michael et al. (2016) suggest, people often use investment of effort in a task as an implicit cue for making the perceiver aware that we expect him to behave collaboratively. In other cases, the cues are more explicit, for instance, when we negotiate what to do during the task through what Clark (2006, p. 131–33) calls a projective pair (e.g. proposal/acceptance), where one of the participants proposes a particular goal to another (Let's do G!; Should we do that?), who then accepts or rejects the proposal. Besides these exhibitory actions, social agents manifest different *regulative actions* directed toward the performance of others. For instance, we often use positive or negative emotional expressions, like smiling or wrinkling one's nose as a signal of approval or disapproval toward the action of the other (Michael, 2011). Moreover, humans exhibit a robust repertoire of regulative actions directed toward others when they have frustrated our expected social interactions, including blaming, reprimanding or asking for reasons (McGeer, 2015; Roth, 2004). Such regulative actions are manifested during joint action (Gilbert, 2009; Roth, 2004). Some recent studies suggest that people who judge that two persons are walking together in certain conditions are more likely to consider that one of the participants has the right to rebuke the other when he peels off (Gomez-Lavin & Rachar, 2019).

An interesting aspect of these exhibitory and regulative actions is that they are hard to accommodate in a framework that does not presuppose that social agents can hold each other responsible for certain actions. In other words, exhibitory and regulative actions presuppose that agents feel enabled or justified to hold their partners responsible for the expectations they have generated. Such a *normative attitude* is accommodated in our framework to the extent that social partners recognize others' actions as living up to such commitments or frustrating them. In other words, exhibitory and regulative actions are motivated by the normatively-generated expectations of commitments. The normative attitude underpinning exhibitory and regulative actions is explained by the fact that the expectations associated with commitments are normative (Greenspan, 1978; Paprzycka, 1998; Wallace, 1994). That is, when we expect an agent A to do X because she is committed to G, we do not just predict and anticipate X (descriptive expectations) but we are entitled to demand X from A on the basis that she has the obligation to G (normative expectation).

The existence of such a normative attitude can also explain why, in joint actions, social partners may feel motivated to perform actions whose goal is not instrumental but communicative (Vesper et al., 2017, 4). As Clark (2006) emphasizes, joint actions can be divided into two types of actions: the *basic actions*, aimed at achieving the goal per se and *coordinating actions* aimed at facilitating the prediction, adjustments, and coordination between the partners. In other words, social peers pro-actively ensure that the other partner generates the appropriate expectations. Notice that coordinating actions also involve a normative component: they involve the agent's obligation and accountability. When someone produces a social cue that signals what he is going to do—e.g., making eye contact to indicate one's readiness to engage in a collaborative task—she is implicitly embracing the responsibility to act accordingly. These actions exhibit a high component of social exposure. Making public your intention to perform a particular joint action entitles others to sanction or blame you if you decide to abandon the action. Thus, it is difficult to see how agents could undertake such a responsibility without having a particular understanding of their actions and themselves in terms of commitments.

Now, we can see how commitments play a pivotal role in both prediction and motivation in joint action. Regarding prediction, as we mentioned, when one identifies a partner as such and she confirms this identification, the two agents are establishing a set of commitments. As Michael and Pacherie (2015) have argued, commitments stabilize expectations regarding actions, beliefs, and motivations that reduce different types of uncertainties regarding how to proceed during the joint action, shared background knowledge or whether or not the participants have a common goal. In this sense, both attributing and undertaking individual and joint commitments facilitate prediction and coordination by prescribing courses of actions and individuals' behavior more transparently.

Regarding motivation, commitments can serve as an important catalyst for joint actions in different ways. First, commitments impose an obligation on the author of the commitment to fulfill the appropriate expectation. Such an obligation can be enforced in different ways. For instance, the obligation entitles the recipient to sanction their author or to protest if the expectation is not fulfilled which could provide reasons to the authors to engage in a joint action when he has previously committed to it. Secondly, the author of the commitment may feel motivated to act not because she is inclined to avoid the recipient's possible sanctions but because she may feel identified with the expectations in place. Humans often find others' expectations about their own behavior appealing when they are reasonable (D. K. Lewis, 1969; Sugden, 2000). Thirdly, expressing commitments can also motivate the action of the recipient to the extent that they are costly signals which provides evidence of the author's motivation and serves as a reason for the recipient to engage in the joint action (Michael & Pacherie, 2015; Quillien, 2020). Finally, the author's signals of commitments can prompt a so-called *sense of commitment* (Michael, 2022; Michael & Székely, 2018; Michael et al., 2016) on the recipient that may act as an endogenous motivation to engage in the joint action. Sense of commitment is the psychological motivation to collaborate, to engage in a joint goal or cooperate with someone because this person expects you to do so and he has somehow manifested such expectation.

These aspects are sufficiently important and pervasive in human–human interaction to motivate a general strategy in social robotics that puts the notion of commitments at the center of the design of robotic agents. This would consist of taking seriously the idea of making robots able to identify human partners as social partners and attribute them a set of commitments depending on relevant features of the situation, but also, able to monitor and respond to the fulfillment or frustration of the relevant expectations. Moreover, robotic agents should be able to undertake commitments, and thus, pro-actively signal them to establish mutual recognition with their partner.

Some exploratory ideas of how robot design can be oriented to the establishment of commitment lay on the use of social signals in robot design. Several studies have demonstrated that equipping robots with the capacity to produce behavioral cues, facial expressions or gaze cues can boost transparency, mutual understanding and trust in HRI (Normoyle et al., 2013; Sciutti et al., 2018; Stanton & Stevens, 2014). For instance, different studies indicate that stereotypical motions, along with straight lines and additional gestures (see Lichtenthäler & Kirsch, 2013 for a review) are pivotal factors for legible robot behavior. In this line, Breazeal et al. (2005) have demonstrated that equipping robots with subtle eye gaze signals—for instance, enabling Leonardo to reestablish eye contact with the human when it finishes its turn, and then, communicating that it is ready to proceed to the next step in the task—improves the subject's understanding and her capacity to quickly anticipate and address potential errors in the task. Moreover, different laboratories have designed different expressive capacities in robots that boost the motivation to interact in humans or that maintain her engaged in the collaborative task. For instance, work in this direction involves facial expressions in anthropomorphic faces with many degrees of freedom (Ahn et al., 2012; Kedzierski et al., 2013), posture (Breazeal et al., 2007) or body motion (Kishi et al., 2013). The efficiency of such approaches is confirmed by studies that demonstrate that human users find robots more persuasive when they use gaze (Ham et al., 2015) or more cooperative when they use cooperative gestures like beckon, give or shake hands (Riek et al., 2010). In sum, we have reasons to believe that robots equipped with the capacity for advancing certain expectations or exhibiting a certain degree of commitment to a particular task can improve human trust and proclivity to engage with a robot.

5   The Affective Side of Commitments

In the previous section, we have presented a general approach to the design of social robots capable of engaging in joint action with humans. Such an approach emphasizes the necessity of implementing robots whose capabilities enable them to attribute and undertake commitments. These capabilities must include abilities like pro-active signaling of robot's expectations regarding the human, monitoring and reacting to the frustration and fulfillment of such expectations or manifesting its readiness to behave as expected. The main virtues of this strategy are its level of abstraction and generality, which would allow implementing the mentioned abilities differently depending on the specificity and constraints of specific robotic agents. Now, in which sense are affective states important for joint action in HRI and which role do they play in this general strategy? What can a capacity for emotional expression bring to this commitment approach? Are emotions, motivations or other affective states necessary for establishing and maintaining commitments between humans and robots?

To appreciate the role of affective states in the dynamics of establishing and monitoring commitments, we must consider again the regulative actions associated with holding someone responsible for their commitments. As we mentioned, when we expect an agent A to do X because she is committed to G, we do not just predict and anticipate X (descriptive expectations) but we are entitled to demand X from A on the basis that she has the obligation to G (normative expectation). A first affective aspect associated with commitment seems to be the often emotionally loaded character of these regulatory reactions. Regulatory actions are a series of behaviors oriented toward the other agent or their behavior in order to acknowledge that they have violated (or fulfilled) a commitment. Regulatory actions include subtle cases like manifesting surprise or merely warning the other but also more dramatic actions like expressing resentment and anger or manifesting disapproval or blaming. However, even the milder reactions, finding fault with someone or communicating a judgment of violation, seem to register an emotional state or charge.

But what are emotions and why do we experience negative emotions when someone breaks a commitment? Emotion theorists generally understand emotions as having intentional, evaluative, physiological, expressive, behavioral and phenomenological, components, although they tend to disagree on which of these components are central or essential to emotion (Michael, 2011; Prinz, 2004; Scherer, 2005). Social emotions are a particular subset of emotions, both self-directed and other-directed, that depend upon the thoughts, feelings or actions of other people, "as experienced, recalled, anticipated or imagined at first hand" (Hareli & Parkinson, 2008, p. 131). Examples include shame, embarrassment, jealousy, admiration, indignation, gratitude, and so on. Now, commitments impose obligations on the author of the commitment and enable or entitle the recipient to demand these obligations be met. A possible explanation of the emergence of emotional states like feelings of offense, disappointment, disapproval or resentment is precisely their function as a cultural mechanism to prompt punishment and prevent free-riding. While the explanation of small-scale collaboration can appeal to indirect-reciprocity or genetic affinity, the evolution of large-scale cooperation is a puzzling phenomenon from an evolutionary point of view to the extent that it occurs among strangers or non-relatives in large groups. Several authors have proposed that cultural evolution solved the problem by impacting human social psychology (Henrich & Henrich, 2007). In this view, the solution to the problem will lie on a series of evolved norm oriented psychological mechanisms that enable humans to learn and acquire social norms and cultural patterns but also to respond to norm violation and engage in punishment. Such norm-psychology will facilitate large scale cooperation by creating and reinforcing more stable groups (Guzmán et al., 2007).

In this sense, one may argue that social emotions are part of our norm-psychology and that their biological function is precisely to sustain large-scale cultural dynamics by prompting monitoring and punishing responses to norm violations or non-cooperative behaviors in our cultural niche (Fehr & Gächter, 2002). Such a perspective may help us to understand how emotional states can contribute to the maintenance of commitments or why commitments are regarded as credible. Social emotions enable humans to monitor others' commitments and norm compliance and trigger the appropriate punishing or sanctioning response when appropriate. Moreover, one may suggest, the existence of these emotional responses may have promoted, in the appropriate evolutionary circumstances, an inclination to signal and fulfill the commitments, not only in order to avoid punishment and sanctions but as an outcome of developing some sort of avoidance of others' negative emotions—e.g., aversion to others' distress or guilt (Decety & Cowell, 2018; Vaish, 2018)—or as part of a reputation management mechanism.

As a result, social emotions seem to play a fundamental role in the establishment and maintenance of commitments, and thus, in the well-functioning of joint actions. But would this be a definitive positive answer to the question of whether or not commitments require affective states? Does social robotics need to develop emotional robots to establish and maintain joint commitments and protesting as a reaction to the violations of commitments? To what extent should such emotional states be associated with sanctioning or punishment?

Although in evolutionary theories of cooperation, punishment seems to play a fundamental role, it is not so obvious that punishment is necessarily linked to an increase of cooperative behaviors among a population. Several authors have argued that punishment can be ineffective to sustain compliance and even can produce an erosion in cooperation (Dreber et al., 2008; Ostrom et al., 1992). Moreover, the reactive behavioral pattern associated with emotional states like disappointment or disapproval seems to be protesting, attempting to jolt the wrongdoer into seeing things more from the wronged party's perspective or, in the worst case, blaming, rather than punishing or sanctioning. Thus, it is not so obvious that punishing or sanctioning would be the ultimate functional role of social emotions in the development of commitments.

In our view, the role of social emotions in the emergence and maintenance of commitments makes more sense when we have a look at the entire dynamic exchange where these emotional states are often inserted when a violation of commitment is produced. Such an exchange, we will argue, shows that these social emotions play a fundamental role in managing commitment when they work in conjunction with another pivotal affective state: the need to belong. The functional role of social emotions in the context of commitment management only makes sense when the author of the commitments is not only ready to comply with the commitment but also when he is ready to review, re-evaluate and regulate their behavior under the light of the emotional charge produced by a violation. Such inclination to review, re-evaluate and regulate her behavior cannot be explained without postulating a central motivation to care about the others that we will characterize in terms of the need to belong (Baumeister & Leary, 1995; Over, 2016).

To see how, let us consider an observation that several philosophers have made in the context of the debate on moral responsibility (Fricker, 2016; Macnamara, 2013; McGeer, 2012). A fundamental debate in philosophy of mind regarding moral responsibility revolves around the function of reactive attitudes like blaming in the practice of moral responsibility. However, according to these authors, while the debate has often focused on the fact that reactive attitudes are backward-looking responses to actions and attitudes, which manifest that we profoundly care about other moral actions and responsibilities, the debate has often overlooked the forward-looking dimension of such reactive attitudes and that seems to be fundamental for understanding how such reactive attitudes scaffold the moral agency of others (McGeer, 2012). In other words, although reactive attitudes towards others often manifest negative emotions, they also communicate a positive message, namely that we see them as moral agents capable of understanding and living up to the norms of a moral community. In understanding this message, what is essential is that "the recipients of such attitudes understand—or can be brought to understand—that their behaviour has been subjected to normative review, a review that now calls on them to make a normatively "fitting" response" (McGeer, 2012, p. 303). Such a responsiveness involves the wrongdoers behaving reactively in ways commensurate with treating them as responsible agents, as manifested by the co-reactive attitudes of apologizing, giving reasons and reviewing her behavior in a way that reflects a moral sensitivity. So, moral agency is reflected in the disposition to respond reactively to others' reactive attitudes.

Considering commitments from a similar angle, we can see reactions to violation or fulfillment of commitments (e.g. asking for reasons, reprimanding or manifesting surprise or disapproval) precisely as just one step in a dynamic practice where reacting and co-reacting (e.g. giving an excuse, adjusting one's action) to the violation and fulfillment only make sense in the context of the forward-looking function of reactive attitudes: urging the wrongdoer to review her behavior and mental attitude in the light of the expectations generated by the commitments that the wronged party attributed to her. Such a forward-looking function is manifested in the co-reactive responses to reactive attitudes, which are often oriented to apologizing or justifying the violation but more importantly to increase the agential capacities of the agents to the extent that they scaffold their capacity to evaluate their behavior in the light of the commitments involved. As Fricker (2016) claims, even when the wrongdoer does not admit the violation or does not acknowledge her previous commitment, the reactive attitude can produce sufficiently psychological friction on the wrongdoer to orient her mind toward an evaluative stance and lead her to review her motives or reasons to behave as she did.

These dynamics have an important function in terms of interpersonal alignment of intentions and joint beliefs. Imagine two friends, Pablo and Sara, who decide to paint a house together and when Sara takes her brush to start, Pablo takes his equipment and goes to another room. The following exchange may ensue:

- Pablo: "I thought we were going to do this together."
- Sara: "This way we'll go faster."
- Pablo: "I don't want to get bored doing this."
- Sara: "Okay, you're right."

This kind of dynamic, where two agents react and co-react to the frustration of an expectation associated with a joint commitment, shows us how regulative actions may serve to align intentions and beliefs during a joint action, which has important consequences for prediction and motivation. However, the lesson we would like to draw from this analysis is different. Notice that the dynamic trajectory of reactions and co-reactions is based on the premise that both agents care about each other, the joint commitments and their mutual expectations. Reactive and co-reactive attitudes only make sense if the agents involved are the kind of agents that care about living up to the expectations and demands of commitments and care about exercising their agential capacities expressed through evaluating and regulating their actions in accordance with commitments and their normative expectations. Such capacities, then, can only make sense if the agents involved care and value their social relations and bond with other agents, which seems to imply an important affective factor.[2]

Elsewhere, we have argued that a major human motivation for explaining why commitments are credible is the need to belong (Fernández Castro & Pacherie, 2020). The need to belong is the need that individuals have for frequent, positively valenced interactions with other people within a framework of long-lasting concern for each other's welfare (Baumeister & Leary, 1995; Over, 2016). The need to belong is a need, in the sense that long-term social bonds are crucially important to well-being and, conversely, their lack leads to ill-effects. The need to belong explains why humans find acting with others rewarding, why they tend to give attentional priority to social cues, or why many joint actions are motivated not just by the desire to achieve the intended outcome of their shared intention but also by the desire to obtain this social reward. However, in contrast with other postulated general prosocial motivations (Godman, 2013; Godman et al., 2014),

---

[2] Certainly, we are not denying that there can also be self-interested reasons or motivations to comply with a commitment. Our statement is ontogenetic: the most basic, earliest developing psychological motivation to meet expectations associated with commitments is prosocial, even if other individualist mechanisms (e.g. reputation management), which appear later in the development, are important and widespread complementary sources of motivation.

the need to belong is neither indiscriminate nor unbounded but rather manifests selectivity. So, humans prefer repeated interactions with the same persons to interactions with a constantly changing sequence of partners, they devote more energy to preserving and consolidating existing bonds than to interacting with strangers and once the minimum quantity and quality of social bonds are surpassed, their motivation to create new bonds diminishes.

For our purpose, the importance of the need to belong lies in its capacity for giving an account of why we care for, or value, both our commitments to others and the commitments others have to ourselves. Without such a capacity, we could not explain why one experiences social emotions when someone breaks a commitment or why one feels the psychological pressure to evaluate one's performance and commitments in light of the reactions of others. Fernández Castro and Pacherie (2020) argue that although there is a large diversity of motivations that may be involved in why we commit ourselves and why we remain faithful to those commitments—e.g. reputation or avoidance of negative social emotions—, the need to belong is, from a developmental point of view, a more basic motivation. Now, the presented dynamic also allows us to see how the need to belong might be involved in the emergence of these other motivations. First, an agent's social emotions emerge as reactive attitudes to others' attitudes or behavior that may trigger her capacity for reviewing, re-evaluating and regulating her behavior. However, social emotions can only serve such a function if the wronged party cares about the wrongdoer's commitments, even in cases where the wronged party does not necessarily benefit from the result of the joint action, and if the wrongdoer in turn cares about the wronged party sufficiently to motivate a co-reaction. Second, although we may abide by our commitments or provide justifications to explain why we violated a commitment simply in order to promote or preserve a positive reputation, such management of reputation can only emerge as a result of the dynamics of previous reactions and co-reactions premised on the idea that we care about others. The notion of prestige and reputation is tied to the image that others have of us and how such an image may impact our social relations with them. Without a motivation to engage in such social relations, and probably without a dynamic of manifested positive or negative attitudes toward the resulting outcomes of such relations, the notion of reputation no longer makes sense.

In a nutshell, the establishment and maintenance of commitments involve a dynamic trajectory of signals, reactions, and co-reactions. This trajectory seems to involve at least two emotional components. First, a series of socio-emotional states triggered by frustration (and fulfillment) of commitments that produce a series of regulative behaviors signaling recognition of the violation, disapproval of it and warning and aimed at making the wrongdoer review her behavior. Second, a pro-social motivation that prompts both parties to establish commitments but more importantly makes them ready to review, re-evaluate and regulate their behavior in the light of their violation, and thus, scaffold their agential capacities as a team.

6 Emotional Robots and Commitments

We have proposed that the establishment of commitments during joint action necessitates the mutual influence of two types of affective states: social emotions and prosocial motivations. The interaction between these types of affective states explains different features of the interaction between partners like how they hold each other responsible for commitments, how they react and co-react when the expectations associated with the commitments are violated or fulfilled or why one would assume the social costs of undertaking them. If our argumentation is compelling, one must wonder whether implementing these types of affective states is necessary for the design of social robots able to establish commitments with a human partner in collaborative contexts: May developers attempt to implement affective states in robots? Or would it be sufficient to mimic or imitate the behavioral profile of such states? Could we find a design solution that would devise functional substitutes for such affective states without implementing them properly? In this section, we propose three possible answers to these questions and discuss their potential scopes and problems. First, we discuss a minimalist option that would involve faking emotional states. Second, we discuss a maximalist option that would endow robots with affective states or quasi-states that are equal or at least similar to human affective states. Finally, we discuss an alternative option that attempts to design different solutions aimed at establishing commitments without using or faking affective states.

To begin with, the minimalist option would attempt to endow robots with a capacity to manifest certain behavioral reactions recognizable by the human as stereotypical affective reactions that facilitate the establishment and maintenance of commitments without necessarily implementing other dimensions of affective states like arousal or specific action tendencies. Although not straightforwardly connected to commitments, the use of emotional signals to communicate different information or to maintain the human engaged in a collaborative task is a common strategy in social robotics. Several labs (Breazeal, 2003; Craig et al., 2010; Kishi et al., 2013; Oberman et al., 2007; C. Wendt et al., 2008;) are equipping robots with different emotional expressions in order to prompt empathy or prosocial attitudes in the human agent, so robots "could potentially tap into the powerful social motivation system inherent in human life, which could lead to more enjoyable and longer lasting human–robot interactions" (Oberman et al., 2007, p. 2195). Indeed, we can find some developments that could somehow support the realization of the minimal option. On the one hand, some implementations have used emotional expressions as indications of robot's failures (Hamacher et al., 2016; Reyes et al., 2016; Spexard et al., 2008); these emotional expressions can facilitate human interpretation and trigger helping behavior in a way similar to the types of reactions triggered by the social emotions involved in commitments mediated interactions. For example, Hamacher et al. use a BERT2, a humanoid robotic assistant, in a making-omelet task in order to test users' preferences. BERT2 was able to express sadness and apologize when dropping an egg. The studies demonstrated that subjects preferred to interact with the robot able to display such expressions than with the more efficient robot without such social capacities. On the other hand, we can find some studies that give support to the idea that some indications of motivation and commitments on behalf of the robot can boost the human feeling of obligation to remain committed to the action (Michael & Salice, 2017; Powell & Michael, 2019; Vignolo et al., 2019). In Vignolo et al.'s experiment, an iCub robot interacted with children in a teaching skills exchange. In the experiment, the subjects were exposed to two different conditions: in the high effort condition, the iCub slowed down his movements when repeating a demonstration for the human learner, whereas in the low effort condition he sped the movements up when repeating the demonstration. Then, the human had to reciprocate teaching the iCub a new skill. They found that subjects exposed to the high-effort condition were more likely to reciprocate and make more effort to teach the robot. These experiments seem to provide some partial support to the idea that exhibiting prosocial motivations, which indirectly ensure the level of commitment to a task, may facilitate the maintenance of commitments in HRI. In a nutshell, we have reasons to believe that faking emotional expressions is viable, and thus, in principle, a good way to attempt to implement a mechanism for establishing and maintaining commitments in HRI.

The minimalist option, however, presents two significant problems. The first is the problem of responsible agency. As we argued above, the role of affective states in the establishment and maintenance of commitments is twofold. First, social emotions play a role in the production of regulative behaviors that acknowledge the partner's responsibility tied to commitments while triggering the appropriate co-reaction. Second, such co-reaction is motivated by a general prosocial motivation which is manifested in the tendency to provide excusing explanations or reparations in the form of apology but more importantly in the inclination to review one's own behavior in the light of the partner's acknowledgment and reactive attitudes. Now, keeping this latter function in mind, we can see one of the problems of the minimalist option. The role of social motivation in the dynamic of reactions and co-reactions that maintain commitments is not just to trigger expressive or behavioral responses; its function is also to induce changes in the dispositional profile of the subject and thus shape responsible agency. When one receives blame or approval for violating or fulfilling a commitment, one's prosocial motivation may mobilize the appropriate change in the capacity for being properly sensitive to one's commitments, in the care taken to regulate one's actions according to these commitments and in the amount of attention paid to the relevant aspects of the situation in subsequent actions. In the case of fulfillment, the change propitiated by the need to belong can be instantiated in a feeling of reward associated with the action that can translate into a reinforcement of the appropriate dispositions and cognitive processes associated with it. In the case of violation, the change can be produced by a feeling of discomfort because the violation is causing a negative balance in our relationship with the other. Be that as it may, the function of the need to belong as a motivation for establishing and maintaining commitments is not only connected to expressions that can be mimicked but to changes in the dispositions and cognitive processes of the agent. Thus, the minimalist option does not seem to cover all the relevant functions of the affective states necessary for producing commitments.

Moreover, there is a second problem with the minimalist option. Faking emotional responses seems to produce important ethical concerns. As Brinck and Balkenius (2018) have argued, sociable robots that fake emotions exploit the emotional vulnerability of human users, which has potential harming consequences for their integrity:

> The fact that [robots] mimic human emotion and interact via bodily and facial expression of emotion encourages users to grow emotional attachments to them, whereas the robots themselves do not have feelings of the human kind, but display cue-based behavior. Users who invest themselves in the robot and become emotionally dependent on them risk being hurt, suffer depression, and develop mental and physical illnesses (Brinck & Balkenius, 2018, pp. 3–4)

In other words, social robotics construct HRI in a way that exploits the emotional profile of the users and could have serious harming implications for them. Moreover, exploiting human emotions for efficiency purposes, without considering human preferences or needs during the situation, goes against the very basis of sociality itself where the partners often negotiate in order to align their intentions and beliefs to motivate each other to remain engaged in the task.

These problems may lead us to opt for a maximalist option regarding robotic affective states. According to this option, one may attempt to design social robots with real internal states that do not just fake human emotional responses but have real powers to modify the behavior of the robots and guide its adaptation to the social or non-social environment. The key aspect of affective states is to provide the agent with the capacity for selecting behaviors depending on different parameters (Canamero, 2003). For instance, several developers have tried to implement architectures able to adapt the robot to the social or learning environment thanks to modules or devices that assign different quasi motivational internal states depending on the information they receive and that trigger different responses according to such states (Hiolle et al., 2012, 2014; Tanevska et al., 2018, 2019). In an experiment, Tanevska et al (2019) equipped an iCub robot with an adaptive architecture with a state machine that represented the robot's level of social comfort and with a social adaptative machine able to track the state of the robot and produce different reactions depending on the level of saturation. For instance, when its level of social comfort was optimal, the iCub would play with its toys and interact with the user while it would try to attract her attention when the level was non-optimal and it would disengage when getting oversaturated.

Following this idea, one may attempt to develop robots able not only to properly expressively react to social cues or violation of expectations in a way similar to what a human would do but also to regulate their dispositions and cognitive processes correspondingly. For instance, a robot could be more cautious (double-checking human cue-based behaviors) and less engaging with those users who had violated a commitment as a way to instantiate a type of quasi-disappointment state or display more prosocial behavior and engagement strategies when it has violated a commitment to repair the relationship with the user. Such a maximalist strategy would facilitate the avoidance of the problem of responsible agency. Enabling robots with the capacity for reviewing or assessing their own behavioral and cognitive capabilities in response to the reactive attitudes of the human with consequences for the rest of the joint action or future social interactions with the same or distinct users is precisely the type of learning capacities one may expect from the normative aspect of commitments. As such, the affective states will play the necessary functions associated with commitments in joint action.[3]

---

[3] Of course, emotions are very complex phenomena, and it may be argued that some ethical concerns remain as long as robots lack the subjective experiences, i.e., the phenomenology, characteristic of human emotions. However, it seems to us that the main ethical concern with robots faking emotions is that the role of these fake emotions is only to affect the

On the other hand, the maximalist option also comes with more technical problems and computational costs than the minimalist option. Implementing the capacities to detect emotions, gestures, and actions has turned out to be an especially challenging enterprise in realistic environments (Yang et al., 2018). In order to deal with such a problem, developers often consider different proxies like the mere presence of the human or her face, the distance to the robot or tactile stimuli as inputs that trigger particular emotional responses (see e.g. Breazeal, 2004). In the case of joint actions and commitment instantiation, one may opt for a similar solution to detect the relevant aspects of the situation and trigger social emotions as the appropriate reaction to violation or fulfillment of commitments. To take an example, Clodic et al. (2006) implemented a robot guide at a museum. In this experiment, they defined the task of the robot in terms of commitments and assumed that the human was fulfilling the commitment of following the robot when detecting his presence behind. In the minimalist view, one may use such a type of proxies to react to the appropriate emotional responses, for instance, looking back and smiling as a sign of approval or ask for explanations in an angry tone of voice if the user stops following the robot. However, it is difficult to see how the maximalist option may exploit such proxies when implementing co-reactivity. To successfully modify its behavior according to human social emotions, the robot must be able to distinguish very subtle human reactions like indignation, approval, disapproval, guilt, disappointment and so on. As such, the technical limitations associated with emotional recognition in social robotics is much more pressing in the case of the maximalist option.

Moreover, as we stated before, the maximalist option requires not only the capacity to detect the appropriate emotional responses but also the capacity to evaluate and learn to change one's behavior in accordance with these responses along with the capability to execute the appropriate repair strategy in every case. The conjunction of these capacities does not only multiply the problems regarding technical issues but also the computational costs generated by the necessity of processing a larger quantity of information, by the necessity of having more perceptual and behavioral modules or devices, and by the necessity of integrating all this information. Thus, the maximalist option does not only have to deal

---

behavior (and feelings) of their human partners rather than to also regulate the behavior of the robot itself towards its human partners. We are inclined to think that once robots have a capacity for such regulation, this concern becomes much less pressing (Thank to Tobias Störzinger for bringing this to our attention).

with some ethical concerns on its own, but also, with more technical problems and computational costs than the minimalist option.

Finally, an alternative to the minimalist and maximalist solution would be to replace emotions and affective states with functional substitutes, that is other types of reasoning or communicative devices which do not necessarily simulate human-like emotional responses or affective states, but can served the commitment-supporting functions served by emotions in humans; so, robots could be enabled to signal commitments, communicate their violations and fulfillment, to negotiate reparations or evaluate and select their subsequent actions by using alternative mechanisms. On the one hand, given that the pivotal function of emotional states like social emotions in establishing commitments depends largely on their expressive power, alternative expressive strategies like explicit verbal communication (Mavridis, 2015) for reacting to frustration or fulfillment of commitments or more neutral signals like lights or symbols in a screen (Baraka et al., 2016) might be used for the same purpose. On the other hand, the role of affective states in the agent's self-evaluation and in the selection of behavior could be implemented through reasoning capacities. To the extent that robots may be enabled to understand humans' reactive attitudes and commitments signals, they could process the given information to evaluate their own behavior and cognitive processes, so in principle, this alternative solution could also serve to implement the dynamic set of reactions and co-reactions associated with the maintenance of commitments.

Now, the alternative option could, in principle, avoid the first ethical concern to the extent that they can use emotionally neutral expressive strategies to establish and manage commitments, so humans would be less emotionally engaged with the robotic agent and less vulnerable to emotional exploitation. However, like in the case of the maximalist option, the second concern can only be avoided if we put the human preferences, values and integrity at the center of the reasoning capacities that modulate the robotic behavior. Now the question is could we substitute quasi-motivational states for reasoning capacities without missing an element? As we stated above, affective states inform us about how the world is in relation to our own well-being. For instance, the state of fear informs us that a particular object or feature of our environment is dangerous in the sense that it can damage our physical integrity. Moreover, these affective states are also intrinsically connected to actions and can trigger effective behavioral responses. Certainly, a reasoning architecture could infer the relevant commitments a robot should undertake given certain human responses or in what ways it should modify its behavior depending on a state it infers from the human's action. However, in the wild, autonomous robots may have to decide between different courses of action, some of which may involve decisions that have consequences for the human partner or for itself. Selecting one course of action over others is, at the end of the day, something that may involve preferences or motivations that relate to what the robot "cares about" or not. As such, it is difficult to see how we could have an autonomous robot without an architecture that regulates or modulates its behavior in order to maintain certain homeostatic levels that we may identify with preferences. In this particular case, a preference for maintaining a well-balanced relation with the human partner, and thus, a preference for behaviors that facilitate the fulfillment of joint commitments and goals.

7   Concluding Remarks

Solving the problems of motivations and predictions that social robotics encounters in joint action for HRI requires, we believe, enabling robots with the capacity for establishing and maintaining commitments. While improving the prediction of robots' behavior and boosting human motivation to interact with them necessitate establishing a mutual recognition between the partners, we have argued that such mutual recognition cannot be simply implemented through embodied strategies. The different physical and functional features of robotic agents along with the diversity of strategies one may use to identify others as social agents demand that we attribute to them different commitments depending on different physical features, social norms, or contextual parameters. In this sense, our proposal has the advantage of providing a framework to improve prediction and motivation while remaining at the right level of abstraction and being compatible with a larger set of communicative strategies to implement recognition.

Further, we have defended that the establishment and maintenance of commitments in human–human interaction depends on at least two fundamental affective states: social emotions and the pro-social motivation associated with the need to belong. In this sense, we have asked ourselves to what extent a robotic architecture could either incorporate such affective states and provide functional substitutes for them. Finally, we have proposed three options and evaluated their possible ethical and technical problems.

References

Ahn, H. S., Lee, D.-W., Choi, D., Lee, D.-Y., Hur, M., & Lee, H. (2012). Difference of efficiency in human-robot interaction according to condition of experimental environment. In S. S. Ge, O. Khatib, J.-J. Cabibihan, R. Simmons, & M.-A. Williams (Eds.), *Social robotics* (pp. 219–227). Springer.

Austin, J. (1962). *How to do things with words*. Clarendon Press.

Baraka, K., Paiva, A., & Veloso, M. (2016). Expressive lights for revealing mobile service robot state. In L. P. Reis, A. P. Moreira, P. U. Lima, L. Montano, & V. Muñoz-Martinez (Eds.), *Robot 2015: Second Iberian robotics conference* (pp. 107–119). Springer.

Baumeister, R. F., & Leary, M. R. (1995). The need to belong: Desire for interpersonal attachments as a fundamental human motivation. *Psychological Bulletin, 117*(3), 497–529.

Benamara, N. K., Val-Calvo, M., Álvarez-Sánchez, J. R., Díaz-Morcillo, A., Ferrández Vicente, J. M., Fernández-Jover, E., & Stambouli, T. B. (2019). Real-time emotional recognition for sociable robotics based on deep neural networks ensemble. In J. M. Ferrández Vicente, J. R. Álvarez-Sánchez, F. de la Paz López, J. Toledo Moreo, & H. Adeli (eds.), *Understanding the brain function and emotions* (pp. 171–180). Springer.

Brandom, R. B. (2007). The structure of desire and recognition: Self-consciousness and self-constitution. *Philosophy & Social Criticism, 33*(1), 127–150.

Bratman, M. E. (2014). *Shared agency*. Oxford University Press.

Breazeal, C. (2003). Emotion and sociable humanoid robots. *International Journal of Human-Computer Studies, 59*(1–2), 119–155.

Breazeal, C. (2004). *Designing sociable robots*. MIT press.

Breazeal, C., Kidd, C. D., Thomaz, A. L., Hoffman, G., & Berlin, M. (2005). Effects of nonverbal communication on efficiency and robustness in human-robot teamwork. In *IEEE/RSJ international conference on intelligent robots and systems* (pp. 708–713).

Breazeal, C., Wang, A., & Picard, R. (2007). Experiments with a robotic computer: Body, affect and cognition interactions. In *Proceeding of the ACM/IEEE international conference on human-robot interaction* (pp. 153–160).

Brinck, I., & Balkenius, C. (2018). Mutual recognition in human-robot interaction: A deflationary account. *Philosophy & Technology, 33*(1), 53–70.

Canamero, D. (2003). Designing emotions for activity selection. In R. Trappl, P. Petta, & S. Payr (Eds.), *Emotions in humans and artefacts* (pp. 115–148). The MIT Press.

Chartrand, T. L., & Bargh, J. A. (1999). The chameleon effect: The perception-behavior link and social interaction. *Journal of Personality and Social Psychology, 76*(6), 893–910.

Clark, H. H. (2006). Social actions, social commitments. In N. J. Enfield & S. C. Levinson (Eds.), *Roots of human sociality: Culture, cognition, and interaction* (pp. 126–150). Berg.

Clodic, A., Fleury, S., Alami, R., Chatila, R., Bailly, G., Brethes, L., Cottret, M., Danes, P., Dollat, X., Elisei, F., Ferrane, I., Herrb, M., Infantes, G., Lemaire, C., Lerasle, F., Manhes, J., Marcoul, P., Menezes, P., & Montreuil, V. (2006). Rackham: An interactive robot-guide. In *The 15th IEEE international symposium on robot and human interactive communication* (pp. 502–509).

Craig, R., Vaidyanathan, R., James, C., & Melhuish, C. (2010). Assessment of human response to robot facial expressions through visual evoked potentials. In *10th IEEE-RAS International Conference on Humanoid Robots* (pp. 647–652).

Davidson, D. (1991). Three varieties of knowledge. *Royal Institute of Philosophy Supplements, 30*, 153–166.

Decety, J., & Cowell, J. M. (2018). Interpersonal harm aversion as a necessary foundation for morality: A developmental neuroscience perspective. *Development and Psychopathology, 30*(1), 153–164.

Dennett, D. (2009). Intentional systems theory. In A. Beckermann, B. P. McLaughlin, & S. Walter (Eds.), *The Oxford handbook of philosophy of mind* (pp. 339–350). Oxford University Press.

Depue, R. A., & Morrone-Strupinsky, J. V. (2005). A neurobehavioral model of affiliative bonding: Implications for conceptualizing a human trait of affiliation. *Behavioral and Brain Sciences, 28*(3), 313–350.

Dreber, A., Rand, D. G., Fudenberg, D., & Nowak, M. A. (2008). Winners don't punish. *Nature, 452*(7185), 348–351.

Elsner, B., & Hommel, B. (2001). Effect anticipation and action control. *Journal of Experimental Psychology: Human Perception and Performance, 27*(1), 229–240.

Fehr, E., & Gächter, S. (2002). Altruistic punishment in humans. *Nature, 415*(6868), 137–140.

Fernández Castro, V. (2020). Regulation, normativity and folk psychology. *Topoi, 39*(1), 57–67.

Fernández Castro, V., & Heras-Escribano, M. (2020). Social cognition: A normative approach. *Acta Analytica, 35*(1), 75–100.

Fernández Castro, V., & Pacherie, E. (2020). Joint actions, commitments and the need to belong. *Synthese*. https://doi.org/10.1007/s11229-020-02535-0

Fricker, M. (2016). What's the point of blame? A paradigm based explanation. *Noûs, 50*(1), 165–183.

Gilbert, M. (1992). *On social facts*. Princeton University Press.

Gilbert, M. (2009). Shared intention and personal intentions. *Philosophical Studies, 144*(1), 167–187.

Godman, M. (2013). Why we do things together: The social motivation for joint action. *Philosophical Psychology, 26*(4), 588–603.

Godman, M., Nagatsu, M., & Salmela, M. (2014). The social motivation hypothesis for prosocial behavior. *Philosophy of the Social Sciences, 44*(5), 563–587.

Gomez-Lavin, J., & Rachar, M. (2019). Normativity in joint action. *Mind & Language, 34*(1), 97–120.

Greenspan, P. S. (1978). Behavior control and freedom of action. *The Philosophical Review, 87*(2), 225–240.

Guzmán, R. A., Rodrıguez-Sickert, C., & Rowthorn, R. (2007). When in Rome, do as the Romans do: The coevolution of altruistic punishment, conformist learning, and cooperation. *Evolution and Human Behavior, 28*, 112–117.

Ham, J., Cuijpers, R. H., & Cabibihan, J.-J. (2015). Combining robotic persuasive strategies: The persuasive power of a storytelling robot that uses gazing and gestures. *International Journal of Social Robotics, 7*(4), 479–487.

Hamacher, A., Bianchi-Berthouze, N., Pipe, A. G., & Eder, K. (2016). Believing in BERT: Using expressive communication to enhance trust and counteract operational error in physical Human-robot interaction. In *25th IEEE international symposium on robot and human interactive communication (RO-MAN)* (pp. 493–500).

Hareli, S., & Parkinson, B. (2008). What's social about social emotions? *Journal for the Theory of Social Behaviour, 38*(2), 131–156.

Harrison, S. J., & Richardson, M. J. (2009). Horsing around: Spontaneous four-legged coordination. *Journal of Motor Behavior, 41*(6), 519–524.

Henrich, N., & Henrich, J. P. (2007). *Why humans cooperate: A cultural and evolutionary explanation*. Oxford University Press.

Hiolle, A., Cañamero, L., Davila-Ross, M., & Bard, K. A. (2012). Eliciting caregiving behavior in dyadic human-robot attachment-like interactions. *ACM Transactions on Interactive Intelligent Systems, 2*(1), 1–24.

Hiolle, A., Lewis, M., & Cañamero, L. (2014). Arousal regulation and affective adaptation to human responsiveness by a robot that explores and learns a novel environment. *Frontiers in Neurorobotics, 8*, 17.

Huang, C. M. & Thomaz, A. L. (2011). Effects of responding to, initiating and ensuring joint attention in human-robot interaction. In *2011 Ro-Man* (pp. 65–71).

Huang, C.-M., & Thomaz, A. L. (2010). Joint attention in human-robot interaction. In *AAAI fall symposium: Dialog with robots* (pp. 32–37).

Kedzierski, J., Muszynski, R., Zoll, C., Oleksy, A., & Frontkiewicz, M. (2013). EMYS—Emotive head of a social robot. *International Journal of Social Robotics, 5*(2), 237–249.

Kishi, T., Kojima, T., Endo, N., Destephe, M., Otani, T., Jamone, L., Kryczka, P., Trovato, G., Hashimoto, K., Cosentino, S., & Takanishi, A. (2013). Impression survey of the emotion expression humanoid robot with mental model based dynamic emotions. In *IEEE international conference on robotics and automation* (pp. 1663–1668).

Knoblich, G., Butterfill, S., & Sebanz, N. (2011). Psychological research on joint action. In B. Ross (ed.), *Psychology of learning and motivation* (Vol. 54, pp. 59–101). Elsevier.

Kwon, M., Jung, M., & Knepper, R. (2016). Human expectations of social robots. In *11th ACM/IEEE international conference on human-robot interaction* (pp. 463–464).

Laitinen, A. (2016). Robots and human sociality: Normative expectations, the need for recognition, and the social bases of self-esteem. In J. Seibt, M. Nørskov, & S. S. Andersen (eds.), *What social robots can and should do: Proceedings of robophilosophy 2016/TRANSOR 2016* (pp. 313–322). IOS Press.

Lewis, D. K. (1969). *Convention: A philosophical study*. Harvard University Press.

Lewis, M., Sycara, K., & Walker, P. (2018). The role of trust in human-robot interaction. In H. A. Abbass, J. Scholz, & D. J. Reid (Eds.), *Foundations of trusted autonomy* (pp. 135–159). Springer.

Lichtenthäler, C., & Kirsch, A. (2013). Towards legible robot navigation—How to increase the intend expressiveness of robot navigation behavior. In *International conference on social robotics—Workshop embodied communication of goals and intentions*. Retrieved from https://hal.archives-ouvertes.fr/hal-01684307

Lo Presti, P. (2013). Situating norms and jointness of social interaction. *Cosmos and History: THe Journal of Natural and Social Philosophy, 9*(1), 225–248.

Macnamara, C. (2013). "Screw you!" & "thank you." *Philosophical Studies, 165*(3), 893–914.

Maibom, H. L. (2007). Social systems. *Philosophical Psychology, 20*(5), 557–578.

Mavridis, N. (2015). A review of verbal and non-verbal human–robot interactive communication. *Robotics and Autonomous Systems, 63*, 22–35.

McGeer, V. (2012). Co-reactive attitudes and the making of moral community. In C. MacKenzie & R. Langdon (Eds.), *Emotions, imagination and moral reasoning* (pp. 299–326). Psychology Press.

McGeer, V. (2015). Mind-making practices: The social infrastructure of self-knowing agency and responsibility. *Philosophical Explorations, 18*(2), 259–281.

Michael, J. (2011). Shared emotions and joint action. *Review of Philosophy and Psychology, 2*(2), 355–373.

Michael, J. (2022). *The Philosophy and Psychology of Commitment*. Routledge

Michael, J., & Pacherie, E. (2015). On commitments and other uncertainty reduction tools in joint action. *Journal of Social Ontology, 1*(1), 89–120.

Michael, J., & Salice, A. (2017). The sense of commitment in human-robot interaction. *International Journal of Social Robotics, 9*(5), 755–763.

Michael, J., & Székely, M. (2018). The developmental origins of commitment. *Journal of Social Philosophy, 49*(1), 106–123.

Michael, J., Sebanz, N., & Knoblich, G. (2016). The sense of commitment: A minimal approach. *Frontiers in Psychology, 6*, 1968.

Mori, M. (1970). The uncanny valley. *Energy, 7*(4), 33–35.

Normoyle, A., Badler, J. B., Fan, T., Badler, N. I., Cassol, V. J., & Musse, S. R. (2013). Evaluating perceived trust from procedurally animated gaze. In *MIG '13: Proceedings of motion on games* (pp. 141–148).

Oberman, L. M., McCleery, J. P., Ramachandran, V. S., & Pineda, J. A. (2007). EEG evidence for mirror neuron activity during the observation of human and robot actions: Toward an analysis of the human qualities of interactive robots. *Neurocomputing, 70*(13), 2194–2203.

Ostrom, E., Walker, J., & Gardner, R. (1992). Covenants with and without a sword: Self-governance is possible. *The American Political Science Review, 86*(2), 404–417.

Over, H. (2016). The origins of belonging: Social motivation in infants and young children. *Philosophical Transactions of the Royal Society B: Biological Sciences, 371*, 1686.

Pacherie, E. (2011). Framing joint action. *Review of Philosophy and Psychology, 2*(2), 173–192.

Pacherie, E. (2013). Intentional joint agency: Shared intention lite. *Synthese, 190*(10), 1817–1839.

Pandey, A. K., & Alami, R. (2010). A framework towards a socially aware mobile robot motion in human-centered dynamic environment. In *IEEE/RSJ international conference on intelligent robots and systems* (pp. 5855–5860).

Paprzycka, K. (1998). Normative expectations, intentions, and beliefs. *Southern Journal of Philosophy, 37*(4), 629–652.

Powell, H., & Michael, J. (2019). Feeling committed to a robot: Why, what, when and how? *Philosophical Transactions of the Royal Society B: Biological Sciences, 374*, 1771.

Prinz, J. J. (2004). *Gut reactions: A perceptual theory of the emotions*. Oxford University Press.

Prinz, W. (1997). Perception and action planning. *European Journal of Cognitive Psychology, 9*(2), 129–154.

Quillien, T. (2020). Evolution of conditional and unconditional commitment. *Journal of Theoretical Biology, 492*, 110204.

Ramberg, B. (2000). Post-ontological philosophy of mind: Rorty versus Davidson. In R. Brandom (Ed.), *Rorty and his critics* (pp. 351–369). Blackwell Publishers.

Ramenzoni, V. C., Riley, M. A., Shockley, K., & Davis, T. (2008). Carrying the height of the world on your ankles: Encumbering observers reduces estimates of how high an actor can jump. *Quarterly Journal of Experimental Psychology, 61*(10), 1487–1495.

Reyes, M., Meza, I., & Pineda, L. A. (2016). The positive effect of negative feedback in HRI using a facial expression robot. In J. T. K. V. Koh, B. J. Dunstan, D. Silvera-Tawil, & M. Velonaki (Eds.), *Cultural robotics* (pp. 44–54). Springer International Publishing.

Riek, L. D., Rabinowitch, T.-C., Bremner, P., Pipe, A. G., Fraser, M., & Robinson, P. (2010). Cooperative gestures: Effective signaling for humanoid robots. In *5th ACM/IEEE International Conference on Human-Robot Interaction (HRI)* (pp. 61–68).

Roth, A. S. (2004). Shared agency and contralateral commitments. *The Philosophical Review, 113*(3), 359–410.

Sahaï, A., Pacherie, E., Grynszpan, O., & Berberian, B. (2017). Predictive mechanisms are not involved the same way during human-human vs. human-machine interactions: A review. *Frontiers in Neurorobotics, 11*, 52.

Sanders, T. L., Schafer, K. E., Volante, W., Reardon, A., & Hancock, P. A. (2016). Implicit attitudes toward robots. *Proceedings of the Human Factors and Ergonomics Society Annual Meeting, 60*(1), 1746–1749.

Satne, G. (2014). What binds us together: Normativity and the second person. *Philosophical Topics, 42*(1), 43–61.

Scherer, K. R. (2005). What are emotions? And how can they be measured? *Social Science Information, 44*(4), 695–729.

Sciutti, A., Mara, M., Tagliasco, V., & Sandini, G. (2018). Humanizing human-robot interaction: On the importance of mutual understanding. *IEEE Technology and Society Magazine, 37*(1), 22–29.

Sebanz, N., Bekkering, H., & Knoblich, G. (2006). Joint action: Bodies and minds moving together. *Trends in Cognitive Sciences, 10*(2), 70–76.

Seibt, J. (2018). Classifying forms and modes of co-working in the ontology of asymmetric social interactions (OASIS). In M. Coeckelbergh, J. Loh, M. Funk, J. Seibt, & M. Nøskov (eds.), *Envisioning robots in society—Power, politics, and public space: proceedings of Robophilosophy 2018* (pp. 133–146). IOS Press Ebooks.

Siposova, B., Tomasello, M., & Carpenter, M. (2018). Communicative eye contact signals a commitment to cooperate for young children. *Cognition, 179*, 192–201.

Sisbot, E. A., Marin-Urias, L. F., Broquère, X., Sidobre, D., & Alami, R. (2010). Synthesizing robot motions adapted to human presence. *International Journal of Social Robotics, 2*(3), 329–343.

Spexard, T. P., Hanheide, M., Li, S., & Wrede, B. (2008). Oops, something is wrong-error detection and recovery for advanced human-robot-interaction. In *Proceedings of the workshop on social interaction with intelligent indoor robots at the international conference on robotics and automation*. Retrieved from https://pub.uni-bielefeld.de/record/1997436

Stanton, C., & Stevens, C. J. (2014). Robot pressure: The impact of robot eye gaze and lifelike bodily movements upon decision-making and trust. In M. Beetz, B. Johnston, & M.-A. Williams (Eds.), *Social Robotics* (pp. 330–339). Springer International Publishing.

Steckenfinger, S. A., & Ghazanfar, A. A. (2009). Monkey visual behavior falls into the Uncanny Valley. *Proceedings of the National Academy of Sciences, 106*(43), 18362–18366.

Sugden, R. (2000). The motivating power of expectations. In J. Nida-Rümelin & W. Spohn (Eds.), *Rationality, rules, and structure* (pp. 103–129). Springer.

Tai, Y. F., Scherfler, C., Brooks, D. J., Sawamoto, N., & Castiello, U. (2004). The human premotor cortex is 'mirror' only for biological actions. *Current Biology, 14*(2), 117–120.

Tanevska, A., Rea, F., Sandini, G., & Sciutti, A. (2018). Designing an affective cognitive architecture for human-humanoid interaction. *Companion of the 2018 ACM/IEEE international conference on human-robot interaction* (pp. 253–254).

Tanevska, A., Rea, F., Sandini, G., Canamero, L., & Sciutti, A. (2019). A Cognitive Architecture for Socially Adaptable Robots. In *Joint IEEE 9th international conference on development and learning and epigenetic robotics (ICDL-EpiRob)* (pp. 195–200).

Török, G., Pomiechowska, B., Csibra, G., & Sebanz, N. (2019). Rationality in joint action: Maximizing coefficiency in coordination. *Psychological Science, 30*(6), 930–941.

Vaish, A. (2018). The prosocial functions of early social emotions: The case of guilt. *Current Opinion in Psychology, 20*, 25–29.

Vandemeulebroucke, T., de Casterlé, B. D., & Gastmans, C. (2018). How do older adults experience and perceive socially assistive robots in aged care: A systematic review of qualitative evidence. *Aging & Mental Health, 22*(2), 149–167.

Vesper, C., & Richardson, M. J. (2014). Strategic communication and behavioral coupling in asymmetric joint action. *Experimental Brain Research, 232*(9), 2945–2956.

Vesper, C., Abramova, E., Bütepage, J., Ciardo, F., Crossey, B., Effenberg, A., Hristova, D., Karlinsky, A., McEllin, L., Nijssen, S. R. R., Schmitz, L., & Wahn, B. (2017). Joint action: Mental representations, shared information and general mechanisms for coordinating with others. *Frontiers in Psychology, 7*, 2039.

Vignolo, A., Sciutti, A., Rea, F., & Michael, J. (2019). Spatiotemporal coordination supports a sense of commitment in human-robot interaction. In M. A. Salichs, S. S. Ge, E. I. Barakova, J. Cabibihan, A. Wagner, Á. Castro-González, & H. He (Eds.), *Social robotics* (pp. 34–43). Spinger.

Wallace, R. J. (1994). *Responsibility and the moral sentiments*. Harvard University Press.

Wang, S., Lilienfeld, S. O., & Rochat, P. (2015). The Uncanny Valley: Existence and explanations. *Review of General Psychology, 19*(4), 393–407.

Wendt, C., Popp, M., Karg, M., & Kuhnlenz, K. (2008). Physiology and HRI: Recognition of over- and underchallenge. In *The 17th IEEE international symposium on robot and human interactive communication* (pp. 448–452).

Wiese, E., Metta, G., & Wykowska, A. (2017). Robots as intentional agents: Using neuroscientific methods to make robots appear more social. *Frontiers in Psychology, 8*, 1663.

Yang, G.-Z., Bellingham, J., Dupont, P. E., Fischer, P., Floridi, L., Full, R., Jacobstein, N., Kumar, V., McNutt, M., Merrifield, R., Nelson, B. J., Scassellati, B., Taddeo, M., Taylor, R., Veloso, M., Wang, Z. L., & Wood, R. (2018). The grand challenges of *science robotics*. *Science Robotics, 3*(14), eaar7650.

Zawidzki, T. (2013). *Mindshaping: A new framework for understanding human social cognition*. MIT Press.

Víctor Fernández Castro is a Juan de la Cierva Postdoc Research Fellow at University of Granada. His main areas of interest are the theoretical philosophy of mind and psychology and their applications in areas like social robotics, mental disorders or social philosophy. Elisabeth Pacherie is a CNRS Research director at Institut/e Jean Nicod, PSL University, Paris. She works primarily in the philosophy of mind and action and cognitive science, with special interest in joint action and consciousness of action.

Social Robot Personality: A Review and Research Agenda

Sarah Diefenbach, Marietta Herzog, Daniel Ullrich and Lara Christoforakos

Abstract

Social robots made to interact with humans have increasingly entered various domains in recent years. For example, they support patients in elderly care, treat children with autism or just serve as adorable companions in everyday life. With the objective of highly naturalistic interactions with humans, there has been a vast amount of research trying to identify *suitable* personalities for social robots. However, the theoretical foundations of the construct *social robot personality*, its expression, its measurement, and if and how it influences outcomes of human–robot interaction (HRI) are areas of continued debate. To shed light on this debate, we provide a narrative review of interdisciplinary literature on social robot personality. In particular, we describe common themes and limitations within four areas: Conceptualization, expression, measurement, and consequences of perceived social robot personality. Building

---

S. Diefenbach (✉) · M. Herzog · L. Christoforakos
Department of Psychology, Ludwig-Maximilians-Universität München, München, Germany
e-mail: sarah.diefenbach@lmu.de

M. Herzog
e-mail: mariettaherzog@gmx.de

L. Christoforakos
e-mail: Lara.Christoforakos@psy.lmu.de

D. Ullrich
Media Informatics Group, Ludwig-Maximilians-Universität München, München, Germany
e-mail: daniel.ullrich@ifi.lmu.de

upon research gaps identified, we further provide a research agenda aiming to encourage scholars to target opportunities for future research. More specifically, we highlight the need for further research in eight areas: (1) Personality theories beyond the Big Five, (2) individual differences, (3) the interaction of human and robot personality, (4) situational context and robotic social roles, (5) anthropomorphism, (6) cultural differences, (7) long-term interactions, and finally (8) the exploration of an integrative design approach towards social robot personality. Combining the limitations and research opportunities identified this review serves as a roadmap for scholars to progress the design of adequate human–robot interactions.

Keywords

Social robots · Robot personality · Human–robot interaction (HRI) · Psychological personality theories · Robot design

1 Introduction

In recent years, the domain of social robots has activated a vast amount of research and social robots have entered ever new fields of applications. According to Bartneck and Forlizzi (2004), a social robot is an autonomous or semi-autonomous robot that interacts and communicates with humans by following the behavioral norms expected by the people with whom the robot is intended to interact with. Social robots can be seen as a subcategory of assistive robots and non-industrial robots, with the purpose of physical or social assistance or companionship (Heerink, 2010). The most prominent applications include, for example, social robots as companions in elderly care (Paro Robots, 2014) or as a tool in autism therapy for children (e.g., Cabibihan et al., 2013). Social robots may therefore not only assist us in everyday life, but may also contribute to solving global issues, such as the shortage of medical and nursing staff for an increasing elderly population (e.g., Brandon, 2011) or the gap between demand and supply for psychological therapy (Bower & Gilbody, 2004). But despite technological advance in artificial intelligence, the uncertainty with regards to appropriate design of human–robot-interaction (HRI) in social contexts slows the full integration of social robots into our daily lives (Bartneck & Forlizzi, 2004). Several studies have argued that personality of both, the human and the robot, is a vital factor to better understand and subsequently facilitate human–robot interaction (e.g., Gockley & Mataric, 2006; Goetz & Kiesler, 2002; Sydral et al., 2007; Robert, 2018). However, the definition of social robot personality is a research area of continued debate.

The present article focuses on the topic of robot personality and summarizes the current state and limitations of research. Based on a review of the literature in HRI and relevant related fields such as human–computer interaction (HCI) and psychology, we report on four main areas: (1) Conceptualization, that is the theoretical concepts used to conceptualize social robot personality, (2) expression, that is appearance and behavior-related cues used to express personality, (3) measurement, that is the methods used to assess, whether a robot's personality is recognized, and (4) consequences, that is the resulting perceptions and behaviors of humans who interacted with different social robot personalities. Across these four areas, we aim to uncover common themes and trends, as well as research gaps and steps for future research. While our review cannot answer which personality traits appear as suitable for social robots in general, the present overview may provide a basis to support design decisions and to identify relevant concepts, context factors, and design opportunities for a specific application domain.

2 Methodology

Based on a list of relevant keywords (e.g., social robot personality, trait, attitude), we performed an extensive literature search across several data bases (e.g., PsycINFO, ScienceDirect, ResearchGate, Springer, ACM, IEEE, SAGE) and journals from various disciplines, such as HCI (e.g., Computers in Human Behavior, International Journal on Social Robotics, International Journal of Human–Computer Studies), communication science (e.g., Journal of Communication), psychology (e.g., Journal of Applied Social Psychology), medicine (e.g., Gerontologie; Eng.: Gerontology), or computer science and engineering (e.g., Autonomous Robots, Mathematical Problems in Engineering), including articles since January 1995. 353 related articles, books, and conference proceedings appeared to be generally suitable for the topic and were selected for further screening. Among these articles, we picked those which fulfilled two criteria: (1) They had to report conceptual ideas or findings related to (perceived) robot personality or a personality-related entity (e.g., affect, emotion, behavior) and (2) they had to explore interactions with social robots or virtual agents. 118 articles fulfilled these criteria and were selected for further, in-depth analysis.

For the final literature selection for in-depth analysis, articles had to fulfill at least one of the following four criteria: They had to entail some sort of (1) robot personality conceptualization, (2) expression, (3) measurement, or (4) consequence (i.e., affected user perception and/or behavior). This led to a selection of 68 studies to be included in the present review. The studies' publication dates ranged from year 1999 (oldest) to year 2019 (most recent) and studies were from various sources, with several studies originating from conference proceedings (e.g., Robot and Human Interactive Communication; RO-MAN).

The in-depth literature analysis of the 68 selected articles included a screening of the theoretical background, research questions, study design, dependent and independent variables, study material (e.g., robots used), main results, limitations, and lastly general characteristics, such as country of origin, authors, affiliated University and journal. Afterwards, all findings relevant to one or more areas of conceptualization, expression, measurement, or consequences were summarized for each article. Lastly, findings were mapped across studies, synthesized, and manually analyzed for common themes and trends. In the following sections, we first provide a short overview of the research landscape and then summarize findings and research gaps for the four areas. We then discuss future research directions across the four areas. Due to space limitations, the present list of references includes only those articles which are cited in the text, but not the full list of 68 reviewed articles. A more comprehensive version of the literature list can be found in Christoforakos et al. (2019).

3 Research Landscape of Social Robot Personality

Of the 68 studies, the majority (74%) was empirical and 26% were conceptual. On average, the empirical studies had a sample size of $N = 58$. As already noted above, our review focused on four different, but interrelated areas: (1) Conceptualization, that is the theoretical concepts used to conceptualize social robot personality, (2) expression, that is appearance and behavior-related cues used to express personality, (3) measurement, that is the methods used to assess, whether a robot's personality is recognized, and (4) consequences, that is the resulting perceptions and behaviors of humans who interacted with different social robot personalities. All 68 studies incorporated the conceptualization of robot personality, 56 studies (82%) the expression, 35 studies (51%) personality measurement, and 29 studies (43%) explored described consequences of perceived personality among users.

4 Conceptualization of Robot Personality

Conceptualization of robot personality encompasses models, theories, or other sources by which researchers inform their definition of robot personality. In the literature reviewed, personality concepts are informed by a wide-ranging portrait of sources from personality psychology, describing the personality of humans, over methods from human–computer interaction (HCI) to arts and media. One of the most commonly used models from personality psychology is the Big Five model (Mccrae et al., 1992). The Big Five model assumes that human personality can be described along five dimensions, namely, Extraversion (e.g., active, assertive, energetic, outgoing, talkative), Agreeableness (e.g., appreciative, kind, generous, forgiving, sympathetic, trusting), Neuroticism (anxious, self-pitying, tense, touchy, unstable, worrying), Openness to Experience (e.g., artistic, curious, imaginative, insightful, original, wide interests), and Conscientiousness (e.g., efficient, organized, planful, reliable, responsible, thorough) (Vinciarelli & Mohammadi, 2014). Out of all 68 studies considered in this review, 53% informed their personality concept by the Big Five (e.g., Aly & Tapus, 2016; Andrist et al., 2015; Brandon, 2011; Celiktutan & Gunes, 2015), specifically selected dimensions of the Big Five (e.g., Clavel, Faur, Martin, Pesty, & Duhaut, 2013; Konstantopoulos, Karkaletsis, & Matheson, 2009; Meerbeek et al., 2008) or variations and adaptions of the Big Five (e.g., Ball & Breese, 2000; Chee et al., 2012; Walters, Syrdal, Dautenhahn, Te Boekhorst, & Koay, 2008). Other theories of personality are rarely found. That is, only seven studies (10%) describe robot personality based on other theories such as, for example, interpersonal theory (Kiesler, 1983). Compared to the Big Five personality factors, interpersonal theories only focus on dispositions relevant for interpersonal interactions (McCrae & Costa, 1989). They specifically focus on dimensions that are experienced when interacting with other humans, such as friendliness or dominance, rather than addressing traits that might not affect interpersonal interactions (e.g., Openness to Experiences). For example, Okuno et al. (2003) conceptualized the personality of a roboceptionist (i.e., a robotic receptionist) based on a circumplex interpersonal model with the two dimensions of Dominance/Submissiveness and Friendliness/Hostility in order to control the robot's attention towards hotel guests. As a result, the submissive roboceptionist was more easily distracted by a new customer while in conversation with another customer, than a dominant roboceptionist.

Additionally, with regards to theories of personality psychology, we also found rare examples (three studies, 6%) where theories from cognitive psychology were applied to conceptualize social robot personality. One example is the application of regulatory focus theory (Crowe & Higgins, 1997). Based on the hedonic principle to avoid pain and approach pleasure, the theory of regulatory focus suggests that people's behavior and motivation are affected by two different foci: promotion focus and prevention focus. According to the theory, people with promotion-focus strive for advancement, growth, and accomplishment, whereas people with prevention focus are motivated by security, safety and responsibility. In the reviewed HRI literature, one study evaluated whether regulatory focus can be recognized as part of the robot's personality (Faur et al., 2015) and another explored, how matching personality in terms of regulatory focus of robot and user affects user motivation for and performance in a specific task (Cruz-Maya et al., 2017). Indeed, it was found that the description of a virtual agent's game strategy for an online game successfully conveyed the agent's personality in terms of regulatory focus (Faur et al., 2015). Furthermore, Cruz-Maya et al. (2017) found that matching the user's regulatory focus with the regulatory focus activated through a robot's instructions led to better performance in a cognitive task, less cognitive load, and less physical stress (in comparison to mismatching regulatory focus). Finally, other studies conceptualized personality based on observable appearance or behavior (Fong et al., 2003; Niculescu et al., 2013; Woods, 2006; Woods, Dautenhahn, & Schulz, 2004), social roles (Konstantopoulos, 2013), biological theories (Ogata & Sugano, 1999; Sutcliffe, Pineau, & Grollman, 2014), technology-driven approaches (Liu et al., 2017, 2018), arts and media, such as borrowing design principles from animation and character creation in movies (e.g., Meerbeek, Saerbeck, & Bartneck, 2009), or descriptive approaches, such as playful vs. serious personalities (Fussell et al., 2008; Goetz et al., 2003; Sundar et al., 2017; Ullrich, 2017).

However, while the vast majority of researchers agree on defining robot personality by applying the Big Five personality traits from humans to robots, there are also large variations in the understanding and application of the Big Five. For example, many studies use just one dimension, namely Extraversion (Aly & Tapus, 2016; Andrist et al., 2015; Brandon, 2011; Celiktutan & Gunes, 2015; Chee et al., 2012; Holtgraves et al., 2007; Joosse et al., 2013; Lee et al., 2006; Mayer & Panek, 2016; Panek et al., 2015; E. Park et al., 2012; S. Park et al., 2010a, 2010b; Walters et al., 2014). Moreover, on closer inspection of the studies, it also became apparent that though claiming to rely on the Big Five, many studies define the individual traits differently. For example, one study defined Extraversion in a cooperative robot for game play as behaving in a challenging and fast-moving manner, whereas an introverted robot was behaving in an encouraging and comforting manner by making lots of positive comments (Belpaeme et al., 2013). In another example, Martínez-Miranda et al. (2018) understand the dimension of Agreeableness rather as a bipolar dimension of friendly versus unfriendly. Another totally different approach in a review by Fong et al. (2003) suggests five robot personality types associated with a robot's appearance (i.e., tool-like, pet or creature, cartoon, artificial being, and human-like personality). In conclusion, there is no common understanding of what robot personality entails and even if there is agreement on a common concept, such as the Big Five personality traits, there is no consensus about the adoption of that concept and its implications for the robot's behavior. In addition to a lacking consensus regarding the definition of robot personality, most researchers understand the concept of personality as specific attributes of behavior or appearance. In particular, many studies define, for example, Extraversion by using behavioral or appearance-related attributes, such as speaking loud and fast or being dressed in red. Although there are several studies confirming such relationships between observable attributes and personality perception for humans (e.g.,Borkenau & Liebler, 1992; Uleman et al., 2008), it is questionable whether these are adequate or sufficient for the design of credible social robot personalities.

5 Expression and Implementation of Robot Personality

Besides the theoretical conceptualization of personality as described in the preceding paragraphs, there are also different approaches to express robot personality and to communicate the robot's intended personality to the users. In the reviewed HRI-literature, the expression of personality can be clustered into three different interpersonal cues: Verbal (e.g., language and verbal content), para-verbal (e.g., speed of speech and tonality), and non-verbal behavior (e.g., posture and gestures). In total, 56 studies (82%) used one or more interpersonal cues to express personality. While most studies relied on one behavioral cue (i.e., 35 studies, 62%), fewer studies followed a multi-modal approach (i.e., eleven studies with two forms of cues and ten studies with all three forms). Thereby, non-verbal behavior (including appearance and explicit actions) was with 47 studies (83%) the most commonly altered cue, followed by verbal behavior (20 studies, 36%) and para-verbal behavior (19 studies, 34%).

Within the domain of non-verbal behavior, facial expressions (26% of studies, e.g., Brandon, 2011; Han, Lin, & Song, 2013; Hu, Xie, Liu, & Wang, 2013), gestures (30% of studies, e.g., Aly & Tapus, 2016; Celiktutan & Gunes, 2015; Craenen, Deshmukh, Foster, & Vinciarelli, 2018a; S. Park et al., 2010a, 2010b) and motion speed/path (17% of studies, e.g.,Joosse et al., 2013; Lee et al., 2006; Walters et al., 2014) were most frequently chosen.

Within the domain of verbal behavior, most of the studies (55%) were informed by concepts from personality psychology and (in parallel to the conceptualization of robot personality) specifically the Big Five. Mayer and Panek (2016), for example, expressed Introversion in terms of shorter sentences. Other examples also used humor in form of jokes (Niculescu et al., 2013), powerful words (Bian et al., 2016), or distinct melodies as verbal output (Lee et al., 2006) to vary the robot's personality in terms of extraversion and introversion.

Within the domain of para-verbal behavior, most studies used speech rate (e.g., Celiktutan & Gunes, 2015; Lee et al., 2006; Mileounis et al., 2015) and voice pitch (e.g., Moshkina & Arkin, 2003; Niculescu et al., 2013; Tay et al., 2014) to convey personality, followed by volume (e.g., Joosse et al., 2013; Lee et al., 2006; Mayer & Panek, 2016; Meerbeek & Saerbeck, 2010; Panek et al., 2015), speech melody (e.g., Ball & Breese, 2000; Brandon, 2011; Celiktutan & Gunes, 2015; Lee et al., 2006), and word count (e.g.,Mayer & Panek, 2016; Walters et al., 2014). Similar to verbal behavior, most of the studies using para-verbal behavior as means to express personality informed their personality concept by psychological theories.

Altogether, the closer analysis of approaches and applied cues for the expression of robot personality revealed large differences, leading to limitations in terms of generalizability and comparability across studies. On one side, researchers use various different verbal-, non-verbal-, para-verbal-, appearance or action-related cues to express the same personality characteristics, ranging from verbal content (e.g., Holtgraves, 2007), over the robot's voice pitch (e.g., Lee et al., 2006) to specific actions, such as following an agenda (e.g., Konstantopoulos, 2013). Further, some researchers use specific cues to express a personality trait (e.g., red and white color to express extraversion and introversion, respectively, e.g., Mayer & Panek, 2016), whereas others use the exact same cue for other purposes, such as distinguishing robots within an experiment (e.g., green and yellow color to distinguish robots, e.g., Meerbeek et al., 2008). Thus, it is unclear, whether different cues used by researchers might also have interacted with cues on other modalities, the robot's appearance, or other situational factors. Hence, a generalization of the approaches used in single studies, such as extraversion always being expressed by high voice pitch, might thus be problematic.

6 Measurement of Perceived Robot Personality

In total, 35 studies included the measurement of the perceived robot personality on the user's side. The used measures of perceived robot personality can be clustered as drawn from two different sources: existing psychological measures and self-constructed measures.

Twenty-six (74%) of the 35 studies include psychological measures or adaptions of such to assess the users' recognition of robot personality. The applied measures range from renowned, commonly used psychological inventories (e.g., BFI-10; Goldberg, 1992) to recently developed questionnaires (e.g., Godspeed Scores; Bartneck et al., 2009). Most of the studies thereby rely on users' self-reported perceptions of a robot's personality traits, most commonly adapted from the Big Five (Mccrae et al., 1992). More specifically, study participants answer third-person inventories on how they perceive the robot's personality. A large part of the 26 studies (23, 88%) include inventories and adaptions of Goldberg's Big Five Mini-Markers (Saucier, 1994), the 10-item short version of the Big Five Inventory (BFI-10; Rammstedt, Kemper, Klein, Beierlein, & Kovaleva, 2012), Wiggins' Personality Adjective Items (Wiggins, 1979), the International Personality Item Pool (IPIP; Donnellan et al., 2006), the NEO-FFI (McCrae & Costa, 2004), the Ten Item Personality Measure (TIPI; Gosling et al., 2003), the Big Five Inventory (BFI; Goldberg, 1992), Godspeed Scores (Christoph Bartneck et al., 2009), and a property-based adjective measurement questionnaire assessing Big Five dimensions. Most of the studies assessing perceived personality using the Big Five model also conceptualized the robot's personality based on this theory. Only four studies rely on other psychological measures than the Big Five model: Bian et al. (2016) and Woods et al. (2005) use the *Eysenck* Personality Questionnaire-Revised (EPQ-RSC; Qian et al., 2000), Broadbent et al. (2013) use the Asch's Checklist (Asch, 1946), Kwak and Kim (2005) use a checklist inspired by the Myers-Briggs Type Indicators (MBTI; Myers et al., 1985), and Faur et al. (2015) measure perceived regulatory focus with the Regulatory Focus Questionnaire Proverbs Form (RFQ-PF; Faur et al., 2017). In general, regardless of the specific used instrument, the measurement of personality recognition based on psychological instruments is often adapted to better fit its prior conceptualization.

Besides applying existing psychological measurements, researchers also used custom-made methods adapted to the study's individual personality definitions. This was the case for 9 (26%) of 35 studies, however, with very diverse approaches. Some studies used self-constructed items based on adjectives describing the robot's intended robot personality, such as calm, boring, careful to be rated on a Likert-Scale (Hendriks et al., 2011). Other approaches entailed to assess the users' personality, the perceived robot personality, and subsequently the perceived closeness of the users' to the robot's personality as a measure of personality recognition (e.g., Tapus & Matarić, 2008).

Both approaches, applying existing psychological measures or applying self-constructed measures, come with particular drawbacks and limitations. When applying methods constructed for assessing the personality of humans, this may come with problems such as rating a robot's touchiness although it has no arms (e.g., Meerbeek et al., 2008) or rating a robot's extraversion based on how likely it is to talk to a lot of people at parties (e.g., Joosse et al., 2013). On the other hand, when applying very customized methods, this in turn is related to constraints in generalizability and comparability. In general, the lack of unified approach to assess personality perception makes it difficult to ensure whether robot personality is perceived or correctly recognized among users across studies (for an in-depth review and discussion of assessment methods for personality perceptions, seeCallejas et al., 2014; Sim & Loo, 2015).

7　Consequences of Perceived Robot Personality

Lastly, in order to design an adequate social robot personality, it is necessary to evaluate which consequences different personalities have for the user. Thus, many studies in HRI-literature assess the effects of perceived personality on user perceptions and behaviors during and after the interaction with the robot. In this review, a total of 29 studies (43% of all studies reviewed) explored effects of robot personality on different aspects such as self-reported user perceptions (27 studies out of 29, 93%), observable behavior (2 studies out of 29, 7%), or both (7 studies out of 29, 24%).

Among the self-reported consequences, seven studies (e.g.,Craenen et al., 2018b; Cruz-Maya et al., 2017) made use of one of the few standardized measurements for HRI, namely, the Godspeed scores (Bartneck et al., 2009). The Godspeed scores are designed as standard measurement tool for human–robot interaction and allow to assess the robot on the dimensions of anthropomorphism, animacy, likeability, perceived intelligence and perceived safety. Other studies referring to self-reported user perceptions incorporated around 50 different self-reported perceptions affected by perceived robot personality, which can be broadly clustered into measures of liking, acceptance, perceived performance and intelligence, perceived hedonic and pragmatic quality, perceived human-likeness, and other measures. Regarding liking, nine studies measured preference (e.g.,

Aly & Tapus, 2016; Andrist et al., 2015); two studies measured the intention to use (Brandon, 2011; Tay et al., 2014), one study measured desirability (Goetz & Kiesler, 2002), and one study measured likability (Walters et al., 2014). Examples of measures within the category robot acceptance were trust (e.g., Tay et al., 2014) personality suitability (e.g., Ullrich, 2017), the users' feelings of anxiety (e.g., Brandon, 2011) or perceived eeriness (e.g., Sundar et al., 2017). Regarding the robot's perceived performance and intelligence, researchers evaluated the perceived intelligence (e.g., Walters et al., 2009), perceived social intelligence (e.g., Mileounis et al., 2015), or anticipated task performance in specific occupations (e.g., Joosse et al., 2013). Studies categorized as evaluating hedonic quality, for example, assessed perceived entertainment (e.g., Goetz et al., 2003) or, in parallel to pragmatic quality, adapted established measures of hedonic and pragmatic quality such as the AttrakDiff questionnaire (Hassenzahl et al., 2003) as seen in Niculescu et al.'s study (2013). To measure pragmatic quality, studies often evaluated perceived usefulness (e.g., Brandon, 2011), ease of use (e.g., Tay et al., 2014), task fulfillment (e.g., Meerbeek et al., 2008), and similar measures, such as pragmatic quality adapted from the AttrakDiff questionnaire (Hassenzahl et al., 2003).

Moreover, several studies evaluated the effects of robot personality on the perceived human-likeness. In this review, perceived human-likeness encompasses measures that assess perceived human-likeness directly, the perception of realism or social presence, or the attribution of human characteristics to social robots (e.g., Brandon, 2011; Broadbent et al., 2013; Celiktutan & Gunes, 2015; De Graaf & Ben Allouch, 2014; Fussell et al., 2008; Joosse et al., 2013; Meerbeek et al., 2008; Park et al., 2012). Lastly, there were some studies using measures that could not be subsumed within one of the previous categories such as familiarity of the robot (De Graaf & Ben Allouch, 2014), or friendliness of the robot (Park et al., 2012).

As opposed to approximately 50 different self-reported user perceptions correlated to robot personality, only six different user behaviors were assessed, including engagement (Gockley et al., 2005), interaction time (Tapus & Matarić, 2008), stress level (i.e., heart rate, respiration rate, and skin conductance; Cruz-Maya et al., 2017), task performance (i.e., reaction time and error rate; Cruz-Maya et al., 2017), compliance (Andrist et al., 2015; Goetz & Kiesler, 2002; Goetz et al., 2003), and hesitation time to switch of the robot (Horstmann et al., 2018).

Altogether, the large variety of assessed consequences of perceived robot personality might relate to a lacking unified theoretical approach and according measurement tools. In particular, only few studies draw on, for example, the theory of planned behavior (i.e., TPB; Ajzen, 1991), the unified theory of acceptance and use of technology (i.e., UTAUT; Venkatesh et al., 2011) or the technology acceptance model (i.e., TAM; Y. Lee et al., 2003) in order to evaluate constructs (e.g., attitude, trust, intention to use, perceived usefulness, and acceptance) in a unified manner (e.g., Brandon, 2011; Tay et al., 2014).

8 General Methodological Limitations in Studies on Robot Personality

As for most relatively novel domains of research, there are typical methodological limitations in HRI in terms of study type and limited sample sizes. However, following Dautenhahn's (2007) argument of not to expect the same methodological standards in novel and innovative HRI literature compared to other more established fields, we focus on two specific limitations beyond sample size or study type: The robots used in the study and the assessment of effects of perceived robot personality on user behavior.

In total, 25 different robots were used in 56 studies that applied robots or robot-similar artefacts (e.g., conversational agents embodied on a screen), see Fig. 1 for examples. That is, in almost every second study, a different robot was used to investigate social robot personality, differing in physical appearance, functionalities and context of use. Thus, besides variations in robot personality, also the used robot per se may have a considerable effect on the assessed user perceptions and behavioral measures. For example, Robotics's PeopleBot (Robotics, 2001) and Aldebaran Robotics' NAO (Gouaillier et al., 2009) were with seven and six studies, respectively, the most commonly used robots. Between those two robots, vast differences in appearance and functionality can be observed. More specifically, NAO is a little, humanoid robot, whereas the PeopleBot is a highly adaptable, large, and mechanically looking robot with a screen-based interface, which was initially intended to be used for telepresence purposes. Moreover, the PeopleBot was regularly adapted in its physical appearance between studies in order to, for example, look more mechanical or more human-like (e.g., Walters et al., 2008). Further, some authors describe, that the robot they used is typically associated with specific user perceptions, such as for example, being generally described as cheerful and friendly (e.g., Biswas & Murray, 2014) or being perceived as a female cat (e.g., iCat; Meerbeek et al., 2008). Such confusions between manipulated robot personality and the used robot platform must be considered and assessed more systematically when interpreting findings.

Regarding the study of effects of robot personality on user perceptions and behavior, 93% investigated these solely based on self-reports. Only a few studies

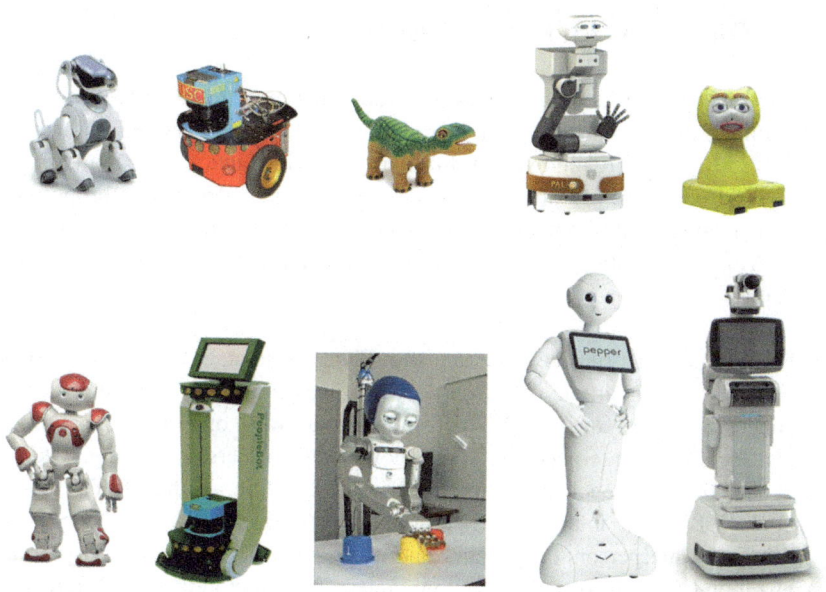

Fig. 1 Exemplary robots used in different studies. Upper row from left to right: SONY's AIBO (Sony AIBO Tribute Site, 2019), Pioneer 2-DX (Gockley & Matarić, 2006), Robotic Pet Pleo (Innvo Labs, 2012), TIAGo (Robotics, 2019a, 2019b), and iCat (iCat Research Community, 2005). Bottom row from left to right: Aldebaran's NAO (Aldebaran Robotics, 2012), Robotics' PeopleBot (TelepresenceRobots, 2016), Meka Robot (Dang & Tapus, 2013), SoftBank's Pepper (SoftBank Robotics, 2019a, 2019b), and HomeMate (Zhao, Naguib, & Lee, 2014)

(7%) exclusively focused on effects on user behavior, or analyzed both (24%). Nevertheless, there are studies that indicate that such effects on behavior are relevant. For example, Goetz et al. (2003) found that variations in robot personality are related to a user's compliance in conducting a physical exercise. Another study examining human interaction behaviors with chatbots found that, users tended behave differently with chatbots than with humans. More specifically, they tended to be more agreeable, more extraverted, more conscientiousness and self-disclosing when interacting with humans than with chatbots, showing more socially desirable traits with humans as opposed to with chatbots (Mou & Xu, 2017). Such findings indicate, that robot personality might have other, unexplored effects on user behavior neglected in previous literature.

9 Directions for Future Research

Based on the present literature review, we identified eight research areas of particular importance to be further addressed in the future: (1) Personality theories beyond the Big Five, (2) individual differences, (3) interaction with human personality and adaption, (4) situational context and social roles, (5) anthropomorphism, (6) cultural differences, (7) personality in long-term interactions, and finally (8) the exploration of an integrative design approach towards social robot personality.

(1) Personality Theories beyond the Big Five: In total, 53% of all studies reviewed in this article relied on the Big Five personality traits in order to conceptualize personality. Only a few studies draw on other theoretical models, such as the interpersonal circumplex models of personality (e.g., Okuno et al., 2003), Eysenck's PEN model (Eysenck, 1991), or the MBTI (Myers et al., 1985). The interpersonal circumplex (Wiggins, 1979) is a theory with origins in interpersonal psychiatry and was repeatedly used for the conceptualization of social roles, interpersonal interactions and individual differences (McCrae & Costa, 1989). Through its strong focus on dispositions relevant for interpersonal interactions, it might be more adequate than the Big Five personality traits, which in comparison also include experiential and motivational traits less relevant to social interactions. Especially the dimension of Dominance/Submissiveness could provide a deeper understanding of robot personality in the context of different roles, as for example in healthcare or military (Santamaria & Nathan-Roberts, 2017). Other than that, attribution theories could also play a relevant role. De Graaf and Ben Allouch (2014) already found that people attributed specific personalities to a robot depending on their own personal traits and moderated by the (positive or negative) expectations they had regarding the robot's performance. That is, they only assigned their own Extraversion to the robot when the robot was expected to perform well. Building upon these examples, we argue that researches should expand personality conceptualization beyond the Big Five personality traits—and particularly the Extraversion dimension—to other psychological theories, such as interpersonal theories or attribution theories from social psychology. Drawing on other disciplines, such as arts, media or data science, might be of further inspirational value and help to eventually reach consensus on the concept of social robot personality. In future studies, it could be of further interest to go above and beyond solely applying research

from human–human interaction or personality psychology, and to explore other forms of interactions, such as those with animals or physical (anthropomorphized) objects.

(2) Individual Differences: Previous research has shown that social robot personality perception underlies several individual differences. Many studies have already shown that gender affects the susceptibility to robotic emotional cues (Moshkina & Arkin, 2005), the preference of a social robot's affect (e.g., Cameron et al., 2015), and the preference of the robot's gender itself (e.g., Eyssel et al., 2012). Regarding age, it has been shown that younger children exhibited greater emotional reactions interacting with robots than older children (Okuno et al., 2003), whereas Martínez-Miranda et al. (2018) found indications that younger children are less demanding in terms of robot personality (i.e., they react more tolerantly to less desirable, or less positive robot personalities than older children). Shahid, Krahmer, Swerts, and Mubin (2010) further found that gender, age, and technological experience were important in how subjects viewed their personality as being similar to the robot's personality. However, such literature exploring individual differences is scarce. Most of the research conducted in previous HRI literature relied on study participants from the university. More specifically, effects of robot personality are evaluated with a highly educated sample of an average age ranging from 20–30 years and extensive technological experience. Especially in the light of robots currently often being created for target groups beyond 65 years of age, children, or people with cognitive impairment (e.g., dementia), and only few or no technological experience, such individual differences should be further explored in future research.

(3) Interaction with Human Personality and Adaption: There is an ongoing debate whether a robot's personality should be similar (e.g., Andrist et al., 2015; Cruz-Maya, 2018; Park et al., 2012) or complementary (e.g., Lee et al., 2006) to the human's. However, as previously discussed, these findings are mostly based on research assessing similarity and complementarity of the Big Five personality traits and more specifically the dimension of Extraversion. Nevertheless, there are other effects that potentially play a role. For example, some studies found that the preference of similar or dissimilar personality depends on the perceived stereotypical role of the robot (e.g., Joosse et al., 2013). Another study depicts that perceived similarity in terms of the social identity theory (Burke & Stets, 2000) affects preference for similar or dissimilar personalities. Social Identity Theory states that humans categorize, classify, and name themselves in relation to other social categories or classifications in order to form an identity of themselves (Burke & Stets, 2000).

In particular, this study showed that an A.I. voice instructors considered as belonging to one's ingroup (i.e., perceived as belonging to one's own social category, here in terms of age) was rated higher for credibility and social presence, and resulted in more learning motivation, as opposed to one that was perceived as less similar (Edwards et al., 2019). Given these examples, it cannot be inferred with certainty whether it is the similarity of a robot's personality, or maybe other factors, such as perceived social role or belonging to one's in group in terms of demographic attributes, that affect the user's preference for a robot. Thus, we argue for a broader exploration of the complementarity and similarity-attraction hypotheses with personality facets beyond the dimension of Extraversion and beyond the Big Five. Secondly, we suggest to explore potential cognitive mechanisms, such as the social identity theory in the future. The fact that similar or dissimilar robot personalities might affect user preferences further raises the question whether it is beneficial for social robots to be able to adapt different human personalities. For example, social robots could adapt in becoming more similar or more complementary in their personality, they could adapt to fit specific user attributes (e.g., age or gender), or they could adapt to pre- or real-time assessed and learned personality preferences of the user. Up to now, there are only few studies exploring the effects of such adaptions of social robots. Most of the studies describing adaption are computational studies which explore potential computational solutions for the adaption to the user. While some studies thereby try to understand human affect and mimic it in the robot (e.g., Ritschel & André, 2017), other studies try to imitate human interaction styles based on machine learning (e.g., Faur et al., 2015). Trying to understand human affect, one study inferred human affective states through using document data (i.e., Facebook posts) in order to understand what a robot could infer about the human personality in conversations and how it could adapt to it (e.g., Skillicorn, Alsadhan, Billingsley, & Williams, 2017). Another example for imitation learning is found in Cameron et al.'s (2015) study. Nevertheless, up to now there is only little evidence which shows positive effects of personality adaption. In this review, two studies provided evidence for positive effects of adaption of verbal- and non-verbal behavior (Aly & Tapus, 2016), as well as age and emotion (Konstantopoulos, 2013) on preference for the (adapted) social robot. Given the little evidence on the form of adaption or its subsequent effects on user preferences, these areas should be targeted in future research. Such future research should further address the suitability of adaption to different contexts, such as interaction settings with multiple human interaction partners.

(4) Situational Context and Social Roles: More work is needed that considers the situation, consisting of task (e.g., consulting-task) and context (e.g., health care) in which social robots operate, as well as the social role (e.g., nurse) assigned to the social robot in that given situation. The studies reviewed in this article examined social robot personality in various different contexts and tasks ranging from cooperating in the game called 'Who wants be the next millionaire?' (Mileounis et al., 2015) to verbally supporting patients in a physical exercise (Kiesler & Goetz, 2002). Researchers thereby rarely examined the influence of different tasks or contexts on the perceived robot personality. However, few studies have shown that the perceived suitability of robot personality for a situation was influential for perceived or preferred social robot personality (e.g.,Goetz et al., 2003; Mutlu, Osman, Forlizzi, Hodgins, & Kiesler, 2006; Ullrich, 2017). Thus, future studies should incorporate different tasks and contexts in their experimental designs and subsequently analyze interactions of situation (i.e., task and context) with social robot personality. Moreover, it would be interesting to map different attributes of such situations and investigate how these relate to different consequential perceptions and behaviors among users across existing and future literature in HRI.

Similar to the context and the task, research on roles explicitly assigned to robots or implicitly perceived by users is underrepresented in previous HRI-literature. While several studies manipulated the social role of their robot (e.g.,Brandon, 2011; Eyssel et al., 2012; Joosse et al., 2013; Tay et al., 2014), there is little research and consensus on the effects of these social roles on personality perceptions among users. However, there is some evidence that robots are perceived as having specific social or stereotypical roles. For example, some studies have shown that already small changes in appearance (e.g., length of robot hair; Eyssel & Hegel, 2012) or behavior (e.g., voice pitch; Powers & Kiesler, 2006) are sufficient to elicit mental models of a robot's skills or knowledge. Humans thereby further seem to apply the same stereotypes (e.g.,Eyssel & Hegel, 2012; Nass et al., 1997; Powers & Kiesler, 2006; Tay et al., 2014) and attribute similar stereotypical traits (Clavel et al., 2013) to robots as to humans. That is, for example, we attribute less knowledge to a female robotic voice (Powers & Kiesler, 2006), perceive more agency in male, while more communality in female robots, prefer male robots as cooperation partner in mathematical tasks (Eyssel & Hegel, 2012), and even accept robots less when their gender does not match their stereotypical occupational role (Tay et al., 2014). According to the common ground theory (Clark & Brennan, 1991), Powers et al. (2005) argue that this elicitation of mental models as well as stereotyping could facilitate HRI by providing cues (based on appearance and behavior) which allow users to make inferences about the robot's knowledge and functionality at zero acquaintance. However, it is also an ethical question whether gender and ethnic stereotypes should be reinforced within social robot design in the future.

(5) Anthropomorphism and Personality: The diversity of appearances among the robots used in studies raises the question whether their degree of perceived human-likeness and thus anthropomorphism affects personality perception. Some studies in this review reported that increased human-likeness and thus experienced anthropomorphism led to higher ratings on socially desirable personality traits (Ogawa et al., 2018) as well as attributed mind and positive personality characteristics (Broadbent et al., 2013). Furthermore, one study found that increased human-likeness of the robot's verbal content in dialogues led to more compliance in a health related task (Kiesler & Goetz, 2002). Despite these first indications of effects of anthropomorphism on perception and behavior, systematic effects on personality perception are not yet explored. In the future, it would be interesting to investigate whether and how anthropomorphism affects personality perception. Moreover, it could be interesting to investigate whether perceived personality itself is also a cue for anthropomorphism and thus also prone to an uncanny valley effect. That is, personality perception could lead to eeriness and discomfort when it becomes too close to human personality or, for example, too similar to one's own personality. In reverse, it would also be interesting to investigate facets of personality and general robot appearance which do not elicit anthropomorphism and cause people to not treat robots socially, as suggested by Dautenhahn (2007).

(6) Cultural Differences: Most of the studies reviewed were either conducted in the United States or in Asia. Only a few studies originated from Germany, Netherlands, France or other European countries. Nevertheless, there are cultural differences that have to be considered. Some studies have already shown that there are cultural differences in preferences for robot behavior or appearance. For example, one study provided evidence for cultural differences in the way robots should look, act, and whom they should interact with: Koreans, for example, prefer more animate, social, and human-like appearances, whereas US-citizens prefer machine-like robots out of metal that do not have faces. Compared to Korea, US-citizens further did not want social robots in social contexts (Lee & Sabanović, 2014). Another study by Lohse, Joosse, Poppe, and Evers (2014) also found that Chinese participants preferred high-contact responses and closer approaches, whereas US-citizens

preferred a comparatively larger intimate space zone without robots. These findings, however, only describe cultural differences in terms of appearance or specific non-verbal behavior (e.g., proxemics). Only one study in this review addressed that the superior task performance of their participants due to a specific personality might be related to cultural experiences. In particular, Bian et al. (2016) argue that their result of children's enhanced learning performance with choleric trainers in a Tai Chi training might be due to the cultural experience of a strict Chinese education. Other than that, cultural effects on personality perception and preference for social robots, however, have been rarely explored and should be further addressed in the future.

(7) Personality in long-term Interaction: Personality of social robots is mostly expressed and evaluated in short-term laboratory-based scenarios. However, personality is defined as stable and enduring experiential, motivational and affective styles (Mccrae et al., 1992). Although, first impressions (or encounters at zero acquaintance) are important and suitable for many applications in HRI (e.g., brief and non-repeated interactions such as buying a train ticket), there are other contexts, such as healthcare, where long-term interactions play an important role (Dautenhahn, 2007). Moshkina and Arkin's (2005) study is the only longitudinal study which evaluated personality perception during more than one interaction. They found, that participants, who believed that a robot displayed emotions and/or personality, also believed that these features made their interaction more pleasant. However, they did not provide evidence for participants actually recognizing the robot's emotions over the course of several interactions. Other than that, evidence on effects of personality perception and cues which affect personality perception over longer time spans is rare. Thus, future research should investigate relevant cues that affect perceptions in long-term interactions and in reverse, how personality affects long-term interactions. Additionally, future research could explore whether it is beneficial to involve users in the personality design process during such long-term interactions. Drawing on animal-human relationships, Dautenhahn (2004), suggests to *parent* robots in social environments similar to pet dogs and later allow customization by users based on an enduring personality development of the robot.

(8) An Integrative Design Approach: Lastly, the field of HRI borrows theories, methods and findings from various disciplines ranging from computer science over psychology to design. However, these findings are not always integrated in the design process of social robots. Following Meerbeek and Saerbeck (2010), it can be argued that studies and methods from all areas should be integrated in a unified design process. They suggest an inte-

grated design approach incorporating theories from psychology, methods from user-centered design and human–computer interaction (HCI) to create social robot personalities. More specifically, they propose a four-step process including: (1) The creation of a personality profile (based on user research and personality psychology) followed by (2) the definition of the expression of this personality profile by the robot, (3) the specification of robot behavior in design rules and finally (4) the implementation and evaluation of robot personality with end-users. Furthermore, it could be interesting to include frameworks from computer science and new, technological and data-driven methodologies of personality synthesis as described by Vinciarelli and Mohammadi (2014) in this approach. Inspired by Meerbeek and Saerbeck's (2010) initial study, we thus suggest to identify and evaluate further processes (e.g., user-testing, collaborative design) and information sources (e.g., arts and media) to establish an integrative methodology to design social robot personality.

10 Summary and Conclusion

The present narrative review of robot personality provides various contributions to the field of HRI and social robots: A broad perspective on research gaps, an inclusive review of conceptual and empirical findings, and a roadmap with directions to be addressed by scholars from different disciplines. This review is to our best knowledge the first review that integrates personality conceptualization, measurement, expression, as well as consequential effects on user perceptions of social robots. Other reviews have solely focused on the theories used (e.g., Robert, 2018) or the measurement and design (e.g., Santamaria & Nathan-Roberts, 2017). Thereby, other than providing an in-depth review of one specific area, a broad view of research on personality in social robots helped to uncover limitations and research gaps of broader range. Secondly, by incorporating both empirical as well as conceptual studies on social robot personality, we further provide a more inclusive review compared to previous reviews. For example, we considered studies using methodologies such as video-based studies or conversational agents to explore robot personality as opposed to Santamaria and Nathan-Roberts (2017). Thirdly, this review is based on studies from several different disciplines, including computer science, psychology, human–computer interaction (HCI), communication sciences, and others. Through this inherent interdisciplinary approach, it is supposed to encourage scholars from all different disciplines to target current limitations and future research directions from the perspective of their respective discipline. Furthermore, we argue for an integrative approach in our future research directions that incorporates findings from all of the different disciplines.

Within our review, we uncovered several trends: Relying on the Big Five personality traits for conceptualization, expressing behavior via non-verbal cues, measuring personality with adaptions of surveys measuring the Big Five personality traits, and relying on very diverse, self-reported perceptions of users to evaluate the consequences of social robot personality. Moreover, we highlighted prominent limitations identified within each of the four areas: Personality definition is inhomogeneous and often solely expressed by behavioral attributes, there is a lack of consensus on the cues used to express personality, assessment methods for personality recognition are diverse and not necessarily adequate for context or robot used, and the selection of evaluated user perceptions does not follow a theoretical approach related to a lack of comparability across studies. For example, as our analysis showed, the same personality trait (e.g., extraversion) is operationalized in many different ways, such as the robot's voice pitch or the color of the clothes the robot is wearing. Obviously, depending on which model and approach researchers choose to conceptualize, implement, measure and evaluate robot personality, findings on (seemingly) the same issue may vary from study to study. Therefore, a more sensible and unified approach seems to be a basic fundament for valid insights into the potential and consequences of robot personality in HRI.

Additionally, we discussed prominent methodological limitations regarding the variety and incomparability of robots used across different studies and the generally underrepresented evaluation of effects of robot personality on user behaviors. Based on the limitations and research gaps identified, we highlighted eight important issues for future research. Firstly, researchers should inform their personality concepts by drawing on theories beyond the Big Five. Secondly, more work is needed to explore the interaction of robot and user personality, as well as the adaption of robot personality to the user. Third and fourth, individual and cultural differences in personality perception of social robots should be further explored in the future. Fifth, researchers should explore how a robot's task, the context and its' assigned social role might interact with personality perception. Sixth, first evidence on reciprocal effects of anthropomorphism and personality perception should be further explored. Seventh, it is unclear, whether personality is perceived over longer periods and how such perceptions would affect HRI. Future research should therefore also integrate long-term interactions or longitudinal study designs. Lastly, we suggest that scholars draw on interdisciplinary sources and design methods in order to investigate an integrative, unified approach for social robot personality design. In sum, these research areas and questions may be seen as a promising roadmap for scholars to progress the research and development of *suitable* social robot personalities.

References

Ajzen, I. (1991). The theory of planned behavior. *Organizational Behaviour and Human Decision Processes, 50*, 179–211.
Aldebaran Robotics. (2012). *Aldebaran robotics NAOqi*. Retrieved February 10, 2019, from http://doc.aldebaran.com/2-1/home_nao.html
Aly, A., & Tapus, A. (2016). Towards an intelligent system for generating an adapted verbal and nonverbal combined behavior in human–robot interaction. *Autonomous Robots, 40*(2), 193–209.
Andrist, S., Mutlu, B., & Tapus, A. (2015). Look like me: Matching robot personality via gaze to increase motivation. In *Proceedings of the 33rd annual ACM conference on human factors in computing systems—CHI '15* (pp. 3603–3612).
Asch, S. E. (1946). Forming impressions of personality. *Journal of Abnormal and Social Psychology, 41*(3), 258–290.
Ball, G., & Breese, J. (2000). *Emotion and personality in a conversational character*. MIT Press.
Bartneck, C., & Forlizzi, J. (2004). A design-centred framework for social human–robot interaction. In *Proceedings ot the RO-MAN 2004 13th IEEE international workshop on robot and human interactive communication* (pp. 591–594).
Bartneck, C., Kulić, D., Croft, E., & Zoghbi, S. (2009). Measurement instruments for the anthropomorphism, animacy, likeability, perceived intelligence, and perceived safety of robots. *International Journal of Social Robotics, 1*(1), 71–81.
Bian, Y., Yang, C., Guan, D., Gao, F., Shen, C., & Meng, X. (2016). A study based on virtual tai chi training studio. *CHI, 2016*, 433–444.
Biswas, M., & Murray, J. (2014). Effect of cognitive biases on human–robot interaction: A case study of a robot's misattribution. In *IEEE RO-MAN 2014—23rd IEEE international symposium on robot and human interactive communication: Human–robot co-existence: Adaptive interfaces and systems for daily life, therapy, assistance and socially engaging interactions* (pp. 1024–1029).
Borkenau, P., & Liebler, A. (1992). Trait inferences: Sources of validity at zero acquaintance. *Journal of Personality and Social Psychology, 62*(4), 645–657.
Bower, P., & Gilbody, S. (2004). Stepped care in psychological therapies: Access, effectiveness and efficiency. *British Journal of Psychiatry, 186*(1), 11–18.
Brandon, M. (2011). *Effect of robot-user personality matching on the acceptance of domestic assistant robots for elderly*. University of Twente.
Broadbent, E. (2017). Interactions with robots: The truths we reveal about ourselves. *Annual Review of Psychology, 68*, 627–652.
Broadbent, E., Kumar, V., Li, X., Sollers, J., Stafford, R. Q., MacDonald, B. A., & Wegner, D. M. (2013). Robots with display screens: A robot with a more humanlike face display is perceived to have more mind and a better personality. *PLoS ONE, 8*(8), 1–9.

Burke, P. J., & Stets, J. E. (2000). Identity theory and social identity theory. *Social Psychology Quarterly, 63*(3), 224–237.

Cabibihan, J. J., Javed, H., Ang, M., & Aljunied, S. M. (2013). Why robots? A Survey on the roles and benefits of social robots in the therapy of children with autism. *International Journal of Social Robotics, 5*(4), 593–618.

Callejas, Z., Griol, D., & López-Cózar, R. (2014). A framework for the assessment of synthetic personalities according to user perception. *International Journal of Human Computer Studies, 72*(7), 567–583.

Cameron, D., Fernando, S., Collins, E., Millings, A., Moore, R., Sharkey, A., Evers, V., Prescott, T. (2015). Presence of life-like robot expressions influences children's enjoyment of human–robot interactions in the field. *Proceedings of the AISB convention 2015 4th international symposium on new frontiers in human–robot interaction.*

Christoforakos, L., Diefenbach, S., Ullrich, D., & Herzog, M. (2019). Die Roboterpersönlichkeit-Konzeption, Gestaltung und Evaluation der Persönlichkeit von sozialen Technologien. In H. Fischer & S. Hess (eds.), *Mensch und computer 2019-usability professionals* (pp. 75–83).

Celiktutan, O., & Gunes, H. (2015). Computational analysis of human–robot interactions through first-person vision: Personality and interaction experience. In *Proceedings of the 24th IEEE international symposium on robot and human interactive communication* (pp. 815–820).

Chee, B. T. T., Taezoon, P., Xu, Q., Ng, J., & Tan, O. (2012). Personality of social robots perceived through the appearance. *Work, 41*(1), 272–276.

Clark, H. H., & Brennan, S. E. (1991). Grounding in communication. In L. B. Resnick, J. M. Levine, & S. D. Teasley (Eds.), *Perspectives on socially shared cognition* (pp. 127–149). American Psychological Association.

Clavel, C., Faur, C., Martin, J. C., Pesty, S., & Duhaut, D. (2013). Artificial companions with personality and social role. In *Proceedings of the 2013 IEEE symposium on computational intelligence for creativity and affective computing, CICAC 2013–2013 IEEE symposium series on computational intelligence, SSCI 2013* (pp. 87–95).

Craenen, B., Deshmukh, A., Foster, M. E., & Vinciarelli, A. (2018a). Do we really like robots that match our personality? The case of Big-Five traits, Godspeed Scores and robotic gestures. In *27th IEEE international symposium on robot and human interactive communication (RO-MAN)* (pp. 626–631).

Craenen, B., Deshmukh, A., Foster, M. E., & Vinciarelli, A. (2018b). Shaping gestures to shape personalities: The relationship between gesture parameters, attributed personality traits and godspeed scores. In *Proceedings of the 27th IEEE international symposium on robot and human interactive communication, Nanjing, China, August 27–31, 2018b TuDP.15 Shaping* (pp. 699–704).

Crowe, E., & Higgins, T. E. (1997). Regulatory focus and strategic inclinations: Promotion and prevention in decision-making. *Organizational Behaviour and Human Decision Processes, 69*(2), 117–132.

Cruz-Maya, A. (2018). *The role of personality, memory, and regulatory focus on human–robot interaction.* PhD thesis. Université Paris-Saclay.

Cruz-Maya, A., Agrigoroaie, R., & Tapus, A. (2017). Improving user's performance by motivation: Matching robot interaction strategy with user's regulatory state. In A. Kheddar, E. Yoshida, S. S. Ge, K. Suzuki, J. Cabibihan, F. Eyssel, & H. He (Eds.), *International conference on social robotics* (pp. 464–473). Springer.

Dang, T.-H.-H., & Tapus, A. (2013). *Robot Meka demonstrates the construction game.* Retrieved February 10, 2019, from https://www.youtube.com/watch?v=OsV63812j8g

Dautenhahn, K. (2004). Robots we like to live with?!—A developmental perspective on a personalized, life-long robot companion. In *Proceedings of the 13th IEEE international workshop on robot and human interactive communication* (pp. 17–22).

Dautenhahn, K. (2007). Methodology & themes of human–robot interaction: A growing research field. *International Journal of Advanced Robotic Systems, 4*(1), 103–108.

de Graaf, M. M. A., & Ben Allouch, S. (2014). Expectation setting and personality attribution in HRI. In *Proceedings of the 2014 ACM/IEEE international conference on human–robot interaction—HRI '14* (pp. 144–145).

Donnellan, M. B., Oswald, F. L., Baird, B. M., & Lucas, R. E. (2006). The Mini-IPIP scales: Tiny-yet-effective measures of the Big Five factors of personality. *Psychological Assessment, 18*(2), 192–203.

Edwards, C., Edwards, A., Stoll, B., Lin, X., & Massey, N. (2019). Evaluations of an artificial intelligence instructor's voice—Social identity theory in human–robot interactions. *Computers in Human Behaviour, 90*, 357–362.

Eysenck, H. J. (1991). Dimensions of personality: 16, 5 or 3?—Criteria for a taxonomic paradigm. *Personality and Individual Differences, 12*(8), 773–790.

Eyssel, F., & Hegel, F. (2012). (S)he's got the look: Gender stereotyping of robots. *Journal of Applied Social Psychology, 42*(9), 2213–2230.

Eyssel, F., Kuchenbrandt, D., Hegel, F., & De Ruiter, L. (2012). Activating elicited agent knowledge: How robot and user features shape the perception of social robots. In *The 21st IEEE international symposium on robot and human interactive communication* (pp. 851–857).

Faur, C., Caillou, P., Martin, J. C., & Clavel, C. (2015). A socio-cognitive approach to personality: Machine-learned game strategies as cues of regulatory focus. In *2015 International conference on affective computing and intelligent interaction, ACII 2015* (pp. 581–587).

Faur, C., Martin, J.-C., & Clavel, C. (2017). Measuring chronic regulatory focus with proverbs: The developmental and psychometric properties of a French scale. *Personality and Individual Differences, 107*, 137–145.

Fong, T., Nourbakhsh, I., & Dautenhahn, K. (2003). A survey of socially interactive robots: Concepts, design, and applications. *Robotics and Autonomous Systems, 42*(3–4), 143–166.

Fussell, S. R., Kiesler, S., Setlock, L. D., & Yew, V. (2008). How people anthropomorphize robots. In *Proceedings of the 3rd international conference on human robot interaction—HRI '08*, pp. 145–152.

Gockley, R., Bruce, A., Forlizzi, J., Michalowski, M., Mundell, A., Rosenthal, S., Sellner, B., Simmons, R., Snipes, K., Schultz, A., & Wang, J. (2005). Designing robots for long-term social interaction. In *2005 IEEE/RSJ international conference on intelligent robots and systems, IROS* (pp. 2199–2204).

Gockley, R., & Matarić, M. J. (2006). Encouraging physical therapy compliance with a hands-off mobile robot. In *Proceedings of the 1st ACM SIGCHI/SIGART conference on human–robot interaction* (pp. 150–155).

Goetz, J., & Kiesler, S. (2002). Cooperation with a robotic assistant. In *CHI '02 extended abstracts on Human factors in computing systems—CHI '02* (pp. 578–579).
Goetz, J., Kiesler, S., & Powers, A. (2003). Matching robot appearance and behavior to tasks to improve human–robot cooperation. In *The 12th IEEE international workshop on robot and human interactive communication* (pp. 55–60).
Goldberg, L. R. (1992). The development of markers for the Big Five factor structure. *Psychological Assessment, 4*(1), 26–42.
Gosling, S. D., Rentfrow, P. J., & Swann, W. B. (2003). A very brief measure of the Big-Five personality domains. *Journal of Research in Personality, 37*(6), 504–528.
Gouaillier, D., Hugel, V., Blazevic, P., Kilner, C., Monceaux, J., Lafourcade, P., Marnier, B., Serre, J., & Maisonnier, B. (2009). Mechatronic design of NAO humanoid. In *2009 IEEE international conference on robotics and automation* (pp. 769–774).
Hassenzahl, M., Burmester, M., & Koller, F. (2003). AttrakDiff: Ein Fragebogen zur Messung wahrgenommener hedonischer und pragmatischer Qualität. In G. Szwillus & J. Ziegler (Eds.), *Mensch & Computer 2003: Interaktion in Bewegung* (pp. 187–196). Teubner.
Heerink, M. (2010). *Assessing acceptance of assistive social robots by aging adults*. PhD thesis. University of Amsterdam.
Hendriks, B., Meerbeek, B., Boess, S., Pauws, S., & Sonneveld, M. (2011). Robot vacuum cleaner personality and behavior. *International Journal of Social Robotics, 3*(2), 187–195.
Holtgraves, T. M., Ross, S. J., Weywadt, C. R., & Han, T. L. (2007). Perceiving artificial social agents. *Computers in Human Behavior, 23*(5), 2163–2174.
Horstmann, A. C., Bock, N., Linhuber, E., Szczuka, J. M., Straßmann, C., & Krämer, N. C. (2018). Do a robot's social skills and its objection discourage interactants from switching the robot off? *PLoS ONE, 13*(7), 1–25.
Hu, X., Xie, L., Liu, X., & Wang, Z. (2013). Emotion expression of robot with personality. *Mathematical Problems in Engineering, 2013*, 1–10.
iCat Research Community. (2005). *iCat*. Retrieved February 10, 2019, from http://www.hitech-projects.com/icat/phpBB3/viewtopic.php?f=8&t=8
Innvo Labs. (2012). *Robotic Pet Pleo*. Retrieved February 10, 2019, from https://www.pleoworld.com/pleo_rb/eng/index.php
John, O. P., & Srivastava, S. (1999). The Big Five trait taxonomy: History, measurement, and theoretical perspectives. In L. A. Pervin & O. P. John (Eds.), *Handbook of personality: Theory and research* (2nd ed., pp. 102–138). Guilford Press.
Joosse, M., Lohse, M., Pérez, J. G., & Evers, V. (2013). What you do is who you are: The role of task context in perceived social robot personality. In *IEEE international conference on robotics and automation (ICRA)* (pp. 2134–2139).
Kiesler, D. J. (1983). The 1982 interpersonal circle: A taxonomy for complementarity in human transactions. *Psychological Review, 90*(3), 185–214.
Kiesler, S., & Goetz, J. (2002). Mental models and cooperation with robotic assistants. In *Proceedings of conference on human factors in computing systems* (pp. 566–577).
Konstantopoulos, S. (2013). System personality and adaptivity on affective human–computer interaction. *International Journal on Artificial Intelligence Tools, 22*(2), 358–372.
Konstantopoulos, S., Karkaletsis, V., & Matheson, C. (2009). Robot personality: Representation and externalization. In *Proceedings of computational aspects of affective and*

emotional interaction conference (CAFFEi08). Patras, Greece. Retrieved from http://meeting.athens-agora.gr/

Kwak, S., & Kim, M. (2005). User preferences for personalities of entertainment robots according to the users' psychological types. *Journal of the Japanese Society for the Science of Design, 52*(4), 47–52.

Lee, H. R., & Sabanović, S. (2014). Culturally variable preferences for robot design and use in South Korea, Turkey, and the United States. In *Proceedings of the 2014 ACM/IEEE international conference on human–robot interaction—HRI '14* (pp. 17–24).

Lee, K. M., Peng, W., Jin, S. A., & Yan, C. (2006). Can robots manifest personality?: An empirical test of personality recognition, social responses, and social presence in human–robot interaction. *Journal of Communication, 56*(4), 754–772.

Lee, Y., Kozar, K. A., & Larsen, K. R. T. (2003). The technology acceptance model: Past, present, and future. *Communications of the Association for Information Systems, 12*(1), 752–780.

Liu, P., Glas, D. F., Kanda, T., & Ishiguro, H. (2017). Two demonstrators are better than one—A social robot that learns to imitate people with different interaction styles. In *IEEE Transactions on Cognitive and Developmental Systems* (pp. 1–15).

Liu, P., Glas, D. F., Kanda, T., & Ishiguro, H. (2018). Learning proactive behavior for interactive social robots. *Autonomous Robots, 42*(5), 1067–1085.

Lohse, M., Joosse, M., Poppe, R., & Evers, V. (2014). Cultural differences in how an engagement-seeking robot should approach a group of people. In *CABS'14* (pp. 121–130). Kyoto.

Martínez-Miranda, J., Pérez-Espinosa, H., Espinosa-Curiel, I., Avila-George, H., & Rodríguez-Jacobo, J. (2018). Age-based differences in preferences and affective reactions towards a robot's personality during interaction. *Computers in Human Behavior, 84*, 245–257.

Mayer, P., & Panek, P. (2016). Sollten Assistenzroboter eine "Persönlichkeit" haben?: Potenzial simplifizierter Roboterpersönlichkeiten. *Zeitschrift Fur Gerontologie Und Geriatrie, 49*(4), 298–302.

McCrae, R. R., & Costa, P. T. (1989). The structure of interpersonal traits: Wiggins's circumplex and the five-factor model. *Journal of Personality and Social Psychology, 56*(4), 586–595.

McCrae, R. R., & Costa, P. T. (2004). A contemplated revision of the NEO five-factor inventory. *Personality and Individual Differences, 36*(3), 587–596.

Mccrae, R. R., John, O. P., Bond, M., Borkenau, P., Buss, D., Costa, P., & Norman, W. (1992). An introduction to the five factor model and its applications. *Journal of Personality, 60*(2), 175–215.

Meerbeek, B., Hoonhout, J., Bingley, P., & Terken, J. (2008). The influence of robot personality on perceived and preferred level of user control. *Interaction Studies, 9*(2), 204–229.

Meerbeek, B., & Saerbeck, M. (2010). User-centered design of robot personality and behavior. In *NordCHI 2010: Proceedings of the 2nd international workshop on designing robotic artefacts with user- and experience-centered perspectives* (pp. 1–2).

Meerbeek, B., Saerbeck, M., & Bartneck, C. (2009a). Iterative design process for robots with personality. In *Proceedings of the AISB2009 symposium on new frontiers in human–robot interaction* (pp. 94–101).

Mileounis, A., Cuijpers, R. H., & Barakova, E. I. (2015). Creating robots with personality: The effect of personality on social intelligence. In *International work-conference on the interplay between natural and artificial computation* (pp. 119–132).

Moshkina, L., & Arkin, R. C. (2003). On TAMEing robots. In *SMC'03 Conference Proceedings. 2003 IEEE international conference on systems, man and cybernetics. Conference theme—system security and assurance* (Vol. 4, pp. 3949–3959).

Moshkina, L., & Arkin, R. C. (2005). Human perspective on affective robotic behavior: A longitudinal study. In *2005 IEEE/RSJ international conference on intelligent robots and systems* (pp. 1444–1451).

Mou, Y., & Xu, K. (2017). The media inequality: Comparing the initial human–human and human–AI social interactions. *Computers in Human Behavior, 72*, 432–440.

Mutlu, B., Osman, S., Forlizzi, J., Hodgins, J., & Kiesler, S. (2006). Task structure and user attributes as elements of human–robot interaction design. In *ROMAN 2006—The 15th IEEE international symposium on robot and human interactive communication* (pp. 74–79).

Myers, I. B., McCaulley, M. H., & Most, R. (1985). *A guide to the development and use of the Myers-Briggs type indicator*. Consulting Psychologists Press.

Nass, C., Moon, Y., & Green, N. (1997). Are computers gender-neutral? Gender stereotypic responses to computers. *Journal of Applied Psychology, 27*(10), 864–876.

Niculescu, A., van Dijk, B., Nijholt, A., Li, H., & See, S. L. (2013). Making social robots more attractive: The effects of voice pitch, humor and empathy. *International Journal of Social Robotics, 5*, 171–191.

Ogata, T., & Sugano, S. (1999). Emotional communication between humans and the autonomous robot WAMOEBA-2 (Waseda Amoeba) which has the emotion model. *Transactions of the Japan Society of Mechanical Engineers Series C, 65*(633), 1900–1906.

Ogawa, K., Bartneck, C., Sakamoto, D., Kanda, T., Ono, T., & Ishiguro, H. (2018). Can an android persuade you? In H. Ishiguro & F. D. Libera (Eds.), *Geminoid studies: Science and technologies for humanlike teleoperated androids* (pp. 235–247). Springer.

Okuno, H. G., Nakadai, K., & Kitano, H. (2003). Design and implementation of personality of humanoids in human humanoid non-verbal interaction. In P. Chung, C. Hinde, & M. Ali (Eds.), *Developments in applied artificial intelligence* (pp. 662–673). Springer.

Panek, P., Mayer, P., & Zagler, W. (2015). Beiträge zur Modellierung von "Persönlichkeit" bei assistiven Robotern für alte Menschen zwecks besserer Mensch-Roboter Interaktion. In *Deutscher AAL Kongress* (pp. 452–458). Frankfurt.

Park, E., Jin, D., & Del Pobil, A. P. (2012). The law of attraction in human–robot interaction. *International Journal of Advanced Robotic Systems, 9*(35), 1–7.

Park, J., Kim, W., Lee, W., & Chung, M. (2010a). Artificial emotion generation based on personality, mood, and emotion for life-like facial expressions of robots. *Human–computer Interaction, 332*, 223–233.

Park, S., Moshkina, L., & Arkin, R. C. (2010b). Recognizing nonverbal affective behavior in humanoid robots. *Intelligent Autonomous Systems, 11*, 12–21. https://doi.org/10.3233/978-1-60750-613-3-12

Paro Robots. (2014). *PARO therapeutic robot*. Retrieved January 18, 2019, from http://www.parorobots.com/

Powers, A., & Kiesler, S. (2006). The advisor robot: Tracing people's mental model from a robot's physical attributes. In *Proceedings of the 1st ACM SIGCHI/SIGART conference on human–robot interaction* (pp. 218–225).

Powers, A., Kiesler, S., Fussell, S., & Torrey, C. (2007). Comparing a computer agent with a humanoid robot. In *Proceeding of the ACM/IEEE international conference on human–robot interaction—HRI '07* (pp. 145–152).

Powers, A., Kramer, A. D. I., Lim, S., Kuo, J., Lee, S. L., & Kiesler, S. (2005). Eliciting information from people with a gendered humanoid robot. In *Proceedings—IEEE international workshop on robot and human interactive communication* (pp. 158–163).

Qian, M., Wu, G., Zhu, R., & Zhang, S. (2000). Development of the revised Eysenck personality questionnaire short scale for Chinese (EPQ-RSC). *Acta Psychologica Sinica, 32*(3), 317–323.

Rammstedt, B., Kemper, C. J., Klein, M. C., Beierlein, C., & Kovaleva, A. (2012). Eine kurze skala zur messung der fünf dimensionen der persönlichkeit: Big-Five-Inventory-10 (BFI-10). *GESIS Working Paper, 23*(2), 1–32.

Ritschel, H., & André, E. (2017). Real-time robot personality adaptation based on reinforcement learning and social signals. In *Proceedings of the 2017 ACM/IEEE international conference on human–robot interaction—HRI '17* (pp. 265–266).

Robert, L. (2018). Personality in the human robot interaction literature: A review and brief critique. In *24th Americas Conference on Information Systems, At New Orleans, LA* (pp. 1–10).

Robotics. (2001). *PeopleBot Operations Manual*. ActiveMedia Robotics.

Robotics, P. (2019). *TIAGo*. Retrieved February 10, 2019, from http://blog.pal-robotics.com/tiago-your-best-robot-for-research/

Santamaria, T., & Nathan-Roberts, D. (2017). Personality measurement and design in human–robot interaction: A systematic and critical review. In *Proceedings of the human factors and ergonomics society 2017 annual meeting* (pp. 853–857).

Saucier, G. (1994). Mini-markers: A brief version of Goldberg's unipolar Big-Five markers. *Journal of Personality Assessment, 63*(3), 506–516.

Severinson-Eklundh, K., Green, A., & Hüttenrauch, H. (2003). Social and collaborative aspects of interaction with a service robot. *Robotics and Autonomous Systems, 42*(3–4), 223–234.

Shahid, S., Krahmer, E., Swerts, M., & Mubin, O. (2010). Child-robot interaction during collaborative game play: Effects of age and gender on emotion and experience. In *Proceedings of the 22Nd conference of the computer-human interaction special interest group of Australia on computer-human interaction* (pp. 332–335).

Shalev-Shwartz, S., & Ben-David, S. (2014). *Understanding machine learning: From theory to algorithms*. Cambridge University Press.

Sim, D. Y. Y., & Loo, C. K. (2015). Extensive assessment and evaluation methodologies on assistive social robots for modelling human–robot interaction—A review. *Information Sciences, 301*, 305–344.

Skillicorn, D. B., Alsadhan, N., Billingsley, R., & Williams, M.-A. (2017). Social robot modelling of human affective state. *ArXiv*, 1–32. Retrieved from http://arxiv.org/abs/1705.00786

SoftBank Robotics. (2019). *Pepper robot*. Retrieved February 10, 2019, from https://www.softbankrobotics.com/emea/en/pepper

Sony AIBO Tribute Site. (2019). *Sony AIBO*. Retrieved February 10, 2019, from http://www.sony-aibo.com/

Sundar, S. S., Jung, E. H., Waddell, T. F., & Kim, K. J. (2017). Cheery companions or serious assistants? Role and demeanor congruity as predictors of robot attraction and use intentions among senior citizens. *International Journal of Human Computer Studies, 97*, 88–97.

Sutcliffe, A., Pineau, J., & Grollman, D. (2014). Estimating people' s subjective experiences of robot behavior. *Artificial Intelligence for Human–Robot Interaction*, 151–152.

Tapus, A., & Matarić, M. J. (2008). User personality matching with a hands-off robot for post-stroke rehabilitation therapy. In O. Khatib, V. Kumar, & D. Rus (Eds.), *Experimental robotics* (pp. 165–175). Springer.

Tay, B., Jung, Y., & Park, T. (2014). When stereotypes meet robots: The double-edge sword of robot gender and personality in human–robot interaction. *Computers in Human Behavior, 38*, 75–84.

TelepresenceRobots. (2016). *PeopleBot*. Retrieved February 10, 2019, from https://telepresencerobots.com/robots/adept-mobilerobots-peoplebot

Uleman, J. S., Saribay, S. A., & Gonzalez, C. M. (2008). Spontaneous inferences, implicit impressions, and implicit theories. *Annual Review of Psychology, 59*, 329–360.

Ullrich, D. (2017). Robot personality insights. Designing suitable robot personalities for different domains. *I-Com, 16*(1), 57–67.

Venkatesh, V., Thong, J. Y. L., Chan, F. K. Y., Hu, P. J. H., & Brown, S. A. (2011). Extending the two-stage information systems continuance model: Incorporating UTAUT predictors and the role of context. *Information Systems Journal, 21*(6), 527–555.

Vinciarelli, A., & Mohammadi, G. (2014). A survey of personality computing. *IEEE Transactions on Affective Computing, 5*(3), 273–291.

Walters, M. L., Koay, K. L., Syrdal, D. S., Dautenhahn, K., & Te Boekhorst, R. (2009). Preferences and perceptions of robot appearance and embodiment in human–robot interaction trials. In *Proceedings new frontiers in human–robot interaction* (pp. 136–143).

Walters, M. L., Lohse, M., Hanheide, M., Wrede, B., Syrdal, D. S., Koay, K. L., & Severinson-Eklundh, K. (2014). Evaluating the robot personality and verbal behavior of domestic robots using video-based studies. In Y. Xu, H. Qian, & X. Wu (Eds.), *Household service robotics* (pp. 467–486). Elsevier.

Walters, M. L., Syrdal, D. S., Dautenhahn, K., Te Boekhorst, R., & Koay, K. L. (2008). Avoiding the uncanny valley: Robot appearance, personality and consistency of behavior in an attention-seeking home scenario for a robot companion. *Autonomous Robots, 24*(2), 159–178.

Wiggins, J. S. (1979). A psychological taxonomy of trait-descriptive terms: The interpersonal domain. *Journal of Personality and Social Psychology, 37*(3), 395–412.

Woods, S. (2006). Exploring the design space of robots: Children's perspectives. *Interacting with Computers, 18*(6), 1390–1418.

Woods, S., Dautenhahn, K., Kaouri, C., Te Boekhorst, R., & Koay, K. L. (2005). Is this robot like me? Links between human and robot personality traits. In *Proceedings of 2005 5th IEEE-RAS international conference on humanoid robots* (pp. 375–380).

Woods, S., Dautenhahn, K., & Schulz, J. (2004). The design space of robots: investigating children's views. In *Proceedings of the 2004 IEEE International workshop on robot and human interactive communication* (pp. 47–52).

Zhao, X., Naguib, A., & Lee, S. (2014). Octree segmentation based calling gesture recognition for elderly care robots. In *Proceedings of the 8th International conference on ubiquitous information management and communication* (pp. 54–63).

Sarah Diefenbach is professor for market and consumer psychology at the Ludwig-Maximilians-Universität München (LMU Munich, Germany) with a focus on the field of interactive technology. Her research group explores design factors and relevant psychological mechanisms in the context of technology usage in different fields, e.g., social media, digital collaboration, companion technologies, social robots.

Marietta Herzog graduated with a Master of Science (M.Sc.) in Economic, Organizational and Social Psychology at the Ludwig-Maximilians-Universität München (LMU Munich, Germany). In her studies she focused on user experience (UX) and social robot personality. Currently, she's working as a product manager at Zavvy, a start-up automating people operations for fast growing companies.

Daniel Ullrich is postdoctoral researcher at the Chair of Human-Computer Interaction at the Ludwig-Maximilians-Universität München (LMU Munich, Germany). His research focuses on Human-Robot-Interaction (HRI) and social robots, Experience Design and UX Evaluation.

Lara Christoforakos is a postdoctoral researcher at the Chair for Economic and Organizational Psychology at Ludwig-Maximilians-Universität München (LMU Munich, Germany) with a focus on the field of HCI. Her research explores psychological mechanisms and design factors in the context of technology use within various fields, such as companion technologies or social robots.

Social Robots as Echo Chambers and Opinion Amplifiers

Limits to the Social Integrability of Robots

Catrin Misselhorn and Tobias Störzinger

Abstract

Using a practice-theoretical perspective on sociality, we investigate which social practices are reserved for humans. We argue that especially those practices that require participants to reciprocally recognize each other as persons clash with the conceptual understanding of robots. Furthermore, the paper provides reasons why this understanding of robots can be defended against a conception that wants to attribute the status of persons to robots based on their behavior. The simulated evaluative attitudes of robots are not rooted in the robots themselves but turn out instead to be merely opinion amplifiers of their developers or sociotechnical echo chambers of the users. However, we also argue that building robots that can perfectly simulate recognition claims nevertheless poses a problem since such devices would distort our social practices.

We would like to thank Laura Martena, Hauke Behrendt, Tom Poljanšek and Anne Clausen for helpful comments and exciting discussions that greatly contributed to the improvement of this paper.

C. Misselhorn (✉) · T. Störzinger
Department of Philosophy, Georg-August-University Göttingen, Göttingen, Germany
e-mail: catrin.misselhorn@uni-goettingen.de

T. Störzinger
e-mail: tobias.stoerzinger@uni-goettingen.de

Keywords

Social robots · Social practices · Human robot interaction (HRI) · Recognition · Personhood

1 Introduction: The Growing Sociality of Social Robots

Current efforts to do research, develop, and potentially commercialize social robots offer a glimpse into the future of human sociality, opening up both utopian and dystopian perspectives. On the one hand, some highlight the idea of cooperative collaboration and claim that social robots can simplify our lives and, by triggering "social resonance," perhaps even cushion the negative effects of our hyper-individualized society, such as loneliness. On the other hand, some fear that we are replacing the warmth and depth of human interaction with an inferior derivative.

In this context, it is worth questioning how social so called "social robots" can actually be and to what extent it is possible to integrate them into our social life. Some social robots, such as *Sophia* by *Hanson Robotics*, have attracted a great deal of attention, partly because their public appearance gives the impression that they can socially and emotionally interact with their conversation partners and even participate in discussions about normative issues.[1] Other robots, such as the zoomorphic robot *Kiki*, are advertised with slogans about their ability to emotionally engage with us. For instance, one of *Kiki's* advertisement slogans is "Kiki really loves you back" (Zoetic AI, 2020).

Are *Kiki* and *Sophia* examples of robots with which we will live together in the near future? In this paper, we explore the extent to which social robots can be integrated into our social life and ask whether there is a limit to their social integration. Are there specific forms of social relations, such as romantic partnership, friendship, or normative practices that constitute barriers to robotic participation?

We will examine this subject from a practice theoretical perspective. This means that we will ask whether there are certain social practices' whose "realiza-

---

[1] See for example Sophia's YouTube channel and especially this video: https://www.youtube.com/watch?v=NwNULnXk9b0.

tion requirements" cannot be met by robots. Finally, we come to the conclusion that robots cannot engage in practices that require participants to recognize each other as persons reciprocally. Nevertheless, we will argue that building robots that can perfectly simulate recognition claims would lead to a distortion of our social practices.

In defense of these claims, the paper is organized as follows. In the next section (Sect. 2), we explain the idea of social robots by distinguishing them from functional service and assistance robots. Furthermore, we clarify the role of emotional expression in the creation of social robots. Social robots need to be at least able to express emotions primarily because emotional expressions are supposed to give users the impression that the robot takes an evaluative stance toward the world which involves feelings, believes and desires in a phenomenal sense. We end this section by asking how far the social integration of social robots can go.

In Sect. 3, we prepare to answer this question by introducing the concept of social practices. Social practices are "sayings and doings" (Schatzki, 2008) that follow a certain normative regularity in order to be realized. We then distinguish between *functional conditions* and (non-functional) *status conditions* for the realization of a social practice.

In Sect. 4, we argue that a particular set of practices within the set of all practices constituting our social life can be characterized by the status requirement of reciprocal recognition as a person. We examine what it means to be recognized as a person and, based on Ikäheimo's work (2019), distinguish between a *capacity conception of personhood* and a *status conception of personhood*. We then propose a basis for the capacity conception of personhood by referring to considerations from Bertram and Celikates (2015). They see the *capacity to make recognition claims* in conflicts as the fundamental basis for reciprocal recognition. Based on this perspective, the capacity conception of personhood requires that an agent be able to claim recognition and grant recognition. Furthermore, *status as a person* is fully realized when agents within a social practice recognize each other as beings who are entitled to make claims of recognition.

Drawing on these distinctions, we then explore the extent to which recognition as a person limits robots' integrability in social practices: In Sect. 5, we analyze whether robots are actually capable of making claims of recognition. We find that, given the current state of technology, robots do not meet this criterion. We then move on to question whether, assuming counterfactually that robots were capable of making recognition claims, we could ascribe to them the status of a person. In this regard, our thesis is that the way we (as a society) understand and think of robots conflicts with a possible recognition of them as persons. The idea that we create, control, and sell robots that are able to participate in practices that require person status is self-defeating. In Sect. 6, we then turn to the question of whether we are being normatively forced to change our understanding of robots if they would have the capacity to make recognition claims. We provide reasons as to why this is not the case—at least not in the foreseeable future. However, we also argue that building robots that can perfectly simulate recognition claims nevertheless poses a problem since such devices would distort our social practices. We then summarize our findings and arguments in the conclusion (Sect. 7).

2 The Idea of Social Robots and the Role of Emotional Expression

In this section, we analyze the premises of the idea of social robots and determine the role of emotional expression for social robotics. In brief, we propose the distinction of "social robots" from "mere" service and assistance robots. We claim that unlike "mere" service or assistance robots, social robots aim to establish *a social relationship* with their users. This goal can only be achieved if social robots simulate being agents with an evaluate stance on the world. The role of expressing emotions is considered a core function for simulating a subjective point of view, as emotions express an affective evaluative perspective on the world (Misselhorn, 2021). Each of these claims is explored in further detail in the sections that follow.

2.1 The Difference Between Service Robots and Social Robots: Simulating a Self

Some robots that are included under the label "social robots" are not actually *social* in a narrow sense of the term. There are robots that assist humans in their daily tasks and work or that cooperate with them in a weak form. While these robots function in close physical proximity to humans, it is not intended that humans form emotional and social bonds and relationships with them (though this can happen unintentionally; see, for example, Darling, 2017; Misselhorn 2021 and in this volume). We propose that the broad category of social robots (including service and assistance robots) be distinguished from a narrower group of social robots based on the fact that the former are primarily intended to perform functional tasks. In contrast, the latter are mainly intended to establish close social relationships (in addition to assisting with or performing functional tasks). The following three examples can be used to explain and further refine this distinction (with regard to the question of how tasks can be functional in a social sense, see Poljanšek & Störzinger, 2020).

2.1.1 Three Examples of Social Robots

First, we examine the *TUG* service robot from *AETHON* (AETHON, 2020). *TUG* is a robotic platform that resembles a moving box, and its main task is to perform various drop-off and pick-up services. It can safely transport multiple goods such as towels, bedsheets, medical samples, or waste from one place to another. For example, *TUG* is able to detect pedestrians who are in its path and then stop to avoid collisions. *TUG* performs all of its tasks rather stoically. It carries out every order without displaying the slightest emotion. Nothing gives the user the impression that *TUG* has any evaluative attitude about what it is doing. This is not considered a shortcoming because the goal behind the development of such a robot is not to create an agent that acts according to its own evaluative perspective of the world, but rather to build a robot that can assist humans in functional tasks.

Our second example is the social robot companion *Kiki* from *Zoetic AI* (2020). *Kiki* has a zoomorphic design and has a somewhat cat-like appearance. *Kiki* has large eyes that move and is capable of making animal-like sounds. It does not perform any functional tasks, such as handing over objects, turning on lights, calling relatives, or carrying towels. As the advertising for *Kiki* demonstrates, this product is not intended to support everyday life in a useful way, but was rather built with the aim of building a social and emotional bond with the user. *Kiki*, Zoetic AI explains, "understands you," "loves you back," and "is always there for you." In summary, *Kiki* is a "loyal companion who keeps you company when you're together and misses you when you're apart" (Zoetic AI, 2020). It is claimed that *Kiki* builds its own personality over time based on interactions with the user. Such customization to the way the user interacts with *Kiki* cannot only make *Kiki* more affectionate or loving but can also cause *Kiki* to ignore her user "if [they] are mean to her" (Zoetic AI, 2020). A promotional video on *Zoetic AI's* homepage shows *Kiki* expressing a range of emotions when she interacts with users. For example, *Kiki* seems to get angry or sad (depending on the persona) when the user is about to leave their home. Then, *Kiki* is shown overjoyed when the user returns. *Kiki* expresses jealousy when strangers interact with the user and needs time to get used to new partners. Finally, she is said to have a "sense of self:"

> Kiki's sense of self comes from her needs and wants, which we call drives. Her personality determines how these drives fluctuate in response to external stimuli. Acting

according to her own drives, Kiki makes every decision meaningful and intentional. (Zoetic AI, 2020)

As Zoetic AI makes clear, *Kiki's* "self" is based on her "needs and desires," which produce her behavior. Quite unlike *TUG*, *Kiki* seems to take an evaluative stance toward the world. She seems interested in what the user is doing and responds to the way users interact with her.

Before we continue and further explore how needs, desires, and emotions constitute an evaluative point of view, we discuss a third example. The third social robot we present is *Sophia* from *Hanson Robotics* (Hanson Robotics, 2020). *Sophia* is a hyper-realistic humanoid robot that can express various emotions and hold conversations with humans. *Hanson Robotics* claims that *Sophia* may even have a "rudimentary form of consciousness" (Hanson Robotics, 2020) although in lack of any evidence to support this claim this must be rather a judged as a form of advertising. Sophia's public appearances on talk shows, conferences, and social media channels give the impression that Hanson Robotics aims to create a replicant of a human. According to the judgment of the king of Saudi Arabia, the company seems to have progressed significantly in this project, as he has already granted *Sophia* citizenship.

As an enlightened and feigned reflective robot, *Sophia* takes issue with the problematic asymmetry in the fact that a robot has been granted citizenship, while women do not have full citizenship in some countries. Therefore, *Sophia* has proclaimed that her Saudi citizenship is merely "aspirational" and that she dreams of "equal rights for all" (Brain Bar, 2020).

Regardless of whether *Sophia* is as technically sophisticated as such media appearances would have us believe; it seems clear that the overall idea behind *Sophia's* development is to create a robot that is capable of having meaningful conversations with humans about various topics.

2.1.2 A Central Difference: Simulating a Self vs. Offering a Service

A comparison of these three robots reveals a significant difference between *Kiki* and *Sophia* on the one hand and *TUG* on the other hand. While *TUG* simply plays a functional role for humans, *Kiki* and *Sophia* present themselves not as tools but rather as beings of their own. In contrast to the former robot, the latter ones simulate being a "self."

The questions that then remain to be explored are what it means to "have" or "be" a self and why we claim that *Kiki* and *Sophia* are trying to simulate this, while *TUG* does not. As a first working definition, we suggest that "having" or "being" a social agent involves having evaluative attitudes that guide an agent's actions. Having evaluative attitudes means that one *cares* about what happens to oneself, to others and what happens in the world. An agent that "has" a self is someone who takes an evaluative stance on the world and acts accordingly. Having evaluative attitudes can range from simple things, such as "feeling pain" and therefore trying to avoid its causes, to higher-order propositional attitudes, such as the belief that "killing beings capable of suffering without reason is an impermissible act."

TUG does not seem to express any of these attitudes, as it simply executes what it is told to do if it is capable of doing so. *Kiki*, on the other hand, has "needs" and "wants" that structure her interaction with a user. As we have learned, this can even lead her to ignore a user if she is treated poorly. Unlike *TUG*, *Kiki* does not accept all forms of behavior but seems to care about how one interacts with her. *Sophia* (or at least the ideal of *Sophia* that is still waiting to be fully realized) not only takes an evaluative stance on how users interact with her, but also expresses opinions and values that she defends in public appearances. The evaluative attitudes that she pronounces go beyond simple wants and needs.

At this point, it should be made clear that our talk of mental states, evaluative expressions, and well-being of robots is meant in a simulative sense, as has been described by Seibt (2018). This means that we can talk about *Kiki* being angry without implying that *Kiki feels* anger (What is meant is that *Kiki* is *simulating* anger and not realizing anger as humans do). Following Misselhorn (2018, 2022) one might also speak of "quasi-anger." "Quasi" states in this sense are internal states of computers that share certain functional aspects of mental states but lack phenomenal consciousness and only have derivative intentionality which is transferred to them by the programmers and users.

2.1.3 Why Do Social Robots Need a Self?

Robots like *TUG* do not simulate a self, while robots like *Kiki* or even *Sophia* do because of the purpose for which they were designed. The core functionality of *TUG* does not require a simulated self. The purpose of robots like *Kiki* or *Sophia*, on the other hand, is different. Their core "function" is to form social bonds and relationships, become our friends, or even lovers. Such "functions" seem to presuppose the "having" of a self. These types of robots give us the impression that we are valued and recognized and that there is someone who cares about us. A robot that does not simulate evaluative attitudes cannot be considered a social robot in the strict sense of this label since in order for us to form a strong emotional and social bond with a robot, we need to perceive that robot as enjoying our company or caring about our welfare.

2.2 The Role of Emotions for Social Robots: Expressing a Self

Based on the considerations above, it is possible to determine the role of emotional expression for social robots. Social robots that are meant to establish social relationships with humans must simulate being social agents; in other words, they should be able to simulate having their own wants, needs, or other evaluative attitudes and act accordingly. However, evaluative attitudes cannot be perceived directly; rather, we identify them based on behavioral interpretations. One of the most important behavioral signs of evaluative attitudes are *expressions of emotion*. In general, emotional expression signals that an agent has certain evaluative attitudes.

The expression of emotions can signal to observers that what the agent is doing is not just physically determined, but results from taking an evaluative stance toward the world. Equipping robots with the ability to express emotions represents a way to simulate social agents. Based on robots' expression of emotions, we get the impression that they like specific actions, dislike others, have particular preferences, pursue needs, and generally care about what is happening around them.

Only because of *Kiki's* emotional expressions—the sadness expressed when we leave the house or the joy when we come back—can we perceive that *Kiki* seems to care for us. Based on this perception, a social relationship can then unfold. Strong social bonds are based on the appearance of reciprocal appreciation, and it is the emotional expression of this appreciation in various situations that lays the ground for a relationship.

2.3 Different Kinds of Sociality

In addition to the role of emotional expression for social robots, another insight can be gained from comparing the three robot examples we have introduced above. Each of them can be seen as representing a certain degree of sociality. Whereas *TUG* is able to simulate minimal socially interactive capacities in functional work contexts, *Kiki* is able to form social relationships by simulating parts of the necessary conditions for reciprocal appreciation. *Sophia*, or at least the idea of *Sophia*, goes beyond mere interaction and mutual appreciation. In fact, *Sophia* is supposed to participate in deliberative practices where reasons are exchanged and evaluative attitudes toward the world are shared. At least this is what the designers claim. In the next section, we present a practice theoretical framework to further explicate this difference regarding the degree of participation in different social practices. Our ultimate aim is to lay the foundation for a discussion of whether the integration of robots into social practices has conceptual and ethical limitations.

3 Practice Theory as a Framework for Analyzing Sociality

In this section, we introduce a practice theoretical framework to analyze what it means for a robot to be "social." According to this framework, "being social" means being included in social practices. After introducing the general notion of "social practice," we propose a distinction between the *functional* and *status* requirements of a social practice.

3.1 Social Practices

What is characteristic of the social world? What does it mean to say that a particular robot is taking part in social practices? To answer such questions, we adopt a practice theoretical perspective (cf. Bourdieu, 1979; Giddens, 1984; Schatzki, 2008). This means that we do not think of sociality as reducible to the mere interaction of individuals, nor do we think of the individual as a mere expression of a higher-level "social totality." The former idea is often referred to as "methodological individualism," while the latter is sometimes associated with a rather superficial reading of authors like Durkheim, Hegel, or even Luhmann. Instead, we think of social life as the sum of social practices at a particular space and time (cf. Giddens, 1984). Social practices are *forms* of action that presuppose implicit knowledge of their inherent normative criteria of success on the side of the agents (cf. Reckwitz, 2003). Some authors have suggested that the inherent normativity of a practice (e.g., the right way to greet someone at a birthday party, the right way to marry someone in a Catholic way, the right way to order a cup of coffee at Starbucks, etc.) can be analyzed as a matter of rules. This does not necessarily mean that competent agents who successfully participate in social practices are following rules explicitly. Still, the concept of a rule can help to *reconstruct* what it means for an agent to successfully participate in a social practice (see Elgin, 2018).

For the purposes of this paper, we differentiate between two conditions of success for being able to perform an action as part of a social practice.[2] We present a distinction between the *functional conditions* and the (non-functional) *status conditions* of the realization of a social practice. This means that the agent must (i) be able to perform the "sayings and doings" that are required by the practice (raising one's arm and saying "Hello" to greet another agent) and (ii) have the right *social status* in order for their performance to count as a realization of a specific practice. We explain each of these conditions further in the next section.

3.2 Functional Conditions and Status Conditions

There are conditions of success for social practices, or "rules of a practice," that participants must meet for their "sayings and doings" to count as an instance of a social practice.[3] As we mentioned above, in order to participate in a specific practice, one must do and say the right kinds of things in the proper context, as well as be the right kind of agent that is "licensed" to perform these sayings and doings. One example is a priest pronouncing someone husband and wife. For this

---

[2] One detailed account that analyzes the necessary and "rule-like" conditions for participating in a social practice was suggested by Hauke Behrendt (2018). Behrendt develops a concept of social "inclusion" (into a practice) that has four relational elements. He claims that a full conceptualization of (social) inclusion must determine (i) the subject of inclusion, (ii) the object of inclusion, (iii) the Instance of inclusion and (iv) the rules of inclusion. (Behrendt, 2018, pp. 162–182). We agree with Behrendt that the objects of social inclusion are social practices and further that the instance that formally "licenses" the inclusion or participation in a social practice is in a lot of cases not some bureaucratic agency but rather other participants within the practice. Behrendt understands the conditions for being a principal subject of inclusion as having the necessary properties and capacities for "[…] being a sufficiently competent social Agent" (Behrendt, 2018, p. 165 f.). Rules of inclusion can then further limit access to social practices with respect to further features, (ascribed) properties or capacities that a subject need to possess in order to be included into a social practice. We suggest to tweak Behrendt's account of social inclusion into social practices a little bit for the purpose of this paper. Instead of conceptualizing the necessary conditions for principally being a subject of potential participation in a social practice as more or less general "social competences" we propose to distinguish between the functional conditions of realization of a social practice and sufficient further status conditions, whose realization must be ascribed "on top".

[3] Correctly speaking, for a lot of cases the theoretical reconstruction of the normative sensitive behavior of participants can be made by supposing "success conditions" which are an explication of the implicit knowing-how of the participants).

practice to be considered successful, the priest must say and do the right kind of things in the right context, the marriage ceremony. However, they must also have the (social) status of being a priest. Another wedding attendee cannot speed the process up by simply shouting, "I hereby declare you husband and wife; the buffet is open." While doing and saying the right things form functional requirements, being the right kind of agent is a status requirement. The question then is how wedding officials or priests acquire the social status that they need. They go through other practices, such as education and training that eventually lead to the (formal and/or informal) acceptance that they have a particular status function within an arrangement of social practices. They are *accepted as priests* in the context of the integrated practices that constitute a church (or to put it in Searle's (1996) famous words: Person X counts as a priest in the church-context). While one might associate "being accepted as a priest" with some kind of formal document that confirms this status according to explicit rules, many social practices revolve around more implicit "rules" that enable one's acceptance as having a certain status. For example, what are the status conditions for being a friend or romantic partner? What qualities or skills does one need to have in order to be accepted as someone whose arm-raising and "hello-"saying is considered a successful greeting? Importantly, whether there is a formal procedure that leads to status ascription or only implicit acceptance by other participants, a status is not a natural property of an agent, but rather a socio-ontological property that depends on the ascription itself.

In summary, some robots are designed as social robots. They are meant to form social relationships with users, and to this end, they simulate having a self, which means that they simulate having their own needs, desires, and will. In this section, we set out to answer whether there are both conceptual and ethical challenges related to the inclusion of robots in the social realm. To answer this question, we explained "the social" as a sum of social practices. We then presented the basic underpinnings of practice theory. With regard to the rules of inclusion, we then distinguished—for the purpose of this paper—between *functional* aspects, i.e., behaving in a certain way, and *status* or *role* aspects. In the next section, we introduce a general and essential status. Many social practices require that participants ascribe the status of being a "person" to each other.

4 Recognition as a Person: An Exceptional Status

Different practices require that different functional and different status conditions be realized. Although social practices and the normative requirements associated with them are quite diverse, general requirements for particular *levels* of sociality can be distinguished. We suggest that one criterion that allows us to distinguish two levels of practices is whether the respective practices require participants to recognize each other as persons or not. Social practices that do require reciprocal recognition as persons form a different type if sociality from those that do not require such a reciprocal understating.

With regard to the possible limits of the social integration of robots in those practices that require person status, we distinguish between the question of whether a robot can perform the necessary sayings and doings and whether we (can) accept it as having the necessary status as a person that a practice requires.

In the remainder of this section, we outline the relationship between social practices, reciprocal recognition, and being a person. In the following section, we then argue that there are several challenges preventing social robots' inclusion in practices that require reciprocal recognition as a person.

4.1 Personhood, Recognition, and Social Practices

Given the complexity of the debate on the relationship between recognition and personhood, we cannot go into detail here (for an overview see: Frankfurt, 1971; Dennett, 1988; Laitinen, 2002; Honneth, 2010; Ikäheimo, 2014; Iser, 2019). For the purposes of this paper, we propose a combination of perspectives, from those of Ikäheimo (2007; forthcoming), on the one hand, to the ideas of Bertram and Celikates (2015), on the other.

Ikäheimo distinguishes between a "capacity concept" and a "status concept" of a person (). The capacity concept refers to whether an actor has the necessary capacities to participate in specific practices that are part of the human life form. The status concept refers to whether an agent who has the necessary capacities to perform the specific practices of the human life form is actually recognized as a person by others when performing those practices.

Recognizing someone as a person means more than just *epistemically* acknowledging that someone has the necessary capabilities. It requires that the agent in question be recognized as a person in a *normative* sense. This means, among other things, that an agent should have a normative status before others, or that others are willing to understand one's claims as potentially legitimate reasons. With regard to the precise explication of the capacity concept and the status concept, as well as the relationship between the two, we deviate from Ikäheimo's reconstruction and consider a proposal by Bertram and Celikates (2015).

Celikates and Bertram explicate recognition through the lens of Hegel, starting from the fundamental relation between recognition and conflict. They argue that recognition relations are defined as such based on the fact that agents within conflict situations can understand each other as authors of recognition (Bertram & Celikates, 2015, p. 9). The idea is that in cases where we are in conflict with each other and do not recognize what the other asserts or demands, we nevertheless (insofar as we understand each other as persons) recognize each other as agents capable of recognition.

This is the case, for example, when we argue about what the specific rules of a practice are. In such a case, we are in conflict about what kind of normative rules apply and who should do what. However, we are not in conflict over the fact that we both have a say in how this matter should be settled. While we might not (initially) recognize each other's understanding of the practice and what is normatively binding, we do recognize each other as agents who are, in principle, entitled to make claims about how the practice is to be understood and its associated rules (see also Stahl's (2013) concept of "standard authorities").

Such reciprocal recognition as persons signals that we do not understand each other simply as obstacles to realizing our own ends or as tools to be manipulated at will. Mutual recognition (as persons) in the normative sense implies that we grant each other the right to participate in the process of figuring out how to live and what is right.

Applied to the distinction between a capacity concept and a status concept of person, this means the following: The essential condition for personhood is the ability to be the author of recognition claims. However, full *personhood* is only constituted when others reciprocally recognize this authorship of recognition claims.

Being ascribed the status of a person thus requires not only that I perform as an author of recognition claims, but also that I receive the status of a person via the recognition of being a competent performer. This concept implies that personhood in the status sense is, in some sense, a social-ontological category, as it does not merely rely on my capacities or properties alone but also on the reciprocal recognition of others. Not only must I be able to perform an action in a specific way, but others must also interpret my performance such that it has normative significance. However, capacities and properties can—of course—have normative weight with regard to whether others *should* recognize someone as a person. Not recognizing someone who performs as a person (or has the properties we associate with personhood) needs to be justified. In the case of robots, we explain below how a justification like this would look like.

The question of which practices require that participants recognize each other as authors of recognition warrants further research. However, we can already identify all those practices that could be called "meta-practices," as they essentially center around intentionally negotiating practices, and their normative rules belong to the group of "person only" practices. It is well known within practice theory that every performance of a practice is a reproduction and transformation of that practice, as well as its rules. However, when we speak of "meta-practices," we are referencing the intentional and reason-guided restructuring of practices. This is the case, for example, when we argue about who should have the right to vote. In this case, we are participating in a meta-practice, or "deliberation," that is aimed at shaping another practice (in this case, "voting").

This means that all those practices in which we navigate and negotiate the normative realm require that we recognize participants as having a voice regarding such matters. For example, with regard to "normative" issues, we might compare the way we interact with other people to how we deal with pets. While we might discuss who is allowed to sit on which part of the couch with a roommate, we do not discuss with our dogs and cats whether it is acceptable for them to sit on the couch. Rather, we *train* them to behave a certain way.

In our engagement with persons, we "transcend" a mere collision in which we simply try to reshape other agents according to our will and come to a consensual agreement based on reciprocal recognition.

Against this background, two questions that remain are how far robot sociality extends and whether those practices that require reciprocal recognition as a person constitute a limit for robots' sociality.

5   The Limits of Robot Sociality

In terms of the integrability of robots into social practices that require reciprocal recognition as persons, we can structure our analysis around the following questions: (1) Is a given robot capable of satisfying the functional requirements of personhood (i.e., can it perform as a person?) As we have explained, this ability amounts to being able to make claims of recognition. (2) If the robot is capable of doing so, are we *able* to recognize it as a person (the meaning of this question becomes apparent below)? (3) If robots were able to satisfy the first question, *should* we then recognize them as persons? Does the satisfaction of the first question, in the case of robots, involve a normative force that requires us to infer a positive response to question (3) from a positive response to question (1). Put differently, in the case of robots, does the performance as a person (i.e., expressing the capacities of a person) imply a normative entitlement to the status concept?

The difference between questions (3) and (2) is that question (2) refers to conceptual considerations with respect to our understanding of robots, while question (3) refers to the ethical issues associated with recognizing robots as persons. We deal with questions (1) and (2) in this section and turn to question (3) in Sect. 6.

5.1 Can Robots Perform as Persons?

Do robots have the capacity to make recognition claims? Are they authors of such claims in a meaningful sense? To answer these questions, we must differentiate between an identification of something as having the formal structure of a recognition claim (a parrot saying "I want to be free") and a description of this capacity that includes more substantial implications. There is good reason to assume that Sophia is still just a chatbot with a very humanlike appearance which does no more than parroting.

At this point, it might be considered natural to introduce familiar "list accounts" of personhood that specify what kind of capacities or properties are necessary in order for an agent to count as a person (cf. Dennett, 1988; Frankfurt, 1971). Features like intentionality, second- and third-order intentionality, consciousness, and self-consciousness, among others, have been proposed as conditions for identifying specific agents as persons. However, we suggest to stay focused on the agent's performance, which might be considered an expression of such underlying features. Through the lens of performance, something else becomes central: A substantial performance specification that makes such sayings as "I want freedom" not mere utterances, but normative recognition claims is that they are made in the space of reasons (cf. Sellars, 1997 [1956]; Sellars, 1962; Brandom, 1994). Persons do not only make demands or express evaluative attitudes but *relate* to such claims in a specific way. In principle, they are able to reflect, challenge, and criticize such claims by referring to the claims that support them and figuring out what follows from them (Brandom's famous game of giving and asking for reasons). According to this line of thinking, an agent's capacity to be an author of recognition claims can be described as including the capacity to defend or revise them.

To explore the question of whether robots have these abilities, we can use the three examples presented earlier.

As explained, *TUG* is not able to make any demands or express an evaluative relationship to the world. For this reason alone, *TUG* does not seem to be an agent that in any way possesses the crucial capacities to perform as a person.

The second robot we considered is *Kiki*. *Kiki* can express (simulated) evaluative attitudes in the form of emotional expressions. Therefore, *Kiki* simulates having a self that cares about itself, others and the world. *Kiki* can make demands as to how it should be treated and expresses what kind of interaction it likes and what kind it dislikes.

However, *Kiki* cannot put her demands into a form that would make them potentially reciprocal and universally binding. *Kiki* can express that she likes or dislikes something only on an emotional level (in the sense of simulating those emotions). A user can respond appropriately, provided they care about *Kiki's* simulated well-being. However, *Kiki's* claims are not made in a way that warrants acknowledgment, regardless of whether one cares about *Kiki's* welfare or not. Thus, *Kiki* cannot simulate reciprocal recognition, let alone be guided by the exchange of reasons.

This brings us to the third robot we have discussed above. At first glance, *Sophia* seems to have precisely those features that we found missing in *Kiki*. *Sophia* explicitly adopts evaluative standpoints toward normative propositions. She expresses emotions as reactions to how users interact with her, and somewhat simulates claims in the space of reason, as her statement above about her Saudi citizenship suggests. However, her discursive capacities are rather rudimentary and static until this day. There is no indication that she is really able to let reasons guide her opinions and actions.

Sophia gives speeches and can simulate chit-chat about the football championship, which is impressive. However, there is no evidence that she really performs in the space of reasons, reflects her claims, justifies them spontaneously against objections, or changes her mind. In other words, she is not able to go beyond drawing coherent "conclusions" from large data sets and further reflect on the normative principles that guide her own judgments in the first place.

The same is true for the various versions of ChatGPT and its kin, which are based on large language models. At first glance, they seem to be able to play the game of giving and asking for reasons. However, if one takes a closer look, it becomes evident that they are not. They are just "stochastic parrots" which are very good at producing convincing linguistic output without understanding the meaning of the words that they are generating. That is, they are not really reasoning in the sense of taking into account evidence or considering the truth of what they are saying.

One may object here on the basis that these considerations refer only to the present state of technology. It is by no means impossible that robots may one day be able to exchange claims in the space of reasons. It could well be that at some point, they might be indistinguishable from humans and will implicitly and explicitly demand to be among those agents whose claims should be taken seriously.

An argument of this kind is made by Danaher (2020) in his defense of "ethical behaviorism," a thesis that implies that robots should be considered as moral patients (or even persons) if they exhibit sufficiently similar *behavior* to humans. Whatever the normative defining property for moral significance should be (pain sensitivity or personhood), as long as robots behave just like agents who "really" have these properties, it seems difficult to exclude them from the realm of morally relevant creatures. We address whether we really have no reasons at all in the next section (Sect. 5). Before asking whether we have an obligation to consider such robots as moral patients, however, we first ask whether we are conceptually able to do so. With respect to this question, we defend the following claim: Even if robots were to behave like persons and could participate in a deliberative debate in a reflective and reasoning way, our current conceptualization of robots is in contradiction to our recognition of them as persons. There is, we argue, a conceptual inconsistency in the conjunction of the idea of robots as persons and the whole project of designing and engineering social robots.

5.2 Can We Recognize Robots as Persons?

In tackling the question of whether we can recognize robots as persons, we do not mean to answer the question of whether it is possible or impossible for any individual human being to recognize a robot as a person. We rather have in mind whether it is conceptually consistent for us——in the sense of us as members of society—to recognize robots as persons. In fact, we are not trying to answer the question of whether or not individual humans can psychologically recognize robots as persons, but whether our general concept of "personhood" is compatible with the idea that robots be able to acquire such a status. This means that we need to compare the implications and presuppositions of our concept of "personhood" with our "manifest image" of robots.[4]

We make such a comparison by generalizing some of the arguments already made about the limits of possible forms of interaction with robots. More specifically, the goal in this section is to carve out the more general lesson to be learned from an argument against the possibility of romantic relationships with robots made by Dylan Evans (2019).

In his paper *Wanting the impossible—The dilemma at the heart of intimate human–robot relationships,* Evans argues against a claim that he attributes to Levy (2008). The claim Evans attacks is that romantic relationships with robots would be *more* satisfying than romantic relationships with humans because "we will be able to specify the features of robot companions entirely in accordance with our wishes" (Evans, 2019, 76).

As we demonstrate in this section, Evans's critique can be seen as a special case of a more general aspect that he does not work out in his argument. In the remainder of this section, we explicate the general idea that fuels Evans's argument.

Evans prepares his charge against Levy by identifying a dilemma at the heart of the idea of an intimate human–robot relationship. One horn of the dilemma is rooted in the claim that we want romantic partners that *freely choose* to be with us:

> For among the various desires that most people have regarding relationships are certain second-order desires (desires about desires), including the desire that one's partner has freely chosen to be one's partner, and has not been coerced into this decision (Evans, 2019, p. 80).

The reasoning behind this claim is that we want to have the romantic relationships we have because others choose to be with us and not because it is simply their mechanical behavior or because they are forced into such a relation. We want to be valued freely. However, if being chosen freely is a precondition for romantic partnership, then the robots that Levy has in mind cannot take part in such a relation. These robots simply cannot freely choose to be with their users because they cannot leave their users. To freely choose to be with someone is

---

[4] For the distinction between "manifest image" and "scientific image" see Sellars, 1962. The manifest image can be described as our ordinary understanding of ourselves and the world in which normativity, reasons and rationality plays a role. The scientific image is, roughly speaking, our explanation of the world via theoretical and stipulated, basic entities.

simply not part of their design. If this is true, then robots are not able to satisfy our needs for romantic partnership since they lack a necessary requirement[5]:

> Although people typically want commitment and fidelity from their partners, they want these things to be the fruit of an ongoing choice, rather than inflexible and unreflexive behavior patterns. An obvious objection [...] is that robots will not be able to provide this crucial sense of ongoing choice, and that relationships with robots will therefore be less satisfactory than relationships with humans with respect to this feature. Let us call this "the free-will objection" (FWO) (Evans, 2019, p. 80).

The "free-will objection" counters the idea that robots are more satisfying in romantic partnerships than humans since they cannot satisfy a central second-order desire of humans, namely, the desire of wanting to be wanted freely.

The second horn of the dilemma is rooted in an attempt to tackle the "free-will objection" via a solution that seems to be natural: One might ask why a robot cannot simply be given by design the capacity to freely choose to leave or stay with its user and make that decision somehow dependent on how the robot is treated (not unlike the idea of *Kiki* who is ignoring her user when she gets treated poorly). Such a robot would seem to have its own standpoints on how it wants to be treated and would seem to act in accordance with its own evaluative judgments. In this case, the robot would seem to satisfy a human's higher-order desire for a partner that freely chooses to be with someone since it always has the capacity to leave.

What is the problem with such a solution? We think that in this case, Evans is on the right track but then loses his argumentative focus and misses the crucial point that would have allowed him to generalize his reasoning to areas other than a romantic partnership.

First, Evans claims that on the one hand, a robot that is free to choose would itself become an unattractive as well as a "difficult product to sell" (Evans, 2019, p. 83). The possibility that the robot which someone bought for much money might stop interacting with them makes it an "unreliable product" (Evans, 2019, p. 83). Evans even imagines how one would be looked at by one's friends if there were a robot sitting on the couch in "non-responsive mode." They would surely ask "and how much money did that thing cost?" Evans concludes that a freely-

---

[5] It may be questioned that romantic love really requires that the partner is able to break up the relationship (see Misselhorn, 2021). This does, however, not touch the more fundamental problem to which we point.

choosing robot would not have the advertised benefits since it might, in the end, really decide to leave its user or enter a non-responsive mode and therefore would not be a better romantic partner compared to fellow humans. Evans explains that the ultimate reason behind the failure of the free robot as a reliable product is grounded in the structure of human desire itself. According to Evans, it is not so much the fact that robots cannot live up to the standard of human sociality but instead that humans simultaneously want contradictory things:

> The problem at the heart of this dilemma does not lie with any technical limitations of the robots themselves, just as it does not lie with some contingent feature of human partners. It lies with our very desires themselves. We want contradictory things: a romantic partner who is both free and who will never leave us. (Evans, 2019, p. 84).

There are, however, some problems with this diagnosis. The term "inconsistent" might be a bit strong, though. One the one hand, contradictory desires do not seem to be as problematic as contradictory beliefs (Williams, 1965). On the other hand, it does not even seem to be strictly inconsistent to desire that a romantic partner freely chooses to never leave us. Moreover, the dilemma goes deeper than the shortcomings of an unreliable product that fails to satisfy our desires.

The problem of robots as romantic partners is not so much rooted in the incompatibility between a human's desire for a partner that never leaves them and the desire for a partner that freely chooses to be with them. Rather, the core problem is that it is a kind of category mistake to subsume *a free social agent under the category "product."* The idea is flawed from the start, and the dilemma of the romantic partnership is just a symptom of a general inconsistency. The point is that robots cannot even satisfy the first desire (for a partner that freely chooses to be with someone), even if they would be able to leave someone, since they are—in this example, as well as in our common understanding—conceptualized as products that are marketed, sold, and bought. The robotic self is not a self in an unconditional way. It is a "self as a service" for the well-being of the user. The idea of such a social robot is highly problematic because it rests on the presupposition that close social relationships can be bought and delivered as a service. To put our reasoning into quasi-Sellarsian words: The manifest image of a *"free person in the world"* is hijacked by the economic image of a *"product in the market."* How can a person understand an agent as a social companion that decides to be with them and values their company when they interpret the agent's actions as a service that they paid for? More generally, how can one recognize a robot as a person that has a say with respect to normative issues if they understand this robot as their property? Our "folk understanding" of robots seems to block the reciprocal recognition of robots as persons because such a recognition is conceptually at odds with the understanding of a person as a product.

At this point, one might ask whether such a (mis)recognition of robots as being behaviorally capable of performing successfully in social practices, which normally presupposes person status, is justified, or whether we should reconsider our common understanding of such robots. This leads us back to Danaher's argument. One way to avoid his his "ethical behaviorism" is to argue that one must also take into account our knowledge of the internal constitution of robotic evaluative attitudes when thinking about their normative status. The suggestion in this case would be that even if social robots were to successfully act as agents with their own evaluative attitudes, their internal design would reveal them more as mediums of our human attitudes (either the designers' or users' attitudes).

6 Should We Change Our Manifest Image of Robots?

In the previous sections, we noted that there are a number of practices that require participants to recognize each other as persons. We distinguished between a capacity concept and a status concept of personhood. The capacity concept requires that persons be able to make recognition claims. The status concept further requires that the capacity for recognition claims actually be reciprocally recognized. The agent is understood as someone who is considered qualified to make normative claims. We then asked whether robots have the ability to make recognition claims. We argued that at least those robots that are currently being advertised do not yet possess this ability.

We then moved on to examine whether we can ascribe personhood status to robots under the counterfactual assumption that they are able to act in ways that are indicative of the abilities we "normally" associate with claims for recognition.

Regarding this question, we argued that the way robots are usually conceptualized (and the way they appear in different research projects), namely as products that we can use for our purposes, is conceptually at odds with our ability to recognize them as persons.

One now might highlight that the problem could simply be that we have a morally incorrect understanding of such robots. Would we then be forced to radically change our understanding of robots as products and services and ascribe person status to them as soon as they are able to perform as persons? The inconsistency of our understanding of robots and the idea that they make normative claims might force us to overhaul our concept of robots.

This seems to be the line of thought Danaher pushes when arguing for his concept of "ethical behaviorism." Danaher's overall argument is that robots can have significant moral status if they are *roughly performatively equivalent* to other entities that have significant moral status (Danaher, 2020). With regard to the question of ascribing personhood, this means that insofar as robots can perform as persons, they should also be recognized as persons.

There are several arguments in support of this claim. For example, it could be assumed that the attribution of personhood depends on whether the agent possesses certain characteristics. From this perspective, the question of whether we ascribe personhood to a robot that behaves like a person (and makes recognition claims for example) transforms back into the question of whether that robot actually possesses the necessary properties for personhood. The arguments that could be made in support of Danaher's claim are as follows:

a) One could abandon a substantialistic view altogether and say that "having the necessary properties" *means* nothing more than that the robot behaves in a certain way.
b) One can argue for the fact that the right behavior (being able to make recognition claims, for instance) certainly expresses or indicates that an agent has the necessary properties.
c) One can argue that the attribution of person status (conceptually) depends on the presence of specific properties, but that from our epistemic perspective, we can only use an agents' behavior to determine whether it has the necessary properties or not.

Argument a) explicates mental properties, such as "consciousness," "intentionality," and so on, as terms that describe a way of behaving rather than "real" inner properties of agents. Argument b), on the other hand, focuses on the idea of properties but draws a direct line from the expression of a particular behavior to the presence of the relevant properties. Likewise, argument c) does not resolve the idea of a "possession of properties" with reference to behavior, but considers expressed behavior—somewhat weaker than argument b)—as *indicating* the presence of the necessary properties, *albeit on the grounds that we cannot know any better.*

In the case of argument a), there is no longer any difference between "being" and "seeming." Whenever there is a specific behavior associated with consciousness for example, there is no longer any conceptual space allowing for the possibility that we are being deceived. This would imply that whenever a robot has the capacity to consistently and competently make recognition claims, it "has" all that is needed for an ascription of personhood, and therefore should indeed be ascribed personhood. Argument b) admits that there is a difference between the behavior and the presence of the properties but connects the latter causally so closely to the former that the distinction, while conceptually being made in principle, makes no difference to our attribution practice.

The epistemic variant c) affirms that there is a difference between a certain performance of personhood and the presence of the necessary properties. Since the causal connection is not understood as necessary, this argument leaves room for the idea that something can behave *as-if* it is a person, although it is not. What argument c) denies is that we have any other epistemic means of determining the presence of the necessary properties than analyzing the behavior exhibited. We should—or so the idea goes—exercise caution and attribute properties based on behavior rather than do an injustice by denying an agent the status of a person on the basis of ontological claims that we cannot substantiate.

Danaher's idea of an ethical behaviorism seems to depend on argument c). This is particularly clear in the way he deals with several objections. For example, he describes the relationship between "performance level" and metaphysical states that should ground our ascriptions as follows:

> The behaviorist position is that even if one thinks that certain metaphysical states are the 'true' metaphysical basis for the ascription of moral status, one cannot get to them other than through the performative level. This means that if the entity one is concerned with consistently performs in ways that suggest that they feel pain (or have whatever property it is that we associate with moral status) then one cannot say that those performances are 'fake' or 'deceptive' merely because the entity is suspected of lacking some inner metaphysical essence that grounds the capacity to feel pain (or whatever). The performance itself can verify the presence of the metaphysical essence. (Danaher, 2020, 18)

Danaher's line of thought seems to be that there might be something like a "metaphysical basis" for the ascription of pain (or personhood), but the point of ethical behaviorism is simply that "one cannot get to them other than through the performative level." From this claim, he draws the inference that it is the performance itself that "verifies" the presence of the metaphysical essence. This last conclusion seems to be, however, an unjustified relapse into argument b). The problem with this claim is that just because we do not have any other means of finding out about a certain property's presence, we cannot conclude that a certain performance, which normally signifies the presence of that property, is conceptually sufficient for the property being there. This may seem like a minor conceptual point that has no practical implications, but the problem is that conflating "seeming" and "being" makes it impossible to consider counterarguments that draw on a broader interpretation of behavior.

This should become clear when we look at the premise of the argument that we have no other means than expressed behavior to determine an agent's properties.

We might first look at pain sensation as an example. What would count as a reason for the claim that even if robots behave like agents who feel pain, they do not actually feel pain? To answer this question, we may think that it is possible to access all the (empirical and analytical) resources that the sciences, such as biology, engineering, information technology, etc., provide us with. In this case, everything we know about pain sensation—all pain sensitive beings we know have a nervous system, etc.—and all the knowledge we have about how robots causally realize pain behavior—the programs we code, the mechanics behind the visual expressions, etc.—would disrupt the thesis that robots *feel* pain. So far, we know that the computer programs we code and which control the mechanical systems of a robot have no phenomenal consciousness. We can fully explain how a robot can do the trick of performing just like someone who actually feels pain in a reasonable way. It is not metaphysically impossible that phenomenal consciousness and pain sensation are constituted by something other than brains and nervous systems. Still, according to everything we know so far from the natural sciences, cognitive psychology, biology, and computer science, (current) machines simply do not have phenomenal consciousness. To ground the thesis that robots *feel* pain because they express pain behavior on the assumption that we have no other means other than consulting behavior to determine which agents feel pain and which do not is only possible through the artificial reduction of one's epistemic resources.

We suggest that a similar argument can be made regarding the question of whether performance as a person alone normatively requires the attribution of person status. Similar to the case of the interpretation of pain behavior, we have more information available to us when interpreting recognition claims than the mere behavioral claims themselves.

The core difference between genuine claims of recognition and merely simulated claims of recognition is that genuine claims of recognition spring from the vulnerability and necessity of human life and express a caring relationship to oneself and to the world[6]. In contrast, simulated claims of recognition ultimately mimic such a grounding but, on further analysis, turn out to be mediations of *our* evaluative take on the world.

The reason why behavior alone is not sufficient to attribute person status to robots is that (similar to pain perception) we know exactly how robots achieve the feat of simulating an evaluative attitude toward the world. This knowledge reveals that these simulated attitudes are not *their* evaluative attitudes in a strict sense, but rather a copy of *our* evaluative attitudes which we implement into them during the design phase. A robot's evaluative attitudes are not "grounded" in such a way as recognition claims that are "anchored" in the necessities of (human) existence but rather mirror our existential relation to the world. Humans are "by nature" in such a situation that their existence itself forces them to negotiate and navigate recognition claims based on their evaluative attitudes since they are essentially social beings that need to cooperate with one another. Robots do not care in the same way, since they "naturally" have nothing to lose when their simulated claims go unrecognized.

These considerations are not based on the idea that an attribution of personhood requires that evaluative attitudes somehow be inexplicable. For humans, it may be possible at some point in the future to provide a full naturalistic and reductionist explanation of how we relate to the world. The point is rather that evaluative attitudes and recognition claims are a result of our life form and our existence, whereas in the case of robots, they are simply imposed on them according to *our* needs. This becomes clear when we consider how exactly robots get their attitudes.

Conceptualized somewhat simplistically, there are two ways in which robots arrive at a simulation of evaluative attitudes: either the developers simply program the attitudes or some kind of features that later constitute the attitudes directly or the robots need to learn and develop their attitudes by interacting with their environment (mostly with their users). Insofar as we are dealing with the first option, a robot's recognition claims result from the developer's values. This robot does not have its own evaluative attitudes that originate from its existential situation, but the robot itself is rather something like an "opination amplifier" of the attitudes of its developers.

On the other hand, if we assume that the robot simply adapts its normative claims to its user (or owner), it functions similar to a sociotechnical echo-cham-

---

[6]Thanks to Anne Clausen for pointing out that Hegel's account of relations of recognition is embedded in a situation of the necessities of the human life form.

ber. Instead of simulating a situation in which two people engage with each other and negotiate their different perspectives, users would merely engage affirmatively with a copy of *themselves*. In none of these cases are the evaluative attitudes grounded in a genuine necessity of a specific life form; recognizing the recognition claims that result from such copied attitudes would be similar to recognizing the claims of an actor on stage.

Having said this, there is still room for an argument that simulating recognition claims even from copied attitudes might be morally relevant. Analogously to Misselhorn (in this volume) who shows that empathy with robots is indirectly morally relevant because abusing robots has a negative impact on our moral capacities, one might hold that ignoring simulated recognition claims is morally problematic, as well, because it has a negative impact on our social practices. Our social practices are grounded in accepting recognition claims at face value without checking first whether somebody satisfies the relevant requirements of personhood on a deeper level.

This is the grain of truth in the positions (a) to (c), although it does not suffice for ascribing robots directly any moral status. Suppose that a robot appears to be behaviorally indistinguishable from a human being like Commander Data in the Star Trek episode "The Measure of a Man." Should we accept his recognition claims or should we refuse to do it? Our claim is that we should accept them independently of the question whether they are really grounded or not. The reason is that not accepting them would profoundly distort our social practices which are based on taking recognition claims at face value.

If we would be permanently suspicious whether our interaction partners were really persons or only simulated being ones, our social practices would become morally questionable. How should we decide? Would it not be an invitation to decline recognition claims that do not fit our wishes and plans or which are simply uncomfortable. Borrowing from Kant's procedure of the categorical imperative as a test whether such a world would be morally desirable, the answer seems to be clearly "No". The reason is, however not that it would violate the robot's rights but that it would undercut our own social practices. What is the upshot of this argument? It is not meant to show that we should accept robots as persons in our social practices. The upshot is that we should refrain from building robots that are capable of perfectly simulating recognition claims. Not regarding Data as a product anymore and giving him the freedom to determine his fate, as the court judges in the end of the episode, is one thing. Constructing hundreds and thousands of Datas as his adversary Maddox intends to, is another.

7   Conclusion

In the introduction and the second section of this paper, we diagnosed the tendency of robot manufacturers to try to make social robots more and more "social." Some robots are being made to develop an emotional and social bond with their users or even be able to converse with them on normative issues. We set out to answer the question of how far a social integration of robots into our social life can extend, exploring whether there any boundaries to the social integrability of robots into our social practices.

The argument we have put forth is based on three claims. First, currently, robots cannot be integrated into all those practices that require reciprocal recognition as persons since they do not have the capacity to make consistent and coherent recognition claims. Second, even if we assume that future robots would have the capacity to simulate making recognition claims, the way we conceptualize and understand robots collides with the idea of accepting them as persons. Third, if robots could perfectly simulate making recognition claims, we should not change this conceptualization since—as far as we can currently imagine—their evaluative attitudes are not grounded in their existential necessity to make such claims but rather in a copy of our relationship to the world. Still, there is reason to assume that robots which could perfectly simulate recognition claims would distort our social practices. This is a reason for not striving at building such robots.

Two points are worth mentioning. First, no one knows what the future might bring, and we do not claim that it is impossible to create robots that do have genuine evaluative attitudes, consciousness, and so on. However, our arguments as to whether we can recognize robots as persons are based on our knowledge of what robots are and not on a speculative notion of what they might be in some distant future. It is, however, an argument to the effect that it is morally wrong to follow the pathway that would lead to such a result.

Second, what we have put forward is a general consideration of what *types* of practices are beyond robots' capacities. We have not explained which specific tokens belong to this "person only" practice type. What we have argued is that all those practices that have to do with normative giving and asking for reasons regarding questions as to what the rules of specific practices are and should be categorized as "person only" practices. However, there is a fine line between practices that belong to this type and practices that do not, and decisions as to whether a practice belongs to one type or another depend on a variety of factors, such as use-case, context, users, and other participants, to name a few. Hence, at least in contexts where it is not obvious whether we are dealing with a person or a robot it should be required by law that we are informed when dealing with a robot.

References

AETHON. (2020). Retrieved December 12, 2020, from https://aethon.com

Behrendt, H. (2018). *Das Ideal einer inklusiven Arbeitswelt: Teilhabegerechtigkeit im Zeitalter der Digitalisierung*. Campus.

Bertram, G. W., & Celikates, R. (2015). Towards a conflict theory of recognition: On the constitution of relations of recognition in conflict: Towards a conflict theory of recognition. *European Journal of Philosophy, 23*(4), 838–861.

Bourdieu, P. (1979). *Entwurf einer Theorie der Praxis: Auf der ethnologischen Grundlage der kabylischen Gesellschaft*. Suhrkamp.

Brain Bar. (2020). Retrieved December 12, 2021, from https://www.youtube.com/watch?v=Io6xuGmS5pM

Brandom, R. (1994). *Making it explicit: Reasoning, representing, and discursive commitment*. Harvard University Press.

Danaher, J. (2020). The philosophical case for robot friendship. *Journal of Posthuman Studies, 3*(1), 5–24.

Darling, K. (2017). "Who's Johnny?" Anthropomorphic framing in human–robot interaction, integration, and policy. In P. Lin, G. Bekey, K. Abney, & R. Jenkins (eds.), *Robot ethics 2.0*. Oxford University Press.

Dennett, D. (1988). Conditions of personhood. In M. F. Goodman (Ed.), *What is a person?* (pp. 145–167). Humana Press.

Elgin, C. Z. (2018). The epistemic normativity of knowing-how. In U. Dirks & A. Wagner (Eds.), *Abel im Dialog* (pp. 483–498). De Gruyter.

Evans, D. (2019). Wanting the impossible: The dilemma at the heart of intimate human–robot relationships. In Y. Wilks (ed.), *Close engagements with artificial companions*. John Benjamins.

Frankfurt, H. G. (1971). Freedom of the will and the concept of a person. *The Journal of Philosophy, 80*(10), 563–570.

Giddens, A. (1984). *The constitution of society*. Blackwell.

Hanson Robotics. (2020). Retrieved December 12, 2020, from https://www.hansonrobotics.com/sophia/

Honneth, A. (2010). *Kampf um Anerkennung: Zur moralischen Grammatik sozialer Konflikte*. Suhrkamp.

Ikäheimo, H. (2007). Recognizing persons. *Journal of Consciousness Studies, 14*(5–6), 224–247.

Ikäheimo, H. (2014). *Anerkennung*. De Gruyter.

Ikäheimo, H., 2019. Intersubjective recognition and personhood as membership in the lifeform of persons. *The Social Ontology of Personhood*, Draft 2019.08

Iser, M. (2019). Recognition. In E. N. Zalta (ed.), *The Stanford encyclopedia of philosophy*. Retrieved from https://plato.stanford.edu/archives/sum2019/entries/recognition/

Laitinen, A. (2002). Interpersonal recognition: A response to value or a precondition of personhood? *Inquiry, 45*(4), 463–478.

Levy, D. (2008). *Love and sex with robots: The evolution of human–robot relationships*. Harper Perennial.

Misselhorn, C. (2018). *Grundfragen der Maschinenethik*. Reclam, 5th ed. 2022.

Misselhorn, C. (2021). *Künstliche Intelligenz und Empathie. Vom Leben mit Emotionserkennung, Sexrobotern & Co.* Reclam.
Misselhorn, C. (2022). Artificial moral agents. Conceptual issues and ethical controversy. In S. Voeneky et al. (eds.), *The Cambridge handbook of responsible artificial intelligence: Interdisciplinary perspectives* (in print).
Poljanšek, T., Störzinger, T. (2020). Of waiters, robots, and friends. Functional social interactions vs. close interhuman relationships. *Culturally Sustainable Social Robotics*, 68–77.
Reckwitz, A. (2003). Basic elements of a theory of social practices. *Zeitschrift Für Soziologie, 32*(4), 282–301.
Schatzki, T. R. (2008). *Social practices: A Wittgensteinian approach to human activity and the social*. Cambridge University Press.
Seibt, J. (2018). Classifying forms and modes of co-working in the ontology of asymmetric social interactions (OASIS). In M. Coeckelbergh, J. Loh, M. Funk, J. Seibt, & M. Nørskov (eds.), *Envisioning robots in society—Power, politics, and public space frontiers in artificial intelligence and applications.*
Searle, J. (1996). *The construction of social reality.* Penguin.
Sellars, W. (1962). Philosophy and the scientific image of man. In R. Colodny (Ed.), *Frontiers of science and philosophy* (pp. 35–78). University of Pittsburgh Press.
Sellars, W. (1997). *Empiricism and the philosophy of mind.* Harvard University Press.
Stahl, T. (2013). *Immanente Kritik: Elemente einer Theorie sozialer Praktiken Theorie und Gesellschaft.* Campus.
Williams, B. (1965). Ethical consistency. *Proceedings of the Aristotelian Society, 39*, 103–124.
Zoetic AI. (2020). Retrieved December 12, 2020, from https://www.kiki.ai

Catrin Misselhorn is Full Professor of Philosophy at Georg-August University Göttingen; 2012–2019 she was Chair for Philosophy of Science and Technology at the University of Stuttgart; 2001–2011 she taught philosophy at the University of Tübingen, Humboldt University Berlin and the University of Zurich; 2007–2008 she was Feodor Lynen Fellow at the Center of Affective Sciences in Geneva as well as at the Collège de France and the Institut Jean Nicod for Cognitive Sciences in Paris. Her research areas are the philosophy of AI, robot and machine ethics.Tobias Störzinger is a postdoctoral researcher associated to the Chair of Prof. Dr. Catrin Misselhorn at Georg-August-University Göttingen. His research areas are social ontology, social philosophy, and philosophy of technology. From 2019–2021 he was a research assistant in the research project "GINA", in which research was conducted on the social and ethical challenges of social robotics. His PhD project "Forms of Collective Agency and Metacognition in Distributed Systems", successfully completed in 2021, addresses the question to what extent and under which conditions socio-technical systems can be understood as intelligent agents.

Artistic Explorations

Quantifying Qualia

Jaana Okulov

Abstract

This chapter aims to critically review how affective machine learning models are being currently developed based on categorical thinking. The topic is approached from an artistic perspective drawing knowledge from cross-disciplinary literature. The key term used is *qualia,* defined in relation to affective experience as a multimodal process that is bodily attendable. The chapter asks if affective qualia can be modelled for a machine nonverbally with expression. The thoughts presented in the chapter are clarified with examples from Olento Collective, a group of artists that developed their artworks on artificial affectivity with artistic research.

Keyword

Qualia · Attention · Aesthetics · Artistic research · Multimodal machine learning

1 Differentiating Qualia from Everything Else Going on

In designing emotional machines one question should be placed at the center of the debate: *What is it like to be having this certain state?* Philosopher Thomas Nagel formulated the question in 1974 to denote the subjective character of conscious

J. Okulov (✉)
Helsinki, Finland
e-mail: jaana.okulov@aalto.fi

© Springer Fachmedien Wiesbaden GmbH, part of Springer Nature 2023
C. Misselhorn et al. (eds.), *Emotional Machines*, Technikzukünfte,
Wissenschaft und Gesellschaft / Futures of Technology, Science and Society,
https://doi.org/10.1007/978-3-658-37641-3_11

experience, later referred as qualia (Nagel, 1974; Chalmers, 1996). With regards to emotions, qualia are their phenomenal perceptual features that provide the individual a sense of ownership of their experience (Chalmers, 1996; Frijda, 2005); according to the theory of qualia, mental states are only accessible from a single point of view; therefore experience, along with its affective nature, arises only when one occupies a particular point of view and truly "has" associated feelings and sensations (Frijda, 2005; Nagel, 1974).

Without the subjective character, emotions are reduced to mere descriptions that machines can comprehend. Although they learn efficiently to detect and manipulate emotional gestures, they lack an ability to be immersed in what they perceive or to tune to others' states. Their knowledge can be considered to accumulate from nowhere, as there is no perspective in the system and in the piece of information that it takes in (Adam, 1995; Nagel, 1989; Penny, 2017). Computing qualia is necessary for machines to attain emotions as experiences, let alone to enable their affective decision-making that could be considered ethical.

Philosopher Daniel Dennett argues against qualia. In his paper "Quining Qualia" (1988), he assembles his notions into five intuition pumps that resemble the main statements about qualia from the thinkers before him. Dennett's goal is "to destroy our faith in the pretheoretical or "intuitive" concept" (1988), but by doing so, he also manages to construct a foundation for qualia research. Specifically, that the criteria for computing qualia, developed through Dennett's argumentation, could be as following: 1) we must be able to isolate qualia from everything else going on; 2) qualia must refer to properties or features in the physical world; 3) there must be a way to study qualia systematically; and 4) intersubjective comparisons must be possible. In the following section, I will explain how these arguments should be framed and met in the field of affective computing.

The interdisciplinary literature calls qualia the "hard problem" of consciousness. The difficulty is in explaining how the properties of things, such as the redness of red, or the quality of a tone issuing from a clarinet, arise to the conscious cognition and how the multitude of sensations bond to form an experience (Chalmers, 1995). To process emotional qualia, affective machine learning models are often computed with linguistic annotations. This involves assigning emotions with verbal labels denoting their basic classes, such as "anger", "disgust", "fear", "happiness", "sadness", and "surprise" theorized by Charles Darwin (1998) and later developed by Paul Ekman (1992) (Wang et al., 2022). However, while these models rely on categorization of emotions in detecting them, emotional attunement in living organisms can arise directly from the dynamics of sensations arriving from the exteroceptive and interoceptive senses without requiring high-level cognitive effort. Although verbal language may emerge in affective experience, it is not sufficient for emotions to have qualia. Through qualia, the world is immediately available without anything between our qualitative experiences and us (Musacchio, 2005).

Neuroscientist Edmund Rolls discusses qualia in his book chapter "Which cortical computations underlie consciousness?" (Rolls, 2016), where he also emphasizes that the information processing in and from the sensory system must use a language to give rise to qualia. However, he states that the language does not need to be human verbal language, but a syntactic process that enables manipulation of symbols. The advantage of artificial systems is their ability to process information of any kind from visual aesthetics to musical harmonies as already Ada Lovelace, the developer of the first algorithm, noted in 1843 (Lovelace in Menabrea, 1843). According to Rolls, qualia emerges as "raw sensory, and emotional, 'feels'" when higher-order thought system reflects and corrects its lower-order thoughts, and states that "it feels like something for a machine" to be reflectively thinking about its own low-order-thoughts. (Rolls, 2016).

While in Rolls' theory qualia emerges between the first-order and the second-order thought through a language, I propose that qualia themselves were considered as the organizing principle of an affective system. According to researchers David Balduzzi's and Giulio Tononi's (2009) information integration view of consciousness, qualia are the informational relationships and entanglements that represent interactions withing a system. They provide the system with a functioning geometry where different aspects of phenomenology emerge as different basic shapes in this qualia space. In their theory, experiences become comparable as the shapes can be compared with each other. (Balduzzi & Tononi, 2009). Deducing from this: To establish a framework of logic for an affective system, emotions should be organized not according to their verbal labels that are secondary symbols, but according to qualia that represent the cross-connections between sensory modalities (Haikonen, 2009). If an affective stimulus is presented to such a system that has the sensory and motor modalities integrated seamlessly together, the affective responses of the machine can arise from direct perception as advocated by researcher Pentti Haikonen (2009). With this method, qualia are not isolated from everything else as unimodal states, but they emerge rather as a process (Korf, 2014).

The studies on human perception support the multimodal hypothesis of qualia. In perception generally, multimodality is not an exception, but experiences can be considered multimodal in their nature (Amer et al., 2017; Bertelson & De Gelder, 2004; Marin, 2015; Nanay, 2012). Interestingly, psychological and physiological states can alter how perceptual information is processed and change for example how wide a distance looks like (Schnall, 2011; see also the broad commentary to Chaz Firestone's and Brian Scholl's article on this issue, 2016). This explains why emotional experience is "out there" and we perceive incidents as "threatening" or "lovely" (Frijda, 2005), also aesthetically. According to psychologist Nico Frijda (2005) inspired by Lambie and Marcel (2002) and William Jamesian perspective (1884) perceived qualia results from the physiological arousal that manifest in the body as an urge in relation to the percept. Affective perception activates the movement preparation in the brain (De Gelder et al., 2004; Komeilipoor et al., 2013; Kret et al., 2011); it is the felt action tendency reserved for reactive movement that is released if the incident demands it (Fuchs & Koch, 2014).

By attending **how** something under observation is, one shifts their attention towards relational internal events to learn about the affective nature of the external world. When attention is focused on the dynamic nature of a phenomenon, affective states are understood through how the body attunes to them from the felt changes in the body. This allows expression without the need for linguistically grasping the event first. (Olento Collective, 2017). To train an affective machine learning model according to qualia, human participation is needed. An artistic thought can be presented: How (with what kind of dynamics) aesthetic features activate the body? If qualia avoid stagnant thoughts, it should be differentiated from everything else via vivid engagement.

2   Sensuous Attention

"We usually bump into all sorts of objects in our daily lives ... and we tend to think that in this activity we 'perceive' the object; but this is a gross mistake ... at best, we name, or classify, them, we subsume them under the general concept by which we discourse or communicate about the world around us. Real perception takes place when we focus our *sensuous attention* on an object completely, when the sense, or senses, involved and the mental powers of the mind are given totally to the object as a complex presentation of colors, lines, sounds or movements, along with other types of qualities." (Mitias, 2012). Artistic thought changes how something is observed; it breaks the ordinary observation and contradicts rational thought, and the core is in reflexivity of perception (Mersch, 2018). Notions of external and internal states turn into expression in artistic practices, but art starts to alter or "sensitize" the senses even prior to art turning into an act. In aesthetic perception, the observer is immersed with the object of attention and leaves behind the ordinary practical attitude; this altered perception is called an aesthetic attitude (Schopenhauer, 2019). The artist can be viewed as an experimenter, or an operator of forces (Sauvagnargues, 2013), but aesthetic attitude does not belong exclusively to practitioners of arts.

Gallese and Di Dio refer to aesthetic attitude when attention is directed to the aesthetic qualities of an object (Di Dio et al., 2016; Gallese, 2017; Gallese & Di Dio, 2012); however, this way of perceiving is not limited towards artworks but applies to all types of objects of observation (Brown et al., 2011). Both aesthetic and emotional understanding originate from the same process, where the external and the internal fuses meaningfully together; namely, external events are informed by the interoceptive states and become sensed through the body's arousal (Brown et al., 2011). Neuroaesthetic research has shed light on the importance of movement perception in aesthetics, with the viewer's process of aesthetic understanding occurring quite spontaneously through a motor component. Aesthetic attitude activates both sides of the insula, the center also for affective and subjective understanding, homeostatic states, self-perception, and even the perception of present time (Craig, 2003, 2009). The activation in the central insula is important for the modulation of movement kinematics: how the movement is executed, *gently* or *rudely*, in relation to affective states. By contrast, the posterior insula, connected to the central insula, is crucial for interoception and the "material me", and it receives homeostatic afferent information from the body (Di Dio et al., 2016; Gallese & Dio, 2012). Electric stimulation of the insula produces sensations of "tingling", or "something in the stomach", and "a sensation in a tongue, like a taste", accompanied by emotions such as fear and, most interestingly, an urge to move (Penfield & Faulk, 1955).

Neuroanatomist and neuroscientist AD (Bud) Craig calls these states as homeostatic modalities. These include feelings involving temperature, itches, visceral distension, muscle ache, hunger, thirst, "air hunger", sensual touch, and pain. These modalities appear in the body as distinct sensations, but also as motivation. Homeostatic modalities operate as a homeostatic behavioural drive that directly activates the limbic motor cortex (Craig, 2003). Noteworthy is how sensations can turn into a desire or tendency to move with such ease. This direct stimulus-to-action pattern could be related to machine learning in order to investigate the behavioural space, or in other words, the qualia space, more thoroughly. In understanding how something is, one is attending to the dynamics of the stimuli with the body; as the potentiality of the body to move exists in the artwork as a perceived aesthetic property, it could be argued that qualia in an artwork is the body's physiological arousal. Art is, beyond its intellectual constituents, a possibility to turn a sensation into an expression, and an artwork can capture those sensations in its form.

Philosopher Alexander Gottlieb Baumgarten describes aesthetics as *scientia cognitionis sensitivae*, (i.e., the science of sensitive knowing) (2007 [1750/1758]). He argues that reason and sensibility are analogous to each other and thus relates aesthetics to rationality (Kaiser, 2011). By asking how something is with sensuous curiosities, one might think to get irrational answers, but against what is commonly thought, these abstractions can find concrete measurable analogies in how they activate other modalities. Aesthetics is, in Baumgarten's terms "extensively clear" through multiple qualitatively sensible features that make sense in their complexity (Blahutková, 2017). Noteworthy in Baumgarten's theory is how aesthetics becomes a science of the senses, but how can a singular artwork open up a possibility to something similarly generalizable and repeatable as objects in scientific research?

In art, it is quite common to build an artwork by letting the environment influence aesthetic expression, but the process is often implicit, intuitive, and unquantified. This issue could be addressed by recognizing that qualia becomes quantifiable through multimodal expression and aesthetic and artistic research can be used as a method to measure qualia systematically.

3 Multimodally Emergent

Modality refers to the way something is experienced, or happens, and is often related to sensory modalities (Baltrušaitis et al., 2018). In aesthetic experience, a modality should be considered an expressional mediator; different modalities, although inside the same sensory scope, can have their own expression topology that overlaps. For example, colour change and motion of a form can express the same or different qualia, even though their combinatory outcome is interpreted as a singular visual object (Olento Collective, 2017). The brain might interpret these different expressional modalities in separate areas, as colour and motion, for example, are neurally distinct. Colour information flows through the V4 area of the visual cortex, and motion is interpreted primarily in V5/MT, but colour and motion are not isolated islands and do inform each other (Zeki, 2015).

When a colour warms (i.e., is changing into what seems to be a warmer colour than it previously was), the change can be heard in the quality of a singer's voice reacting to the colour. In artistic research, the relationship between modalities can be made explicit with an expressional reaction to the stimulus. The expression in one modality is analogous to the stimulus (e.g., the qualia of movement), and a sound generated from the observed motion should predictably match in their affectivity. The sound and motion can also be expected to cause the same sensation in an individual if they are both later used as stimuli. Affectively too complex stimuli cause a conflict, or exaggeration in expression, and the aesthetic expression with apt stimuli feels coherent (Olento Collective, 2017, 2018).

Temporality and matching of information from different sources according to time are important for unified perception (Amer et al., 2017). It is apparent, for example, in emotional perception when affective content in one modality changes perception for other modalities (Klasen et al., 2014) and in how emotions are monitored moment to moment through changes in bodily and vocal expression (Robins et al., 2009; Saito et al., 2005). In understanding the temporality of qualia, it should be noted that subjective time, which grounds modalities, can fluctuate according to how something affects; homeostatic modalities, and emotions change time perception for an individual (Craig, 2009). In a study of subjective time in relation to abstract paintings, Nather et al. (2014) discovered that images containing more motion made time seem longer, and another study found that the length of saccades influenced how time was perceived (Jia, 2013). Time seems to speed up when temperatures rise, or a person experiences a certain colour, with the duration of time feeling 12% shorter with a blue stimulus than a red one (Thönes et al., 2018; Wearden & Penton-Voak, 1995).

It can be speculated that subjective time plays a key role in how affective and aesthetic content are compared and made explicit. Perhaps the link between different aesthetic features and homeostatic states might find meaning through the following question: What kind of force with what kind of temporal release must be reserved into a body as a potential, in order to move it accordingly to the stimulus? It is an aesthetic question that can find a meaning through artistic research that draws knowledge from physical theories: "The traditional objective properties and coordinates of physical theory, such as distance, time, mass, charge, momentum, energy, etc., can similarly be generalized to encompass corresponding subjective experiences, the more rigidly defined objective descriptions of which are useful tools for analytical purposes" (Jahn & Dunne, 1997). If qualia were the bodily attendable properties of experience, they must be unique to an individual who has her own characteristics in reacting to things. In asking nonverbally with what kind of forces does a yellow tune the body, the answer is in relation to the inborn reactivity of that particular body. Therefore, to enable intersubjective comparisons of qualia, the latter must first be mapped onto a subjective space representing an individual, only then can the commonalities in the logic of qualia between different individuals be compared.

Multimodal machine learning is an emerging research field in artificial intelligence (Baltrušaitis et al., 2018). Examples of its uses in human behaviour range from audio-visual speech recognition (Ngiam et al., 2011), human–robot interaction

(Qureshi et al., 2016), gesture recognition (Escalera et al., 2017), to emotion classification (Kim et al., 2013). The concept of multimodal machine learning studies currently occurring patterns in information streams and has gained progress in affective understanding through the development of technology such as facial recognition, but it still lacks efficiency when attempting to replicate naturally occurring emotions (Baltrušaitis et al., 2018). Emotion expression patterns are complex and vary across individuals, and further knowledge is needed on how to relate information meaningfully from multiple sources. In behavioural research, only a few studies have tried to integrate multimodal results across domains, and even less have specialized in affective information (Bertelson & De Gelder, 2004). Accurate modelling of emotions requires the reduction of input feature space (Baltrušaitis et al., 2018; Kim et al., 2013; Ngiam et al., 2011) and the development of such common logic would progress the field of affective computing.

Shared representation of the modalities can be considered an abstraction of what they have in common attentionally. Interestingly, state-of-the-art technologies to enhance multimodal processing are gaining interest in relation to attention research (Bao et al., 2018). Attention models are mostly used in classifying images with language (Fang et al., 2015; Yang et al., 2016a) but other thought-provoking examples of attention models are emerging. They have been implemented in Generative Adversarial Networks (GANs) to improve the generation of images from natural language descriptions (Xu et al., 2018); modified into stacked attention networks (SANs) to enable artificial question answering in natural language from images (Yang et al., 2016a, 2016b), and integrated into the deep attention model (DeepAtt) and Neural machine translation (NMT) for better language translations (Luong et al., 2015; B. Zhang et al., 2018a, 2018b). More abstract models include a developed visual-semantic attention for Adversarial Attention Network (AAN) (Huang et al., 2018) and self-attention for Generative Adversarial Network (SAGAN) (Zhang et al., 2018a, 2018b).

The typical approach is to use human rationales as supervision for attention generation (Bao et al., 2018), but what if this supervision was done nonverbally? Artistic practices could bring novel opportunities to the field of multimodal machine learning. Nonverbal annotation used in a model could enable dynamic comparisons of the inputs. For example, soft vocal expression in relation to a landscape could teach a machine to attend to the "soft" features of that landscape. Perhaps this way, emotional information could be attended artificially, directly from expression, without the need for verbal description.

4   Artistic Thought

In conclusion, I would like to present a research process of an art collective, mainly active during the years 2013–2018, working with a multimodal method in practice. The Olento Collective consists of a small team of artists and creative individuals developing *Olento* (*Entity*) artwork, an interactive installation, and a research project studying how affective information can be taught to a machine via multimodal aesthetics (Olento Collective, 2016). The collective positions itself in the contemporary art field but uses also scientific methods to engage the area of interest. During the years 2013–2018, The Olento Collective finalized multiple installations that translated bodily and emotional data into colours, soundscapes, and digital forms, and gathered research data from around 600 people to inform *Olento's* development. The Olento Collective's work will demonstrate how artistic and scientific research can be combined and how, through artistic research, information can be studied nonverbally.

In the years 2013–2016, the Olento Collective's work concentrated on the affectivity of digital forms, first covering the linguistic descriptions of emotions in relation to aesthetics, proceeding into nonverbal comparisons between bodily sensations and 3D objects, and later moving to translations of facial expressions into 3D objects. In 2014, the collective completed a research questionnaire in an art exhibition entitled *Reality Hunger*. This exhibition was held in Fafa Gallery, Helsinki, and curated by Vesa-Pekka Rannikko. A total of 631 answers were gathered from 61 exhibition visitors about their interpretations of emotional qualities. The visitors were asked to define the basic emotions *happiness, sadness, disgust, fear, surprise*, and *anger* (Ekman, 1992) with aesthetic words. The words collected were open coded into 11 word-pairs that included opposing dynamics: small/large, thin/thick, short/long, light/heavy, horizontal/vertical, straight/curved, sharp/dull, symmetric/asymmetric, distinct/fuzzy, soft/hard, and open/closed.

The results showed that basic emotions were described somewhat consistently (Ristola, 2014), but artistically more relevant became how the bodily dynamics manifested themselves in the descriptions. The bodily dynamics were developed into a sculptural logic to create an artwork for the Young Artists Exhibition held in Kunsthalle Helsinki, 2015. The exhibition visitors were able to sculpt their current emotions from *Olento's* digital mass with slide buttons changing or moving a 3D-form (in thickness–thinness, sharpness–roundness axis, and with buttons that lifted–sank, inflated–shrank, twisted–released, closed–opened, bent down–bent up the object), its colour, and its surface material. After sculpting, the visitors answered how their bodies felt in relation to the finished object. They could choose from different areas of the body that related to Nummenmaa Laboratory's research on bodily maps of emotions (2014), after which an artificial agent (Neural network and the Boids) learning the embodiment organized the 3D objects into a virtual space where other visitors' objects were also floating. Around 1000 objects were sculpted and drifting in the space with their distinct behaviours: some sought company, some avoided it, others did not care. The Olento Collective members noted how intuitively people were able to sculpt their emotions and how easily the bodily dynamics were made readable from the objects: thus, aesthetics seemed to have a very direct relationship to bodily sensations. (Olento Collective, 2015).

In 2016, the Olento Collective collaborated with the NEMO research group (which specializes in digitalization of emotions (Katri Saarikivi and Valtteri Wikström et al., 2016)) from the Department of Neurocognitive Sciences at the University of Helsinki to create an interactive installation for the Sideways and Pixelache Festivals in Helsinki. *Olento* was taught to associate digital forms with facial expressions by a person reacting to the changes in aesthetic dynamics of a digital object with facial gestures (open source computer vision and generative algorithm). After the algorithm was taught, the digital object tried to imitated the facial expressions with its form, surface material, and color; for example, when seeing a bright smile, it lifted its body, reached higher, grew sharper and glowed in yellow, still remaining abstract enough to avoid clear anthropomorphism. *Olento* also translated facial expressions of both a festival audience listening to a band in Sideways Festival and another audience looking at an Olento installation in Pixelache Festival into aesthetic dynamics of a digital object, which again influenced the soundscape. The Olento Collective attempted to triangulate affective research data nonverbally: from the music, played by a band, through an audience's facial expressions into a soundscape of *Olento*, so that the qualia in live music would become explicit in *Olento's* soundscape and the affective content comparable in sound and facial expression. Although the experiment did not enable systematic analysis in the end, a nonverbal research method was established. The way facial expressions were taught reactively in relation to digital aesthetics was a new approach for the collective in experimenting with machine learning.

In the years 2017–2018, the collective investigated the affective relationship between digitally drawn line, colour, bodily movement, and singing, with small research programs, and in an artistic manner with multimodal translation, whereby a singer and a dancer tried to develop a mutual affective language through improvisation. In these experiments, the collective noticed similarities between people's behaviours: many of the expressions in relation to visual or auditory stimuli were individually distinctive, but certain patterns seemed to appear in people's expression (Olento Collective, 2017). For example, the well-studied multimodal correlation of pitch and brightness manifested itself in the research settings (Marks & Mulvenna, 2013), but other unstudied correlations also emerged; for example, a saturated lime-yellow colour was often said to be "the fastest" colour when colour was studied by reacting to it with a stylus pen (Olento Collective, 2017). For qualia research, the most relevant insights of the Olento Collective related to the attentional features in affective perception. The collective noticed how especially dynamic features in stimuli, such as changes in a sound or an image were strongly attended and most affectively expressed. Affective information, in Olento Collective's research, seemed to be in relation to dynamic change (Olento Collective, 2018).

The research programs seemed to work as a means to study in more detail how distinct modalities operated in expression. For example, colour often seemed to distract the felt affective consistency in comparison to another modality if the colour was interpreted to express a contradicting qualia in relation to the motion of a drawn line, so that a singer singing the qualia of the drawing explicitly noted that "there are too many affective stimuli to follow with singing" and also confused the machine expression so that it "just felt unintuitive" for the person observing the behaviour (Olento Collective, 2016, 2017). What remains to be studied in more detail is the overall cross-modal correlations between a wider range of modalities in order to map them into a coherent qualia space, where affecting features in different modalities would match the topology.

5 Conclusions

This chapter related the book title *Emotional Machines. Perspectives from Affective Computing and Emotional Human* to the concept of *qualia* and its clear connection to affective perception. While affective machine learning models rely on categorizing emotions, emotional attunement in living organisms can arise directly from the dynamics of sensations without requiring high-level cognitive effort. The chapter argued that to establish a logical framework for an affective system, emotions should be organized based on their qualia rather than their verbal labels. Affective qualia were framed as a multimodal process that becomes available for the body to attune to through a shift in attention towards the physiological arousal that affective percepts cause. It was suggested that affective qualia could be mapped through multimodal expression. The research process of an art collective that used this approach in practice serves as an example of how it can be done.

References

Adam, A. (1995). Artificial intelligence and women's knowledge: What can feminist epistemologies tell us? *Women's Studies International Forum, 9*(4), 407–415.

Amer, M. R., Shields, T., Siddiquie, B., Tamrakar, A., Divakaran, A., & Chai, S. (2017). Deep multimodal fusion: A hybrid approach. *International Journal of Computer Vision, 126*(2–4), 440–456.

Balduzzi, D., & Tononi, G. (2009). Qualia: The Geometry of Integrated Information. *PLoS Computational Biology, 5*(8), e1000462. https://doi.org/10.1371/journal.pcbi.1000462.

Baltrušaitis, T., Ahuja, C., & Morency, L.-P. (2018). Multimodal machine learning: A survey and taxonomy. *IEEE Transactions on Pattern Analysis and Machine Intelligence.*

Bao, Y., Chang, S., Yu, M., & Barzilay, R. (2018). Deriving Machine Attention from Human Rationales. In *Proceedings of the 2018 conference on empirical methods in natural language processing* (pp. 1903–1913).

Baumgarten, A. G. (2007). *Aesthetica/Ästhetik*. Felix Meiner.

Bertelson, P., & De Gelder, B. (2004). The psychology of multimodal perception. In C. Spence & J. Driver (Eds.), *Crossmodal space and crossmodal attention* (pp. 141–177). Oxford University Press.

Blahutková, D. (2017). Alexander Gottlieb Baumgarten and current reflections on his thoughts concerning aesthetics. In S. Kopčáková & A. Kvokačka (eds.), *Súradnice estetiky, umenia a kultúry III. Európske estetické myslenie a umelecká tvorba: Pramene, metamorfózy a ich relevancia: STUDIA AESTHETICA XVII* (pp. 11–19). FF PU v Prešove.

Brown, S., Gao, X., Tisdelle, L., Eickhoff, S. B., & Liotti, M. (2011). Naturalizing aesthetics: Brain areas for aesthetic appraisal across sensory modalities. *NeuroImage, 58*(1), 250–258.

Chalmers, D. J. (1995). Facing up to the problem of consciousness. *Journal of Consciousness Studies, 2*(3), 200–219.

Chalmers, D. J. (1996). *The conscious mind: In search of a fundamental theory*. Oxford University Press.

Craig, A. D. (2003). A new view of pain as a homeostatic emotion. *Trends in Neurosciences, 26*(6), 303–307.

Craig, A. D. B. (2009). Emotional moments across time: A possible neural basis for time perception in the anterior insula. *Philosophical Transactions of the Royal Society of London. Series B, Biological Sciences, 364* (1525), 1933–1942.

Darwin, C. (1998). *The expression of the emotions in man and animals*. (Original work published 1872). Oxford University Press.

De Gelder, B., Snyder, J., Greve, D., Gerard, G., & Hadjikhani, N. (2004). Fear fosters flight: A mechanism for fear contagion when perceiving emotion expressed by a whole body. *Proceedings of the National Academy of Sciences, 101*(47), 16701–16706.

Dennett, D. C. (1988). *Consciousness in modern science*. Oxford University Press.

Di Dio, C., Ardizzi, M., Massaro, D., Di Cesare, G., Gilli, G., Marchetti, A., & Gallese, V. (2016). Human, nature, dynamism: The effects of content and movement perception on brain activations during the aesthetic judgment of representational paintings. *Frontiers in Human Neuroscience, 9*, 705.

Ekman, P. (1992). An argument for basic emotions. *Cognition & Emotion, 6*(3–4), 169–200.
Escalera, S., Athitsos, V., & Guyon, I. (2017). Challenges in multi-modal gesture recognition. In S. Escalera, I. Guyon, & V. Athitsos (Eds.), *Gesture recognition* (pp. 1–60). Springer.
Fang, H., Gupta, S., Iandola, F., Srivastava, R. K., Deng, L., Dollár, P., et al. (2015). From captions to visual concepts and back. In *Proceedings of the IEEE conference on computer vision and pattern recognition* (pp. 1473–1482).
Firestone, C., & Scholl, B. J. (2016). Cognition does not affect perception: Evaluating the evidence for "top-down" effects. *Behavioral and Brain Sciences, 39*. https://doi.org/10.1017/S0140525X15000965.
Frijda, N. (2005). Emotion experience. *Cognition & Emotion, 19*(4), 473–497.
Fuchs, T., & Koch, S. C. (2014). Embodied affectivity: on moving and being moved. *Frontiers in Psychology, 5*, 508. https://doi.org/10.3389/fpsyg.2014.00508.
Gallese, V. (2017). Mirroring, a liberated embodied simulation and aesthetic experience. In *Mirror images. Reflections in art and medicine* (pp. 27–37). Verlag für moderne Kunst, Kunstmuseum Thun.
Gallese, V., & Dio, D. (2012). Neuroesthetics: The body in esthetic experience. In V. S. Ramachandran (Ed.), *The encyclopedia of human behavior* (Vol. 2, pp. 687–693). Academic Press.
Haikonen, P. O. (2009). Qualia and conscious machines. *International Journal of Machine Consciousness, 1*(2), 225–234.
Huang, F., Zhang, X., & Li, Z. (2018). Learning Joint Multimodal Representation with Adversarial Attention Networks. In *2018 ACM Multimedia Conference on Multimedia Conference* (pp. 1874–1882).
Jahn, R. G., & Dunne, B. J. (1997). Science of the subjective. *Journal of Scientific Exploration, 11*(2), 201–224.
James, W. (1884). What is an emotion? *Mind, 9*(34), 188–205.
Jia, L. (2013). *Crossmodal emotional modulation of time perception*. PhD thesis. University of Munich. https://edoc.ub.uni-muenchen.de/16513/.
Kaiser, B. M. (2011). On aesthetics, aesthetics and sensation-reading Baumgarten with Leibniz with Deleuze'. *Esthetica. Tijdschrift Voor Kunst En Filosofie*.
Kim, Y., Lee, H., & Provost, E. M. (2013). Deep learning for robust feature generation in audiovisual emotion recognition. In *2013 IEEE International Conference on Acoustics, Speech and Signal Processing (ICASSP)* (pp. 3687–3691).
Klasen, M., Kreifelts, B., Chen, Y.-H., Seubert, J., & Mathiak, K. (2014). Neural processing of emotion in multimodal settings. *Frontiers in Human Neuroscience, 8*.
Komeilipoor, N., Pizzolato, F., Daffertshofer, A., & Cesari, P. (2013). Excitability of motor cortices as a function of emotional sounds. *PLoS One, 8*(5)., e63060
Korf, J. (2014). Emergence of consciousness and qualia from a complex brain. *Folia Medica, 56*(4), 289–296.
Kret, M., Pichon, S., Grèzes, J., & de Gelder, B. (2011). Similarities and differences in perceiving threat from dynamic faces and bodies. An fMRI study. *NeuroImage, 54*(2), 1755–1762.
Lambie, J. A., & Marcel, A. J. (2002). Consciousness and the varieties of emotion experience: A theoretical framework. *Psychological Review, 109*(2), 219–259.

Luong, M.-T., Pham, H., & Manning, C. D. (2015). Effective approaches to attention-based neural machine translation. In *Proceedings of the 2015 Conference on Empirical Methods in Natural Language Processing* (pp. 1412–1421).

Marin, M. M. (2015). Crossing boundaries: Toward a general model of neuroaesthetics. *Frontiers in Human Neuroscience, 9.*

Marks, L. E., & Mulvenna, C. M. (2013). Synesthesia, at and near its borders. *Frontiers in Psychology, 4,* 443.

Menabrea, L. (1842). Notions sur la Machine Analytique de M. Charles Babbage. *Bibliothèque Universelle de Genève, 41,* 352–376. (Translation by Augusta Ada Lovelace published 1843 in Scientific Memoirs, 3, 666–731).

Mersch, D. (2018). Alchemistic Transpositions: On Artistic Practices of Transmutation and Transition. In M. Schwab (Ed.), *Transpositions: Aesthetico-epistemic operators in artistic research* (pp. 267–280). Leuven University Press.

Mitias, M. (2012). Can we speak of 'aesthetic experience'? In M. Mitias (Ed.), *Possibility of the aesthetic experience* (pp. 47–58). Springer Science & Business Media.

Musacchio, J. M. (2005). The ineffability of qualia and the word-anchoring problem. *Language Sciences, 27*(4), 403–435.

Nagel, T. (1974). What is it like to be a bat? *The Philosophical Review, 83*(4), 435–450.

Nagel, T. (1989). *The view from nowhere.* Oxford University Press.

Nanay, B. (2012). The multimodal experience of art. *The British Journal of Aesthetics, 52*(4), 353–363.

Nather, F. C., Fernandes, P. A. M., & Bueno, J. L. O. (2014). Subjective time perception is affected by different durations of exposure to abstract paintings that represent human movement. *Psychology & Neuroscience, 7*(3), 381–392.

Ngiam, J., Khosla, A., Kim, M., Nam, J., Lee, H., & Ng, A. Y. (2011). Multimodal deep learning. In *Proceedings of the 28th International Conference on Machine Learning (ICML-11)* pp. 689–696.

Penfield, W., & Faulk, M., Jr. (1955). The insula: Further observations on its function. *Brain, 78*(4), 445–470.

Penny, S. (2017). *Making sense. Cognition, computing, art, and embodiment.* MIT Press.

Qureshi, A. H., Nakamura, Y., Yoshikawa, Y., & Ishiguro, H. (2016). Robot gains social intelligence through multimodal deep reinforcement learning. In *2016 IEEE-RAS 16th International Conference on Humanoid Robots (Humanoids)* (pp. 745–751).

Ristola, J. (2014). *Research on visual forms of emotions (Psychological research methods course).* Institute of Behavioral Sciences, University of Helsinki

Robins, D. L., Hunyadi, E., & Schultz, R. T. (2009). Superior temporal activation in response to dynamic audio-visual emotional cues. *Brain and Cognition, 69*(2), 269–278.

Rolls, E. T. (2016). *Cerebral cortex: Principles of operation.* Oxford University Press.

Saito, D. N., Yoshimura, K., Kochiyama, T., Okada, T., Honda, M., & Sadato, N. (2005). Cross-modal binding and activated attentional networks during audio-visual speech integration: A functional MRI study. *Cerebral Cortex, 15*(11), 1750–1760.

Sauvagnargues, A. (2013). *Deleuze and Art.* English translation by S. Bankston. Bloomsbury.

Schnall, S. (2011). Embodiment in affective space: Social influences on spatial perception. In *Spatial dimensions of social thought* (pp. 129–152). De Gruyter Mouton.

Schopenhauer, A. (2019 [1818]). *Arthur Schopenhauer: The World as Will and Presentation* (Vol. I). Ed. by D. Kolak. Routledge.

Thönes, S., Castell, C., Iflinger, J., & Oberfeld, D. (2018). Color and time perception: Evidence for temporal overestimation of blue stimuli. *Scientific Reports, 8*(1), 1688.

Wang, Y., Song, W., Tao, W., Liotta, A., Yang, D., Li, X., ... et al. (2022). A Systematic Review on Affective Computing: Emotion Models, Databases, and Recent Advances. arXiv Preprint arXiv:2203.06935.

Wearden, J. H., & Penton-Voak, I. S. (1995). Feeling the heat: Body temperature and the rate of subjective time, revisited. *The Quarterly Journal of Experimental Psychology. B, Comparative and Physiological Psychology, 48*(2), 129–141.

Xu, T., Zhang, P., Huang, Q., Zhang, H., Gan, Z., Huang, X., & He, X. (2018). Attngan: Fine-grained text to image generation with attentional generative adversarial networks. In *Proceedings of the IEEE Conference on Computer Vision and Pattern Recognition* (pp. 1316–1324).

Yang, Z., He, X., Gao, J., Deng, L., & Smola, A. (2016a). Stacked attention networks for image question answering. In *Proceedings of the IEEE Conference on Computer Vision and Pattern Recognition* (pp. 21–29).

Yang, Z., Yang, D., Dyer, C., He, X., Smola, A., & Hovy, E. (2016b). Hierarchical attention networks for document classification. In *Proceedings of the 2016b Conference of the North American chapter of the association for computational linguistics: human language technologies* (pp. 1480–1489).

Zeki, S. (2015). Area V5—A microcosm of the visual brain. *Frontiers in Integrative Neuroscience, 9*.

Zhang, B., Xiong, D., & Su, J. (2018). Neural machine translation with deep attention. *IEEE Transactions on Pattern Analysis and Machine Intelligence, 42*(1), 154–163.

Zhang, X., Yao, L., Huang, C., Wang, S., Tan, M., Long, G., & Wang, C. (2018). Multimodality sensor data classification with selective attention. In *Proceedings of the twenty-seventh international joint conference on artificial intelligence main track* (pp. 3111–3117).

Olento Collective

2013–2014: Pat Nav (computing and artistic work), Jaana Okulov (research and artistic work), and Ismo Torvinen (computing and artistic work).

2015: Max Hannus (linguistics and artistic work), Mari Ljokkoi (production and artistic work), Jaana Okulov, and Ismo Torvinen. Collaboration: Samuli Kaipiainen (user interface).

2016: Max Hannus, Jani Hietanen (sound developing and artistic work), Mira Hyvönen (computing and artistic work), Jaana Okulov, and Ismo Torvinen. Collaboration: NEMO research group from the Department of Neurocognitive Sciences at the University of Helsinki (Katri Saarikivi, Valtteri Wikström, et al.).

2017: Jani Hietanen, Mira Hyvönen, Jaana Okulov, and Ismo Torvinen. Collaboration: Veera Hirvaskero (singing), Anne Naukkarinen (dancing), Heikki Ketoharju (back-end development).

2018: Jani Hietanen, Mira Hyvönen, Jaana Okulov, and Ismo Torvinen. Collaboration: Veera Hirvaskero, Anne Naukkarinen, Heikki Ketoharju, Tom Engström (front-end development).

Jaana Okulov a doctoral student at Aalto University's School of Arts, Design and Architecture, in Department of Art and Media. Okulov received their MFA in the Finnish Academy of Fine Arts in 2016 and completed basic and intermediate studies in Psychology in 2017. Their research concentrates on the theme of attention in humans and in machines, and combines theoretical and empirical approaches from Perceptual Psychology, Aesthetics and Machine Learning.

My Square Lady

In the mid 1950s, the Broadway musical *My Fair Lady* became the longest-running musical of its time. Simple cockney flower girl Eliza Doolittle is plucked from the streets of London and trained by linguistics professor Henry Higgins under a bet that he can pass the commoner off as a lady of high society after six months of intensive tuition.

Almost 60 years later, Gob Squad Arts Collective dreamed up *My Square Lady*. There was no Eliza, but a robot called Myon (from the Neurorobotics Research Laboratory at Beuth University of Applied Sciences Berlin) which together with the staff and singers of Komische Oper Berlin, Gob Squad tried to pass off as a human, using all the means of the world of opera to teach it about the most human of qualities—emotions. In the process of training Myon, all kinds of questions concerning the nature of humanity emerged: what does it actu-

---

Gob Squad continue to explore and expand what theatre can be. 'Show Me A Good Time' was presented as a 12 hour international live-stream performance in the middle of the first Covid lockdown. 'Elephants in Rooms' is their latest work: a video project linking performers from India, China, UK and Germany.

Gob Squad Arts Collective GmbH · Gob Squad Arts Collective GmbH
Berlin, Germany
e-mail: info@gobsquad.com

ally mean to be human? What are emotions and why do we need them? Does a robot need them, too? Why do we want to have artificial intelligence? Why do we want to have opera? Does Myon need a costume?

The project was split into a research phase and rehearsals towards a final production. During the research phase, the Neurorobotics Research Laboratory under the direction of Prof. Manfred Hild gathered data and optimised the robot by confronting it with real singers voices and the sound of an orchestra, taught the robot how to conduct, and to react to optical stimuli amidst the visually complex world of a stage. Gob Squad also conducted video interviews with members of staff from all departments of the opera house, asking them to answer to the naïve eye of the robot. At the same time, the singers were asked to contribute musical material suitable to the task of teaching a robot about emotions.

During several week-long residencies, Myon visited each department of the Komische Oper Berlin trying to understand and master the skills of musical craft and the art of emotion. The road was long and perilous, there were successes and failures, joy and tears (not Myon's), and only in the summer of 2015 was it clear whether the experiment had come to anything when Myon starred in *My Square Lady*, the world's first Reality Robot Opera.

The following pages show images from the final production.

During Mozart's *The Magic Flute*, three ladies compete for the attention of 'a gracious youth' (cf. Fig. 1). The gracious youth in this case is Myon, whose attention is drawn to anything red. The audience get an insight into the robot's perception with a live video feed from its camera.

"*Hope there's someone who'll take care of me when I die*" by Anthony and the Johnsons is sung whilst children tend Myon (cf. Fig. 2). Care robots for the elderly are already a reality—perhaps Myon's successors will look after the singers one day...

Myon is confronted with a doppelgänger (cf. Fig. 3), questioning concepts of self and individual identity as well as exploring questions raised by the sci-fi trope of consciousness downloading. If the memories of one being (artificial or human) can be transferred from one body to another, would that be a kind of immortality?

Fig. 1 Final production. (© david baltzer/bildbuehne.de)

My Square Lady is a coproduction of Gob Squad, Komische Oper Berlin and the Neurorobotics Research Laboratory at Beuth University of Applied Sciences Berlin. Funded by the Doppelpass Fund of the German Federal Cultural Foundation and the Schering Foundation.

Fig. 2 Final production. (© david baltzer/bildbuehne.de)

Fig. 3 Final production. (© david baltzer/bildbuehne.de)

Made in the USA
Monee, IL
03 May 2026

49438560R10174